Lecture Notes in Civil Engineering 000

Lecture Notes in Civil Engineering (LNCE) publishes the latest developments in Civil Engineering—quickly, informally and in top quality. Though original research reported in proceedings and post-proceedings represents the core of LNCE, edited volumes of exceptionally high quality and interest may also be considered for publication. Volumes published in LNCE embrace all aspects and subfields of, as well as new challenges in, Civil Engineering. Topics in the series include:

- Construction and Structural Mechanics
- Building Materials
- Concrete, Steel and Timber Structures
- Geotechnical Engineering
- Earthquake Engineering
- Coastal Engineering
- Ocean and Offshore Engineering; Ships and Floating Structures
- Hydraulics, Hydrology and Water Resources Engineering
- Environmental Engineering and Sustainability
- Structural Health and Monitoring
- Surveying and Geographical Information Systems
- Indoor Environments
- Transportation and Traffic
- Risk Analysis
- Safety and Security

To submit a proposal or request further information, please contact the appropriate Springer Editor:

- Pierpaolo Riva at pierpaolo.riva@springer.com (Europe and Americas);
- Swati Meherishi at swati.meherishi@springer.com (Asia—except China, Australia, and New Zealand);
- Wayne Hu at wayne.hu@springer.com (China).

All books in the series now indexed by Scopus and EI Compendex database!

Yuanqing Zhu
Editor

Proceedings of the 5th International Symposium in Environmental Science and Industrial Ecology

Springer

Editor
Yuanqing Zhu
School of Power and Energy Engineering
Harbin Engineering University
Harbin, Heilongjiang, China

ISSN 2366-2557 ISSN 2366-2565 (electronic)
Lecture Notes in Civil Engineering
ISBN 978-981-92-2388-6 ISBN 978-981-92-2389-3 (eBook)
https://doi.org/10.1007/978-981-92-2389-3

This Springer imprint is published by the registered company Springer Nature Singapore Pte Ltd.
The registered company address is: 152 Beach Road, #21-01/04 Gateway East, Singapore 189721, Singapore

Preface

This book presents selected papers from the 5th International Symposium on Environmental Science and Industrial Ecology (ISESIE 2025), held as a hybrid conference in Bangkok, Thailand, on August 1–3, 2025. The symposium brought together researchers, scientists, academics, and practitioners from around the world to share their latest findings and insights.

The contributions in this book cover a broad spectrum of themes, including waste valorization and sustainable construction materials, urban green infrastructure and energy systems, eco-design, environmental health and architecture, energy and emerging energy applications, carbon emissions, water resources and environmental quality assessment, as well as smart cities, digitalization, and environmental monitoring.

This book will provide valuable perspectives for students, researchers, and professionals engaged in environmental science and industrial ecology. We hope it will serve as a useful academic resource, foster scholarly exchange, and contribute to advancing sustainable solutions to environmental challenges.

Harbin, China

Yuanqing Zhu
Book Editor

Contents

Enhancing Environmental Sustainability of Calcium Aluminate Cement: Sustainable Efflorescence Mitigation Through Diatomite-Mediated Dual Pore-Chemical Regulation

Bing Zhou[1], Yong Lai[2(✉)], Qifeng Zhang[3], Daoxun Ma[4], Zhibin Du[5], Weidong Huo[4], Xingang Luan[6], Lei Li[6], and Wen Zhang[2,4]

[1] Shanghai New Era Airport Design & Research Institute Co., Ltd.of CAAC, Shanghai 200131, China
[2] Civil Aviation Research Base (Beijing) Co., Ltd., Beijing 100621, China
cacclaiyong@126.com
[3] Guizhou Civil Aviation Construction Investment Co., Ltd., Guiyang 550002, China
[4] China Civil Aviation Airport Construction Group Co., Ltd., Beijing 101317, China
[5] Beijing Construction Project Management General, Beijing 100621, China
[6] Heilongjiang Airport Group Co., Ltd., Harbin 150000, China

Abstract. This study investigates the eco-friendly efflorescence inhibition mechanism of natural diatomite in calcium aluminate cement (CAC) mortars through multi-scale analyses. Optical microscopy and SEM revealed that diatomite incorporation reduced visible efflorescence by transforming surface deposits ($CaCO_3$ and AFt phases) into submicron-sized microcrystalline or amorphous forms (< 100 μm). Capillary water absorption tests demonstrated a significant 45% reduction in water uptake with 30% diatomite, linked to pore structure refinement and enhanced impermeability, thereby reducing long-term deterioration risks. PH analysis showed a decrease, destabilizing AFt and suppressing $CaCO_3$ crystallization. Synergistic mechanisms included chemical neutralization (SiO_2 reacting with $Ca(OH)_2$ to form C-S-H gel), pore filling (blocking capillary channels), and ion adsorption (hydroxyl groups trapping Ca^{2+}/Na^{+}, reducing ion leaching). These effects collectively reduced ion migration and alkaline dissolution, correlating with a 50% decrease in efflorescence area. The findings highlight diatomite's dual role in pore optimization and pH regulation, offering a sustainable and resource-efficient strategy for enhancing the long-term durability, service life, and aesthetic longevity of CAC-based systems, contributing to reduced environmental impact through minimized maintenance and material replacement needs.

Keywords: Calcium Aluminate Cement (CAC) · Efflorescence · Diatomite · Ph Regulation · Capillary Absorption

Y. Zhu (Ed.): ISESIE 2025, LNCE 000, pp. 1–10, 2026.
https://doi.org/10.1007/978-981-92-2389-3_1

1 Introduction

Efflorescence, characterized by unsightly white deposits (e.g., $CaCO_3$, AFt) on cementitious surfaces, remains a critical durability challenge in calcium aluminate cement (CAC)-based systems [1–3]. This phenomenon arises from the migration and crystallization of soluble ions (Ca^{2+}, Na^+) under alternating wet-dry cycles, exacerbated by the alkalinity and porous structure of CAC mortars [4, 5].

Traditional mitigation strategies, such as polymer emulsions [1, 6, 7] or sulfate additives [1, 2, 8], often face limitations in cost-effectiveness or long-term stability. Diatomite, a natural porous material rich in reactive SiO_2, offers potential as a sustainable alternative due to its dual functionality in pore refinement and alkali regulation [9, 10]. However, the interplay between its physicochemical properties (e.g., ion adsorption, pH modulation) and efflorescence inhibition mechanisms in CAC systems remains underexplored.

This study systematically investigates diatomite's role in suppressing efflorescence through multi-scale analyses, including pore structure characterization, ion leaching tests, and microstructural evolution tracking. The findings aim to establish a mechanistic framework for optimizing diatomite-modified CAC systems, balancing aesthetic durability with eco-efficiency.

2 Experimental

2.1 Materials

Commercially available white calcium aluminate cement (CAC) clinker with a whiteness of no less than 92%. A blended quartz sand composed of two particle size distributions: 40–70 mesh and 70–100 mesh, mixed at a ratio of 3:2. Iron oxide red pigment manufactured by LANXESS, Germany. The diatomite used in this study is Type TS1, produced by Yixing Junlian Diatomite Co., Ltd. It is classified as a fine-grade filler, with a moisture content of 0.3%, a bulk density of 370 kg/m^3, and an average particle size of 6 μm. The chemical composition of the diatomite is shown in Table 1. Deionized water was used for immersion solutions, while tap water was used for all other experimental procedures.

Table 1. The chemical composition of CAC and diatomite.

Raw materials	SiO_2	Al_2O_3	CaO	Fe_2O_3	Na_2O	K_2O	MgO	Others
CAC	0.65	68.76	29.44	0.18	0.15	0.09	–	0.73
Diatomite	94.48	2.72	0.36	0.6	0.22	1.11	0.34	0.17

2.2 Mix Design and Sample Preparation

Diatomite was used to replace calcium aluminate cement (CAC) at substitution levels of 0%, 10%, 20%, and 30% by mass, and the corresponding samples were designated as G0, G10, G20, and G30, respectively. The pigment dosage was 3% of the total mass of the cementitious materials. The water to binder ratio (w/b) of the neat cement paste was 0.5. For mortar specimens, the binder to sand ratio was 1:4, and the water to binder ratio (w/b) was 1.0.

The mortar specimens were primarily used for efflorescence assessment and capillary water absorption tests. The preparation of mortar specimens followed the procedures outlined in JC/T 1024–2007 *Wall Decorative Mortar*. The mixing water temperature was maintained consistent with the curing temperature. For efflorescence testing, mortar specimens were cast in shallow dishes with dimensions of ϕ100 mm × 5 mm and leveled with a trowel. For capillary water absorption testing, specimens were cast into molds measuring 40 mm × 40 mm × 160 mm. All molded specimens were cured under controlled conditions at a temperature of (20 ± 2) °C and a relative humidity of (60% ± 5%) for 24 h before demolding.

Neat cement paste specimens were also cast in shallow dishes (ϕ100 mm × 5 mm) and cured under the same temperature and humidity conditions for 24 h prior to demolding. These specimens were used for optical microscopy, Raman spectroscopy, scanning electron microscopy (SEM), and pH testing of the immersion solution.

2.3 Sample Characterizations

(1) Quantitative characterization of primary and secondary efflorescence

Primary Efflorescence: After 24 h of curing, the mortar specimens were placed in an environment at (40 ± 2) °C for an additional 12 h. The surface condition was then photographed.

Secondary Efflorescence: After photographing the primary efflorescence, the specimens were immersed in 100 mL of deionized water for 12 h, followed by air drying, after which the surface was photographed again.

Efflorescence Degree Characterization: The degree of efflorescence was quantified by calculating the percentage of the efflorescence-covered area relative to the total surface area of the mortar specimen. Surface images of the specimens were captured using a digital camera, and the efflorescence coverage area was analyzed using Image-Pro Plus® 6.0 image analysis software.

(2) Optical microscopy analysis

A VMS 3020 optical microscope with a maximum magnification of 155 × was used to observe the surface of the paste specimens exhibiting secondary efflorescence.

(3) Raman spectroscopy analysis

After accelerating efflorescence in the paste specimens via water immersion, samples were prepared from the central efflorescence region for Raman analysis. Raman spectra were acquired using a LabRAM HR Evolution confocal Raman spectrometer (HORIBA, Japan), equipped with a 532 nm solid-state laser (point source) at a power of no less than 100 mW. Low wavenumber measurements were performed down to 50 cm^{-1}.

(4) Scanning electron microscopy (SEM) analysis

Samples were taken from areas with concentrated secondary efflorescence on the paste specimens for SEM analysis to observe the morphological characteristics of the efflorescence-related phases. A ZEISS Sigma 300 VP field emission scanning electron microscope (Carl Zeiss, Germany) was used with an accelerating voltage of 2 kV.

(5) pH Measurement of the immersion solution

The pH of the efflorescence immersion solution was measured using a METTLER TOLEDO FE28 pH meter. Prior to measurement, the instrument was calibrated using standard buffer solutions with pH values of 4.00, 6.86, and 9.18. Each sample was measured in triplicate, and the arithmetic mean was reported.

(6) Capillary water absorption test

Mortar specimens at the designated curing age were oven-dried to constant mass in a vacuum drying chamber at 20 °C. After cooling, all four lateral faces were sealed with paraffin, leaving only the top and bottom surfaces exposed. The dry mass was recorded as W_0. The specimen's top surface was placed 2–3 mm deep in water, supported by $\phi 5$ mm glass rods to avoid contact with the bottom. The timing began upon immersion, and the specimens were weighed at the following intervals: 5 min, 10 min, 20 min, 30 min, 1 h, 2 h, 3 h, 4 h, 6 h, 8 h, 10 h, 12 h, 18 h, 24 h, 30 h, and 48 h. The mass at time t was recorded as W_t, and the water absorption at each time point was calculated as $\Delta W_t = W_t - W_0$.

Capillary water absorption was characterized by the capillary absorption rate S, calculated as follows:

$$S = \frac{\delta}{\sqrt{t}} \tag{1}$$

$$\delta = \frac{\Delta W_t}{A} \tag{2}$$

where, S denotes Capillary absorption rate, kg/(m^2·h$^{0.5}$); δ denotes cumulative water absorption per unit area, kg/m^2; t denotes time, h; ΔW_t denotes water uptake at time t, kg; A denotes cross-sectional area in contact with water, m^2.

Based on the above, a δ–$t^{1/2}$ plot was constructed to assess the capillary absorption behavior.

3 Results

3.1 Efflorescence Observation

Visual Observation

The effect of diatomite content on mortar efflorescence is shown in Fig. 1. As observed, the amount of primary efflorescence on the mortar surface was minimal, with almost no visible efflorescence. However, after water immersion, pronounced secondary efflorescence appeared on the surface. With increasing diatomite content, the severity of secondary efflorescence on the mortar surface decreased significantly. When the replacement level reached 30%, secondary efflorescence was nearly eliminated.

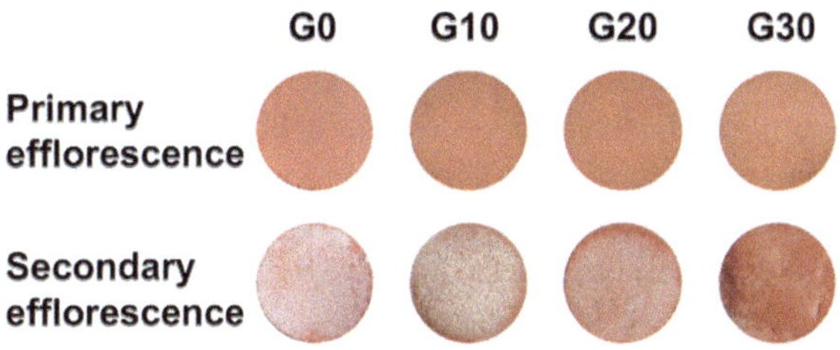

Fig. 1 . Efflorescence of CAC mortars with different dosage of diatomite.

Image Analysis

Image analysis results quantifying the efflorescence area percentage as a function of diatomite content are presented in Fig. 2. As shown, the specimen without diatomite exhibited a primary efflorescence area percentage of 0.72%, while the secondary efflorescence reached 85.9%. With increasing diatomite content from 10% to 30%, the primary efflorescence area percentages were 2.6%, 1.0%, and 3.8%, respectively. In contrast, the secondary efflorescence displayed a clear decreasing trend, with area percentages of 66.7%, 49.1%, and 10.7%, respectively. Notably, the mix containing 30% diatomite showed the most significant reduction in secondary efflorescence area by approximately 75.2%, compared to the control. These results indicate that the incorporation of diatomite effectively mitigates secondary efflorescence in CAC-based mortar systems.

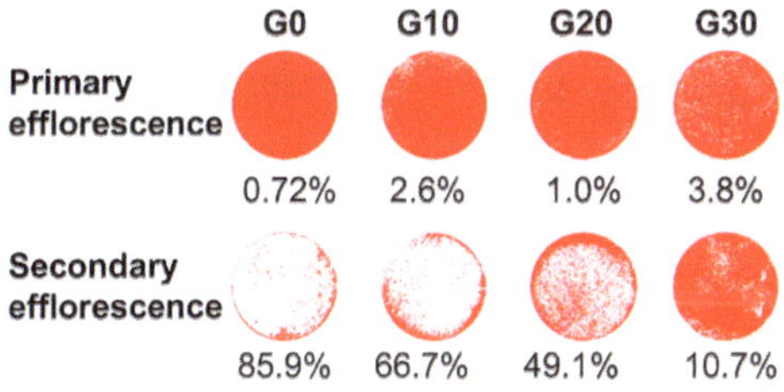

Fig. 2. Efflorescence area percentage of CAC mortars in relation to dosage of diatomite.

3.2 Efflorescence-Causing Substances

Optical Microscopy Observation and In-situ Raman Analysis

Optical microscopy images of the paste surfaces in the calcium aluminate cement–hemihydrate gypsum binary system with varying diatomite contents are shown in Fig. 3. Fig. 4 presents the Raman spectroscopy analysis of the binary system modified with 10% diatomite. As illustrated in Fig. 3, when the diatomite content is 10%, needle- and rod-like white efflorescence products are distributed along pores and cracks at 155 × magnification. At 21 × magnification (Fig. 3b), these efflorescence products are observed to be scattered across the paste surface, with some red-colored areas of the matrix remaining uncovered. When the diatomite content increases to 20%, the complete morphology of the white needle- and rod-like crystals becomes apparent on the paste surface. White efflorescence is still observable in surface pores under 21 × magnification (Fig. 3c). Combined with the Raman spectral analysis (Fig. 4), these products are identified as ettringite, the primary hydration product of the system. In contrast, hexagonal white crystalline phases commonly observed in systems containing limestone powder—typically indicative of calcium carbonate—are much less prominent in the diatomite-modified system, as evidenced by the relatively weak peaks in the Raman spectrum, indicating a lower presence of calcium carbonate. At a content of 30%, needle- and rod-like white crystals approximately 100 μm in length can be seen under 155 × magnification. However, under 21 × magnification (Fig. 3f), only a minimal amount of surface efflorescence is visible, indicating a significant reduction.

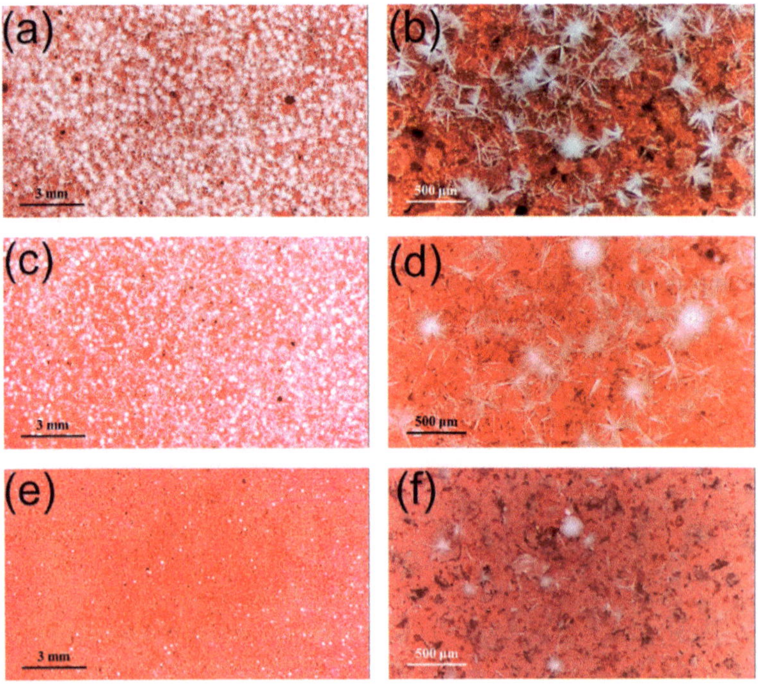

Fig. 3. Optical microscopy images of pastes with different dosage of diatomite.

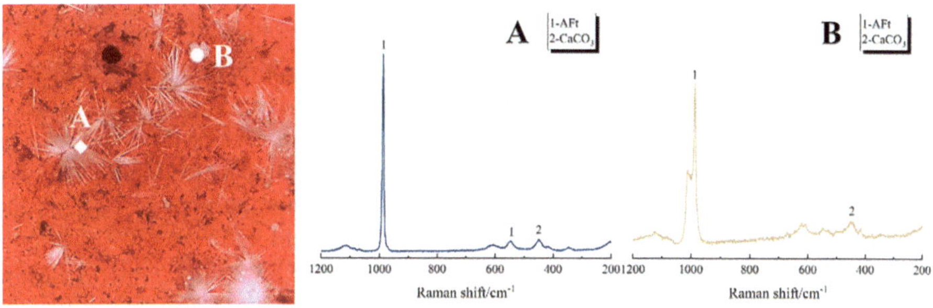

Fig. 4. Raman spectrum of the simples with 10% diatomite.

SEM-EDS Analysis

Fig. 5 presents the SEM-EDS analysis of the specimen containing 10% diatomite. It can be observed that the surface efflorescence consists of long rod-like structures that agglomerate and radiate outward, with dimensions exceeding 100 μm. Additionally, hexagonal crystalline particles are found resting atop the rod-like structures (region corresponding to point B), with particle sizes around 25 μm. EDS analysis reveals that the long rod-like phase (point A) is identified as ettringite, while the cubic-shaped particles are confirmed to be calcium carbonate. In combination with the Raman spectroscopy results, it can be concluded that the efflorescence products in the calcium aluminate cement–hemihydrate gypsum system modified with diatomite are predominantly composed of ettringite, with minor amounts of calcium carbonate present.

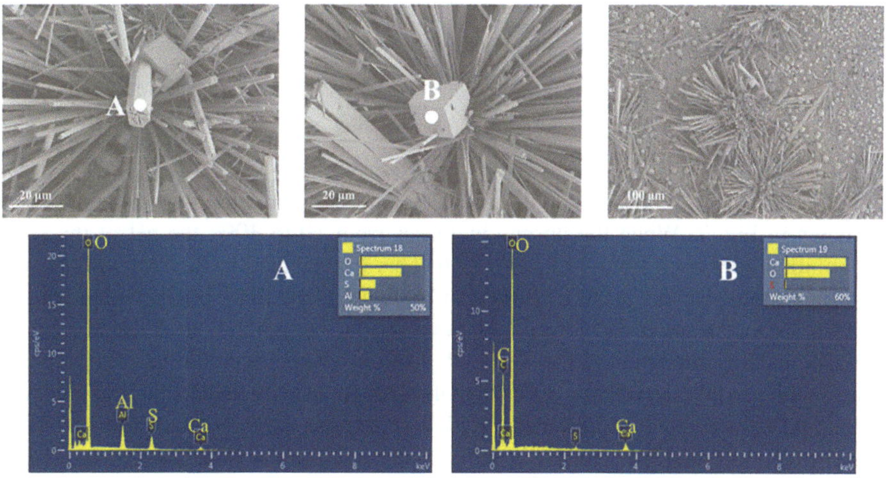

Fig. 5. SEM-EDS analysis of the simples with 10% diatomite.

3.3 Capillary Water Absorption

The effect of diatomite content on the capillary water absorption of calcium aluminate cement mortar is shown in Fig. 6. As illustrated in Fig. 6, the capillary water absorption of mortar with diatomite increases linearly with the square root of time and reaches saturation after 24 h. The diatomite content significantly influences both the capillary water absorption rate and total absorption. When the diatomite content is 10%, the capillary water absorption reaches the maximum value of 6.11 kg/m^2. At 20% diatomite content, the absorption drops to the minimum value of 3.83 kg/m^2, while a further increase to 30% results in a similar absorption of 3.37 kg/m^2. Fig. 6 also reveals that the capillary water absorption rate gradually decreases with increasing diatomite content. Combined with the trend of efflorescence area variation, a positive correlation is observed between capillary water absorption and efflorescence area: as the diatomite content increases, both capillary water absorption and efflorescence area decrease.

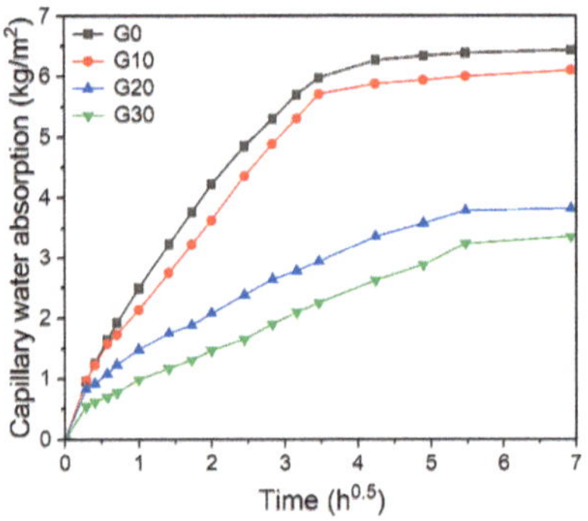

Fig. 6. Capillary pore water absorption rate of mortar under different dosage of diatomite.

3.4 Leachate Analysis

The effect of diatomite content on the pH value of the calcium aluminate cement system is shown in Fig. 7. When the dosage increases from 10 to 30%, the pH of the soaking solution decreases from 11.22 to 10.66. This is attributed to the reaction between SiO_2 in diatomite and $Ca(OH)_2$, forming C-S-H gel ($CaO{\cdot}SiO_2{\cdot}H_2O$), which reduces free Ca^{2+} concentration and lowers pH. As a result, the generation of $CaCO_3$ is suppressed, and any precipitated $CaCO_3$ tends to exist as microcrystalline or amorphous phases with particle sizes below 100 μm, making them visually undetectable and thereby mitigating efflorescence. Additionally, hydroxyl groups (-OH) on the diatomite surface adsorb free Ca^{2+} and Na^+, reducing ion concentration in the pore solution and limiting ion migration to the surface. Furthermore, the porous structure of diatomite fills 10–1000 nm capillary pores, slowing ion migration rates. Combined with pH regulation, this synergy reduces Ca^{2+}/Na^+ leaching. Therefore, diatomite significantly suppresses efflorescence through three mechanisms, chemical neutralization, pore filling, and ion adsorption. These actions collectively lower the system's pH, disrupt the alkaline environment necessary for $CaCO_3$ formation, and reduce ion migration pathways.

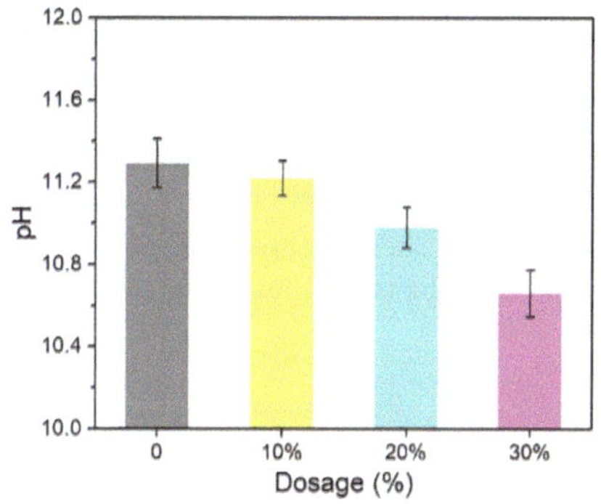

Fig. 7. pH value of the soaking solution of the sample under different dosage of diatomite.

4 Conclusions

This study reveals diatomite's efflorescence inhibition in CAC mortars through triple mechanisms. Microscopy/SEM showed diatomite reduced efflorescence area by transforming $CaCO_3$ and AFt into submicron phases (< 100 μm). Capillary absorption decreased from 6.11 kg/m^2 (10% diatomite) to 3.37 kg/m^2 (30%), correlating with pore refinement (10–1000 nm reduction). pH dropped from 11.22 to 10.66, destabilizing AFt and suppressing $CaCO_3$ crystallization. Synergistic effects included chemical neutralization (SiO_2-$Ca(OH)_2$ → C-S-H), pore filling, and ion adsorption (-OH trapping Ca^{2+}/Na^+). Diatomite's dual role in pore optimization and alkaline regulation offers an eco-friendly solution for CAC-based systems.

References

1. Ma, H., Wang, R., Xi, Z., Gao, H., Fryda, H., Wang, P.: Regulation and inhibition of early whitening of calcium aluminate cement/hemihydrate gypsum decorative mortar by polymer emulsion. Constr. Build. Mater. **393**, 132023 (2023)
2. Sun, Z. et al.Z.: Calcium carbonate and ettringite induced efflorescence in CAC- anhydrite binary systems. J. Wuhan Univ. Technol. Mater Sci Ed. 39, 1129–1137 (2024).
3. Xu, L., Liu, S., Wang, P., Yang, Z.: Influence of Mineral Additives on the Efflorescence of Ettringite-Rich Systems. Materials. **14**, 5464 (2021)
4. Wu, B., Ma, X., Deng, H., Li, Y., Xiang, Y., Zhu, Y.: An efficient approach for mitigation of efflorescence in fly ash-based geopolymer mortars under high-low humidity cycles. Constr. Build. Mater. **317**, 126159 (2022)
5. Liu, X., Gao, H., Fryda, H., Wang, P.: Surface whitening mechanism of ettringite rich decorative mortar. Cem. Concr. Res. **189**, 107786 (2025)
6. Ma, H., Wang, R., Xi, Z., Gao, H., Fryda, H., Wang, P.: Inhibiting effect of styrene-butadiene copolymer on the whitening of calcium aluminate cement mortar. Constr. Build. Mater. **404**, 133030 (2023)
7. Ma, H., Wang, R., Zhang, S., Xu, L., Wang, P.: Synthesis of silane-modified styrene-acrylic ester emulsion and its inhibition mechanism on whitening of calcium aluminate cement mortar. Constr. Build. Mater. **450**, 138600 (2024)

8. Liang, K., Cui, K., Sabri, M.M.S., Huang, J.: Influence factors in the wide application of alkali-activated materials: a critical review about efflorescence. Materials. **15**, 6436 (2022)
9. Alex, A.G., Jose, P.A., Saberian, M., Li, J.: Green pervious concrete containing diatomaceous earth as supplementary cementitous materials for pavement applications. Materials. **16**, 48 (2022)
10. Hu, Z., Wen, X., Zheng, S., Yin, J., Shen, X., Zhang, T.: Regulating pore structure of diatomite with alkali dissolution and its influence on humidity control performance. Ceram. Int. **50**, 42322–42332 (2024)

Exploring an Integrated Spatial Analysis Using a Sampling Survey Model to Accurately Estimate Waste in Mountainous Areas

Chin Hsien Liao(✉)

Department of Land Resources, Chinese Culture University, Taipei, Taiwan
ljx15@ulive.pccu.edu.tw

Abstract. Illegal dumping of waste in mountainous areas seriously threatens our environment. Estimating the volume of waste generated in mountainous areas is not as straightforward as calculating waste volumes in urban areas, which typically rely on population figures. Due to considerable variations in land use development across mountainous terrains, actual waste volumes can differ significantly, making accurate quantification challenging. This study focuses on the Chiayi region of Taiwan, aiming to establish an integrated spatial analysis using a sampling survey model of land use intensity to estimate waste generation in mountainous areas more precisely. This sampling survey model defines "high-intensity land use" based on the seven categories of land use identified in the current land use survey. The findings indicate that it is feasible to identify potential sites for waste generation in mountainous areas, with the volume of waste produced correlating to land area; larger areas tend to generate greater amounts of waste. Meishan Township exhibited the highest waste production in these mountainous areas at 68.78 metric tons daily, while Jhuzaki Township recorded the lowest at 20.58 metric tons daily. Furthermore, an analysis of various high-intensity land use types revealed that residential areas generated the most significant daily waste amount, totaling 78.2 metric tons, followed by recreational facilities, which produced 9.4 metric tons. This method assesses the fixed scale of solid waste in mountainous areas.

Keywords: Illegal Dumping · Spatial Analysis · Sampling Survey Model · Land Use Intensity · Mountainous Areas

1 Introduction

The challenging mountainous terrain, characterized by steep inclines and narrow, winding roads, presents significant obstacles for residents in these areas. Consequently, many individuals resort to indiscriminate waste disposal in remote mountain valleys, neglecting regulations that mandate waste transport to designated collection points for proper disposal by municipal sanitation services. Over time, this careless practice has led to the troubling phenomenon of "illegal waste dumping," where discarded trash accumulates and gradually tumbles down the mountainsides, creating unsightly and hazardous conditions. In response to such pressing issues, the United Nations Environment Program (UNEP) published the "Global Waste Management Outlook" in 2015, presenting a

Y. Zhu (Ed.): ISESIE 2025, LNCE 000, pp. 11–20, 2026.
https://doi.org/10.1007/978-981-92-2389-3_2

comprehensive array of policy recommendations aimed at delivering practical solutions for waste management in specific regions, particularly in mountainous areas [1–3]. The environmental consequences of such waste disposal practices in these unique ecosystems are profound, and thorough research is necessary to understand their full impact. Mountainous communities nestled in mountainous areas often find themselves inadequately equipped with waste removal resources, resulting in a cycle of illegal dumping and open burning of refuse [4–6]. While numerous studies have delved into the ecological ramifications of waste, there remains a significant gap in research that estimates waste generation—particularly concerning the types and intensities of land use. Most waste management discussions tend to concentrate on urban settings, leaving the challenges faced by mountainous areas largely overlooked [7]. Given the absence of reliable statistical data on waste generation in mountainous areas, managing and understanding the volumes of waste produced and removed has proven to be a complex task [8–10]. This study aims to tackle the pressing waste management challenges in mountainous areas by focusing on the Chiayi region of Taiwan. It employs a sampling survey model centered around land use intensity to develop an integrated spatial analysis for more accurate waste generation estimates. Furthermore, by collecting case data, the research enhances the reliability of the model's predictions.

2 Materials and Methods

2.1 Traditional Assessment Methods for Solid Waste

In estimating the demand for waste collection services over the past three years, demographic data from the Social and Economic Data Service Platform was combined with geographic information to spatially analyze the population distribution in urban areas PU_i and the mountainous regions PM_i Multiplying the population distribution ratio by the estimated solid waste generation rate *Wi*, an calculate the waste generation rates for these regions. If this figure is divided by 365 days, it provides the estimated daily solid waste generation volume SW for mountainous and urban areas.

$$\mathrm{SW} = \Sigma_{\mathrm{i}} \left(\mathrm{PU_i} \times \mathrm{W_i}\right) + \left(\mathrm{PM_i} \times \mathrm{W_i}\right) \tag{1}$$

The abovementioned formula fails to account for the differences between urban and mountainous environments, particularly regarding land use types and population distribution. It estimates total waste volume based solely on the total population of urban and mountainous areas, without identifying specific collection locations or total collection volumes for mountainous areas. This oversight affects the frequency of waste truck collections and the allocation of manpower. As a result, inadequate waste collection support in mountainous areas has contributed to the ongoing issue of illegal waste disposal in these regions.

2.2 Identify the Locations of Solid Waste Generated in Mountainous Areas

According to the 2014 Land Use Status Survey data, land use is divided into seven primary categories that contribute to solid waste generation: residential areas, commercial establishments, government agencies, social welfare facilities, recreational facilities, schools, and healthcare facilities. These categories are collectively referred to as "high-intensity land use" in this study. As illustrated in Fig. 1.

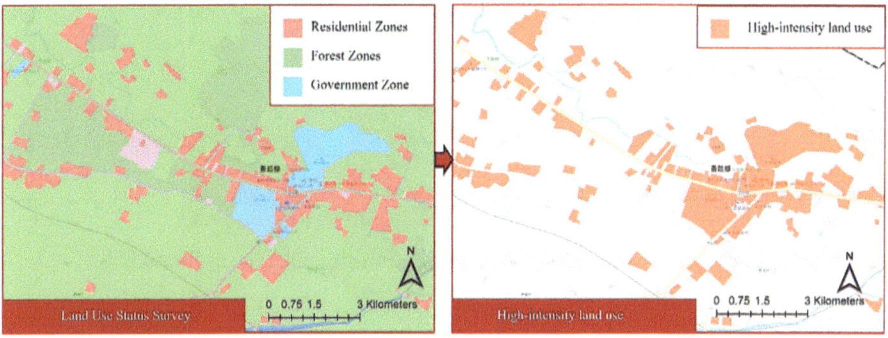

Fig. 1. Illustration of "high-intensity land use" as the primary source of solid waste in Fanlu.

Based on the estimated waste generation quantities for flatland and mountainous areas within each township for 2020–2021, this study classifies the current land use as "high-intensity land use." A corresponding "future waste removal demand per unit area" (measured in kilograms per hectare) has also been assigned.

2.3 Define the Basic Units for Waste Collection and Assign Corresponding Codes

Through interviews, current waste collection planning in mountainous areas primarily uses small-scale settlements as the basic units for establishing waste collection hubs. Therefore, it is important to evaluate whether these "high-intensity land use areas" are suitable to serve as basic waste collection units. This study references the fundamental principles for designating national land conservation areas and agricultural development areas: if the shortest distance between two "high-intensity land use" areas does not exceed 25 m, they will be classified as the same spatial unit using GIS (Aggregate Polygon function). This classification will then be used as the basic unit for waste collection demand in subsequent planning. I successfully transformed 13,703 scattered high-intensity land uses into a streamlined total of 6,709 basic units for waste collection using advanced GIS technology. This significant conversion not only enhances clarity but also optimizes land management efficiency. As illustrated in Fig. 2.

$$SW = \sum_{i \sim j=1}^{n} \left(U_{it} \times W_i + R_{jt} \times W_j \right) \tag{2}$$

SW = basic units for waste collection in hihg intensity land use areas

U_{it}= Area of urban land use within the i unit
R_{jt}= Area of mountainous land use within the j unit
W_i=Rate of Solid Waste by different types of urban land use
W_j= Rate of Solid Waste by different types of mountainous land use

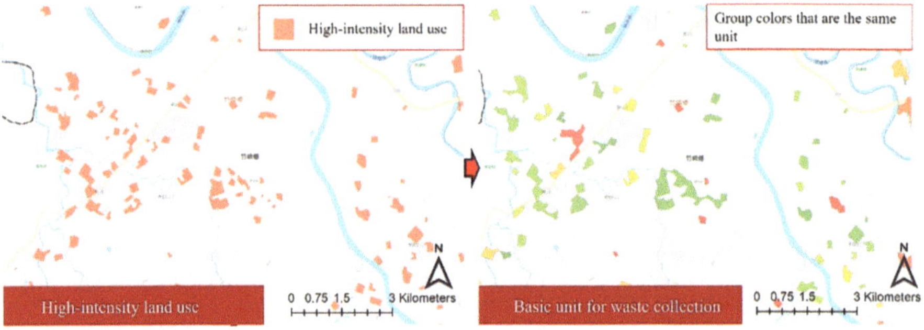

Fig. 2. Illustration of the basic units for waste collection in Jhuzaki.

2.4 Develop a Coding System to Outline High-Intensity Land Use

To estimate waste generation from high-intensity land use, we start by calculating the estimated amount of solid waste produced per unit area for each type of land use. This figure is then multiplied by the area associated with each high-intensity land use type, as identified in the national land use survey, resulting in the total estimated amount of solid waste generated. Next, we will conduct a statistical estimation of waste generation quantities for each basic waste collection unit. Using the "Feature to Point" tool in GIS, we convert the estimated quantities of solid waste for each high-intensity land use area from spatial data into point data. These quantities are then aggregated for each basic waste collection unit. This methodology lays a solid foundation for analyzing the supply and demand dynamics of waste collection services in future evaluations. As illustrated in Fig. 3.

Fig. 3. Basic waste collection demand unit: Estimated volume of solid waste (kg/day) in Jhuzaki.

3 Results and Discussion

3.1 Estimated Volume of Solid Waste Produced in Chiayi County

Over the past decade, data from Chiayi County was used to create a linear function that estimates the potential amount of solid waste the county may generate in the future. It is projected that 194,756 metric tons of solid waste will be generated in 2020, with resource waste being the primary type. The average monthly per capita generation of solid waste is estimated to be 1.073 kg. The estimated results for the period from 2020 to 2022 are summarized in the table below. In 2020, Chiayi County generated 197,386 metric tons of solid waste, ranking 15th in Taiwan. The average daily per capita waste generation was 1.08 kg, placing it in 18th position among counties and cities in the country. This indicates that Chiayi County produces relatively less waste compared to other regions in Taiwan. A closer look at the trends in Chiayi County shows some fluctuations in solid waste generation over the years. However, the overall trend has been upward, with waste increasing from 194,756 metric tons in 2020 to 197,355 metric tons in 2022. Based on current trends, it is anticipated that the amount of solid waste generated in Chiayi County will continue to rise over the next 2–3 years. Additionally, the composition of solid waste in Chiayi County has gradually shifted from mainly general solid waste to a higher proportion of recyclable materials.

3.2 Evaluation of Solid Waste in the Mountainous Areas

Using historical data on solid waste generation and population, this study calculates the average annual per capita waste generation for the six townships in the surveyed area. Zhuzhi Township and Fanlu Township have higher averages, generating approximately 0.32–0.33 metric tons of waste per person annually. Zhongpu Township and Meishan

Township fall in the middle range, with an average of 0.28 metric tons per capita. In contrast, Dapu Township and Alishan Township have lower averages, ranging from 0.18 to 0.22 metric tons per person. The historical data is summarized as follows in Table 1.

Table 1. Annual average amount of solid waste generated per person annually in six townships of Chiayi County (metric tons).

Townships	2016	2017	2018	2019	2020
Zhongpu	0.28	0.27	0.26	0.27	0.28
Jhuzaki	0.33	0.34	0.38	0.36	0.36
Meishan	0.27	0.29	0.32	0.32	0.34
Fanlu	0.35	0.36	0.38	0.36	0.37
Tai Po	0.20	0.20	0.21	0.20	0.20
Alishan	0.20	0.19	0.22	0.27	0.26
Total	0.29	0.30	0.31	0.31	0.32

Between 2016 and 2020, data from six townships in Chiayi County show that the fluctuations in the amount of solid waste generated have remained relatively stable. Overall, there has been a gradual increase in solid waste generation, totaling only 6.4%. This suggests that efforts to control waste generation have been relatively effective, and waste reduction campaigns have seen some success. However, this may also indicate that the township is facing a significant issue with "the problem of illegal dumping of waste."

3.3 A Sampling Survey Model of Land Use Intensity

This study defines "high-intensity land use" based on the seven categories of land use identified in the current land use survey: residential, commercial, government agencies, social welfare facilities, recreational facilities, schools, and healthcare facilities. After analyzing the data, we identified a total of 13,703 plots of high-intensity land use across the six townships, covering an area of 1,939.63 ha. The primary forms of high-intensity land use are residential and commercial. In terms of area, residential use comprises the largest share at 1,259.73 ha, followed by recreational facilities at 463.19 ha. Concerning the number of parcels, residential use holds the highest count with 11,820 parcels, while commercial use has 1,380 parcels. The remaining high-intensity land uses include various government agencies, schools, healthcare facilities, and social welfare facilities. Overall, high-intensity land use is predominantly characterized by residential and commercial categories.

This study estimates the annual average value of solid waste from 2020 to 2022. It is based on the total yearly waste collection volumes for six townships from 2011 to 2020. The study also evaluates the solid waste collection standards for each township's "along-route collection" and "fixed-point collection" volumes from 2021 to 2022. This evaluation relies on the proportion of these two collection methods observed in 2020.

Furthermore, the study aims to estimate the solid waste collection standards for "high-intensity land use areas within urban planning zones" and "high-intensity land use areas outside urban planning zones" for each township, measured in kilograms per day. The estimated solid waste collection requirements per unit area for high-intensity land use areas in each township's urban and non-urban regions are detailed as follows in Table 2.

Table 2. Estimation of solid waste from intensive land use.

Townships	Type	Average estimated daily total amount (kg/day) for 2021–2022	Area of land used intensively (ha)	Estimated waste per unit area of land used intensively (kg/ha/day)
Zhongpu	Urban	22,342.64	142.02	157.32
	Mountainous	8,124.98	394.73	20.58
Jhuzaki	Urban	12,471.12	51.28	243.21
	Mountainous	24,144.37	394.59	61.19
Meishan	Urban	6,162.19	69.92	88.13
	Mountainous	10,871.10	158.05	68.78
Fanlu	Urban	1,343.23	286.20	4.69
	Mountainous	11,284.58	128.83	87.59
Tai Po	Urban	927.22	195.64	4.74
	Mountainous	1,909.61	23.41	81.56
Alishan	Urban	0.00	7.97	37.97*
	Mountainous	3,302.55	86.98	37.97
Total	Urban	43,246.40	753.03	57.43
	Mountainous	59,637.20	1,186.60	50.26

The results indicated that from 2021 to 2022, the estimated daily total amount of general waste generated in the six townships of the mountainous area of Chiayi was 57.43 metric tons per day in the urban planning area and 50.26 metric tons per day in the mountainous area. Among the urban planning areas, Jhuzaki Township produced the highest waste at 243.21 metric tons per day, while Fanlu Township generated the lowest at 4.69 metric tons per day. In the mountainous areas, Meishan Township had the highest waste production at 68.78 metric tons per day, whereas Jhuzaki Township had the lowest at 20.58 metric tons per day. Additionally, when examining various high-intensity land use types, residential areas produced the largest daily waste amount at 78.2 metric tons, followed by recreational facilities at 9.4 metric tons. Government agencies generated the least amount of waste, with only 0.76 metric tons per day.

3.4 Spatial Analysis of Solid Waste from Intensive Land Use

This study employs GIS to assess the total daily waste generated by each basic waste collection unit, categorizing the waste into five ranges: under 1 kg, 1–10 kg, 10–50 kg, 50–100 kg, and over 100 kg. The statistics reveal that 65% (or 4,348 units) of the collection units produce the majority of their waste within the 1–10 kg range, with the 10–50 kg range following closely. The units that generate over 100 kg daily are predominantly located in Jhuzaki and Zhongpu townships. This assessment provides a general overview of waste collection demand by townships (measured in kg/day). As illustrated in Fig. 4.

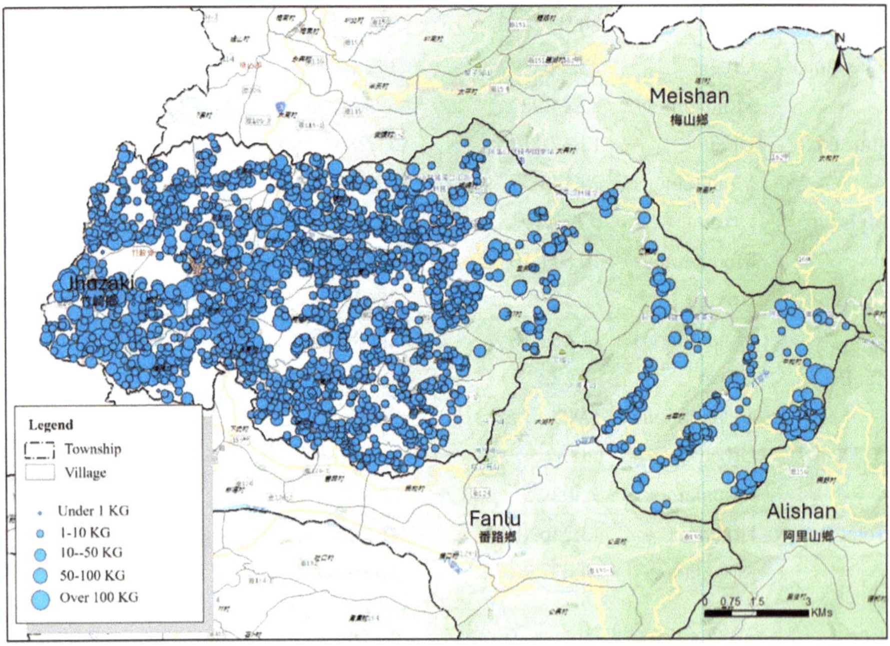

Fig. 4. Estimated volume of solid waste (kg/day) in the Jhuzaki.

4 Conclusion

This study focuses on data collected from the mountainous areas of Chiayi County, Taiwan, over the past decade. By using traditional assessment methods for solid waste, we estimated the potential amount of general waste that may be generated in these mountainous areas in the future. Our projections suggest that 194,756 metric tons of solid waste will be produced in 2021, with resource waste being the primary component. The average monthly per capita generation of general waste is estimated to be 1.073 kg. The results of actual verification indicate that traditional methods are only suitable for certain regions with stable population structures. This analytical model demonstrates better predictive capability for Zhongpu Township, Jhuzaki Township, Fanlu Township, and Dapu Township, while its predictive capacity for Meishan Township and Alishan Township is not significant. To more accurately estimate waste generation in mountainous areas, this study employs a sampling survey model based on land use intensity.

This study estimates potential locations for solid waste and land use areas to calculate the amount of waste produced, operating under the principle that area is proportional to waste generation. This method serves to assess the fixed scale of solid waste in each village. However, it is currently limited by the availability of basic data, which prevents verification of the difference between estimated quantities and actual conditions.

Future recommendations include approaches to establish calibration parameters that reconcile estimated quantities with actual data, thereby improving the accuracy of the estimations. Additionally, this study conducted a sampling survey of land use patterns and waste generation quantities in areas where general waste is generated, identifying solid waste as waste other than industrial waste. Nonetheless, due to this definition, it remains challenging to pinpoint the actual locations of solid waste generation. Therefore, it is recommended that future surveys focus on selected waste collection points to investigate the land use patterns and waste generation quantities of surrounding areas, thus enhancing the accuracy of the estimates.

Conflict of Interest The authors declare no conflict of interest.

Author Contributions Chin Hsien Liao: conceptualization, data collection and analysis, investigation, visualization, writing—original draft preparation.

Acknowledge The author is grateful to the editor and anonymous reviewers whose valuable comments and insightful remarks have led to widespread improvement to the earlier versions of the manuscript.

Funding The present paper results from research project No.112–2410-H-034-046. The National Science and Technology Council financed the research by kindly providing me with information and assistance.

References

1. Alfthan, B. et al.: Waste Management Outlook for Mountain Regions. (2016)
2. Semernya, L., Ramola, A., Alfthan, B., Giacovelli, C.: Waste management outlook for mountain regions: sources and solutions. Waste Manag. Res. **35**(9), 935–939 (2017)
3. Neupane, S., Pinto, H., Pintassilgo, P.: Mountain tourism stakeholders' perspectives on waste management: the case of Everest in Nepal. Tour. Plan. Dev. **22**(2), 207–227 (2025)
4. Dabholkar, A., Muthiyan, B., Srinivasan, S., Ravi, S., Jeon, H., Gao, J.: Smart illegal dumping detection. In: 2017 IEEE Third International Conference on Big Data Computing Service and Applications (BigDataService), pp. 255–260 (2017)
5. Mihai, F.-C., Grozavu, A.: Role of waste collection efficiency in providing a cleaner mountainous environment. Sustainability. **11**(23), 6855 (2019)
6. Du, L., Zuo, J., Vanzo, J., Chang, R., Zillante, G.: Assessing and predicting the illegal dumping risks in relation to road characteristics. Waste Manag. **169**, 332–341 (2023)

7. Liao, C.H.: A method for evaluating spatial characteristics of illegal garbage dumping issues in mountain villages of Taiwan using exploratory special analysis. IJESD. **15**(4), 199–203 (2024)
8. Adhikari, S. et al.: Solid waste management in rural touristic areas in the Himalaya–a case of Ghandruk, Nepal. Habitat Int. **143**, 102994 (2024)
9. Mir, A.A., Mushtaq, J., Dar, A.Q., Patel, M.: A quantitative investigation of methane gas and solid waste management in mountainous Srinagar city-a case study. J. Mater. Cycles Waste Manag. **25**(1), 535–549 (2023)
10. Shahab, S., Anjum, M.: Solid waste management scenario in India and illegal dump detection using deep learning: an AI approach towards the sustainable waste management. Sustainability. **14**(23), 15896 (2022)

Valorization of Calcium Hydroxide-Rich Industrial Solid Wastes as Multifunctional Admixture for Enhanced Performance in Portland Cement-Fly Ash-Slag Systems

Ran Zhuang(✉), Hongbing Chen, Hai Shi, Zhaoqiang Ji, and Jiajun Shi

NO.2 Engineering CO., LTD of CCCC·Third Harbor Engineering CO., LTD, Shanghai 200122, China

zhuangran5269@sina.com

Abstract. To address the slow early-age strength development common in ternary binder systems, this study developed a waste-derived Composite Functional Admixture (CPA) from calcium hydroxide-rich industrial solid wastes to enhance the performance of ternary Portland cement-fly ash-blast furnace slag (PC-FA-BFS) systems. By simultaneously supplying reactive $Ca(OH)_2$ and anhydrite, CPA significantly accelerated early FA/BFS dissolution and hydration, yielding > 6.5% improvement in 1–3 day flexural/compressive strength and refined 3-day paste pore structure. Long-term analyses confirmed CPA maintained 56-day mechanical properties and hydration products, achieving comparable pore refinement (most probable pore size: 17.7–46.2 nm) to controls. Crucially, sulfate resistance exhibited particle-size dependency: Under 5% $MgSO_4$ attack, CPA-D demonstrated 24% higher corrosion resistance than controls, outperforming CPA-B. Mechanistic studies revealed CPA-D's superiority stemmed from its composition/size-modulated formation of stable sulfate phases (e.g., ettringite), enhancing durability. The CPA technology delivers triple sustainable advantages: industrial waste valorization, early-age performance enhancement, and cost efficiency, offering a viable pathway for high-stability green construction materials.

Keywords: Industrial Solid Wastes · Composite Cementitious Materials · Calcium Hydroxide · Mechanical Properties · Sulfate Corrosion Resistance

1 Introduction

The utilization of industrial solid wastes as supplementary cementitious materials (SCMs), such as fly ash (FA) and blast furnace slag (BFS), is a well-established strategy to enhance the sustainability of concrete production by reducing Portland cement (PC) consumption and landfilling burdens [1–3]. However, a significant challenge associated with high-volume FA/BFS incorporation is their inherently low early-age reactivity [4]. These SCMs rely primarily on the portlandite ($Ca(OH)_2$) generated during PC hydration for their pozzolanic/hydraulic activation, leading to delayed strength development

Y. Zhu (Ed.): ISESIE 2025, LNCE 000, pp. 21–29, 2026.
https://doi.org/10.1007/978-981-92-2389-3_3

and potentially compromised early-age performance [1, 5]. However, chemical activators like commercial $Ca(OH)_2$ or gypsum can accelerate SCM reaction [6, 7], but increases cost and environmental footprint, counteracting the sustainability benefits of SCM utilization.

To address this challenge, this study proposes an innovative approach centered on the valorization of calcium hydroxide-containing industrial solid wastes as a novel Composite Functional Admixture (CPA). The core concept is to utilize these specific wastes not merely as inert fillers or partial SCM replacements, but as multi-component activators designed to stimulate the early hydration of FA and BFS within blended cement systems. The CPA employed comprises calcium hydroxide-rich solid wastes with carefully selected, differing particle sizes. Its dual-functional mechanism involves: (1) Providing soluble calcium hydroxide as an internal source of readily available $Ca(OH)_2$ to directly promote the dissolution and pozzolanic reaction of FA and BFS. (2) Incorporating anhydrite ($CaSO_4$), which reacts with aluminate phases to form early ettringite, thereby contributing to early stiffening, refining the pore structure, and enhancing the mechanical properties and sulfate resistance of the cementitious matrix. This strategy offers distinct sustainability advantages over conventional practices. The findings aim to demonstrate the viability and effectiveness of calcium hydroxide-rich industrial solid wastes as a high-performance, multi-functional CPA, providing a sustainable solution for advanced cementitious materials.

2 Experiment

2.1 Materials

Control Group: Portland cement (PC, Type P·I 42.5, supplied by Conch Cement Plant) was blended with fly ash (FA) and blast furnace slag (BFS). Experimental Groups: Two types of calcium hydroxide-containing solid waste (functioning as a composite functional admixture, CPA) with different particle sizes partially replaced both FA and BFS simultaneously. The chemical (XRF) and mineralogical (XRD) compositions of the raw materials are presented in Table 1 and Fig. 1, respectively. Their particle size distributions (determined by laser diffraction) are shown in Fig. 2. Tap water was used in this study and mortar mixtures were prepared using standard sand.

Table 1. Oxide composition of cementitious materials (%).

Oxides	CaO	SiO_2	Al_2O_3	MgO	SO_3	Cl^-	Fe_2O_3	Others
FA	3.15	44.87	44.34	0.58	0.86	0.02	2.80	3.40
PC	61.09	22.67	4.82	2.12	3.58	0.06	4.40	1.26
BFS	36.28	34.39	17.80	7.16	2.20	0.06	0.59	1.53
CPA-B	44.38	26.60	1.03	1.80	24.98	0.03	0.66	0.53
CPA-D	45.30	28.43	0.60	1.51	22.85	0.02	0.81	0.49

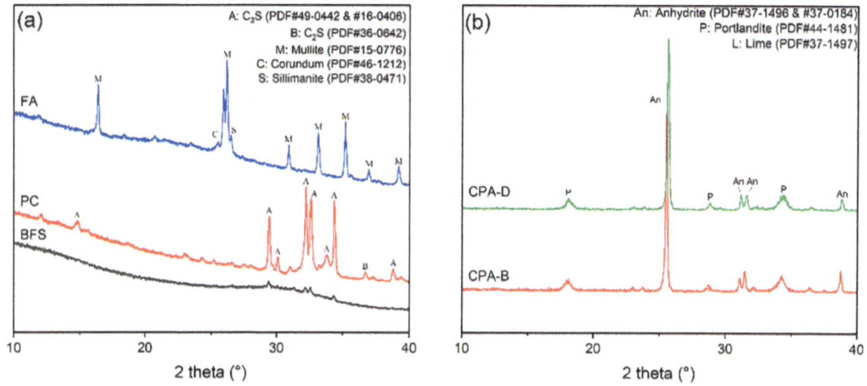

Fig. 1. XRD spectra of raw materials (a) Portland cement, fly ash and blast furnace slag and (b) CPA composite functional admixture.

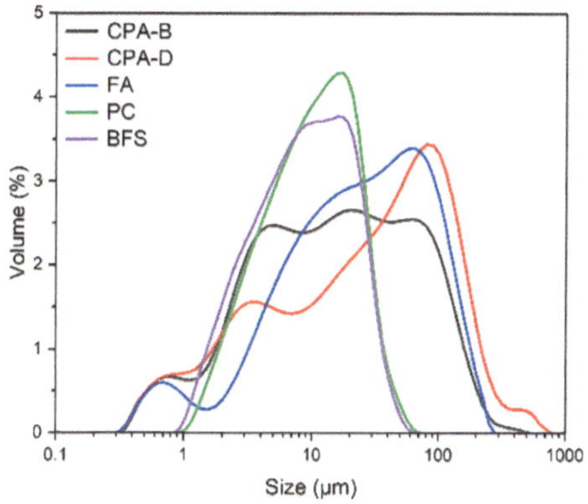

Fig. 2. Particle size distribution of Portland cement, fly ash, blast furnace slag and CPA composite functional admixture materials.

2.2 Mix Ratio and Sample Preparation

Mix Preparation: Paste and mortar mixtures were prepared according to Table 2. Both mixtures employed a constant water-to-binder ratio of 0.5, while the mortar had a sand-to-binder ratio of 3.

Paste Specimen Preparation: Paste mixtures were prepared following the Chinese National Standard GB/T 1346–2011 ("*Test Methods for Water Requirement of Normal Consistency, Setting Time and Soundness of the Portland Cement*"). The mixed paste was immediately cast into six-cube molds (20 × 20 × 20 mm) and subsequently cured 24 h under the same conditions of (20 ± 2) °C and (95 ± 1) % RH. After demolding, the mortar specimens were cured in a constant temperature humidity chamber at 50 °C for 7 days (Fig. 3a).

Curing Regime: After preparation, all mortar specimens were cured under controlled conditions of (20 ± 2) °C and (95 ± 1) % RH until reaching their designated testing ages.

Table 2. Mix ratio (%).

Groups	PC	FA	BFS	CPA-B	CPA-D
R	50	25	25	0	0
B10	50	20	20	10	0
D10	50	20	20	0	10

2.3 Test

Mechanical Testing: Mortar specimens were prepared and tested for mechanical strength strictly following the Chinese National Standard GB/T 17671–2021 ("*Test Methods for Cement Mortar Strength (ISO Method)*").

Sulfate Resistance Testing: Mortar specimens were prepared and tested for sulfate resistance in accordance with the Chinese National Standard GB/T 749–2008 ("*Test Method for Resistance of Cement to Sulfate Attack*"). Subsequently, they were divided into four groups of nine specimens each. Each group was immersed in one of the following solutions at 20 °C for 28 days: deionized water (DW), 3 wt% Na_2SO_4 solution, 5 wt% Na_2SO_4 solution, and 5 wt% $MgSO_4$ solution (Fig. 3b). Phenolphthalein was used as an indicator. Dilute sulfuric acid (H_2SO_4) was added sparingly to the sulfate solutions every other day to maintain a near-neutral pH. After the immersion period, the flexural strength (R) of the mortar specimens was measured using a universal testing machine (UTM). The corrosion resistance coefficient (K) was then calculated.

The oxide composition of the raw materials was determined using a PANalytical® X-ray fluorescence (XRF) spectrometer. The mineralogical composition of both the raw materials and the hydrated pastes was analyzed using a Rigaku® X-ray powder diffractometer (XRD). The particle size distribution of the raw materials was measured via laser diffraction analysis employing an LS230 laser particle size analyzer.

Pore Structure Analysis: Paste samples were cut into fragments approximately 5 mm in diameter using pliers. The pore structure parameters of these fragments were then measured using a Micromeritics® AutoPore V 9600 mercury intrusion porosimeter (MIP).

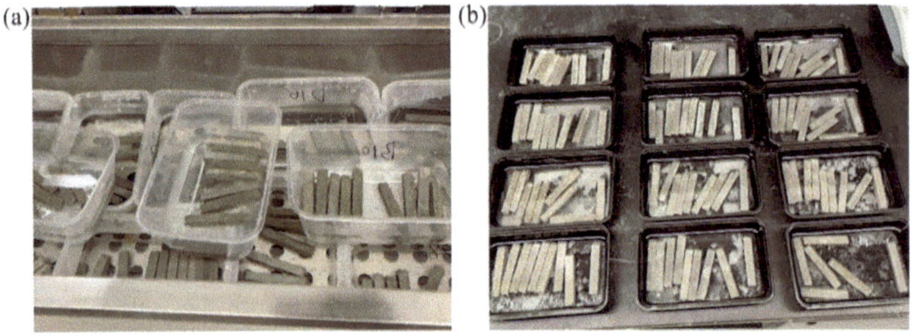

Fig. 3. (a) Illustration of corrosion of a sample cured by damp heat and (b) in sulfate solution.

3 Results and Discussion

3.1 Mechanical Properties

Flexural strength results are shown in Fig. 4 (a). The flexural strength of R (the fly ash-blast furnace slag-Portland cement mortar, FA-BFS-PC mortar) was only 1.8 MPa at 1 d. With curing age increasing, the flexural strength increased, reaching 8.0 MPa at 56 d. This behavior is attributed to the low early hydration reactivity of FA and BFS, which require portlandite generated from PC hydration for activation, leading to accelerated hydration at later ages [8].

Incorporating the two CPA variants with different fineness enhanced the flexural strength of the mortar. The enhancement was more pronounced with CPA-B; specifically, 8% CPA-B increased the flexural strength by 6.5% 28 d, whereas 10% CPA-D yielded only a 1.3% increase. Notably, the strength advantage of CPA-containing mortars over the control diminished by 56 d.

Compressive strength results are shown in Fig. 4 (b). Due to the low hydraulic activity of FA and BFS, the compressive strength of the mortar was merely 5.2 MPa at 1 d, but reached 45 MPa at 56 d. Incorporating CPA improved the compressive strength, with significant enhancements observed at early ages (1 d and 3 d). However, the difference between the CPA-modified mortars and the control became marginal at later ages (28 d and 56 d). These results demonstrate that CPA primarily promotes early-age hydration, thereby boosting mechanical properties during the initial curing period.

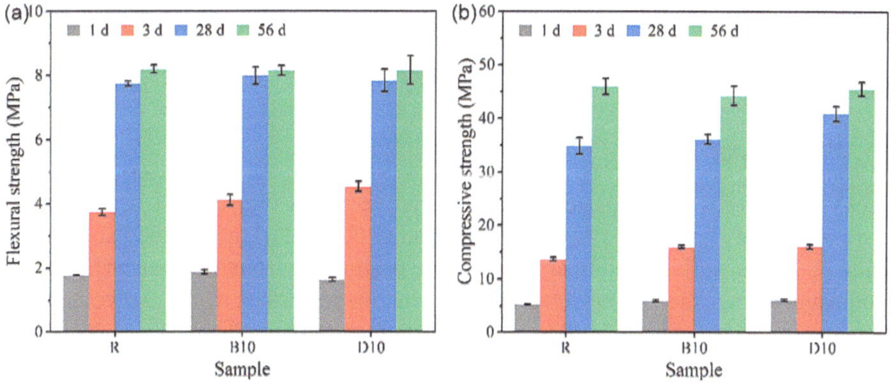

Fig. 4. (a) Flexural strength and (b) compressive strength of the samples at 1, 3, 28, and 56 days.

3.2 Pore Structure

The pore size distributions of the paste samples measured by MIP are shown in Fig. 5. At 1 d, the pore size distributions of the experimental group (with CPA-B) and the control group were similar, with pore sizes predominantly distributed between 1.19 and 1.83 μm. Due to the slow hydration of FA and BFS, the consumption rate of portlandite in the pore solution of the control paste was low, resulting in relatively lower porosity.

Incorporating 10% CPA-B increased the paste porosity above that of the control, indicating that CPA-B promoted the dissolution of FA and BSF. By 3 d, significant differences in pore size distribution emerged. The addition of 10% CPA-B reduced the porosity from 46.2 to 45.1% and decreased the most probable pore size to approximately 0.55 μm, demonstrating that CPA-B refined the pore structure at this stage. At 28 d, the difference in pore size distribution between the two groups substantially diminished, and the pore structure underwent significant further refinement compared to the 3 d state. Both pastes exhibited reduced porosity, and the most probable pore size decreased to the range of 17.7–46.2 nm. The paste containing 10% CPA-B showed a slightly broader pore size distribution than the control, suggesting its pore structure was marginally less refined at later ages. These results confirm that CPA-B promotes the early hydration of the FA-BFS-PC system. Critically, it imposes no adverse effects on the long-term pore structure refinement of the paste.

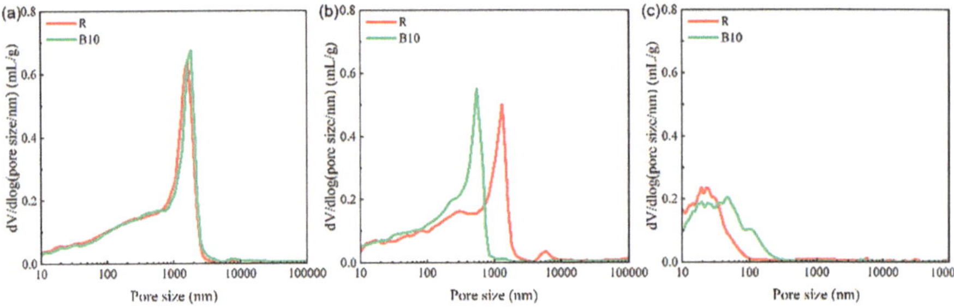

Fig. 5. Pore size distribution of samples cured to (a) 1 d, (b) 3 d, and (c) 28 d.

3.3 Phase Composition

The XRD patterns of the paste samples are presented in Fig. 6. At 1 d, the control paste contained significant amounts of portlandite ($Ca(OH)_2$), unhydrated alite (C_3S), belite (C_2S), as well as mullite and sillimanite originating from the fly ash (FA). With prolonged curing, the content of portlandite increased further, while the contents of alite and belite decreased. Mullite and sillimanite remained chemically stable in the alkaline environment, and thus their contents showed no significant change. The phase composition of the hydration products in the paste incorporating 10% CPA-B exhibited minimal differences compared to the control at all tested ages. This indicates that CPA-B does not alter the intrinsic hydration product assemblage of the FA-BFS-Portland cement system.

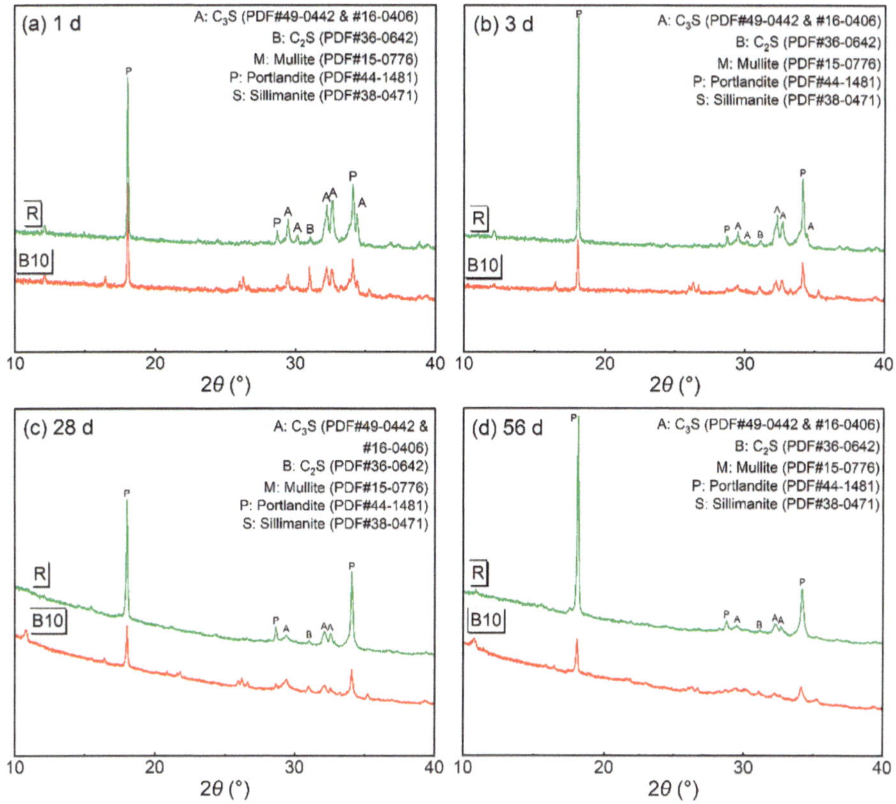

Fig. 6. XRD test results of samples cured to (a) 1 d, (b) 3 d, (c) 28 d, and (d) 56 d.

3.4 Sulfate Corrosion Resistance

The test results of sulfate corrosion resistance of the mortar are shown in Fig. 7. When exposed to 3 wt% Na_2SO_4 solution, the corrosion resistance coefficient (K value) of all mortar specimens increased. The K value of the mortar containing 10% CPA-B was 2% lower than the control. In contrast, the mortar with 10% CPA-D exhibited a 14% increase in sulfate resistance. This behavior can be attributed to the dual effect of sulfate ions in low-concentration solutions: while they disrupt the existing gel structure, they simultaneously promote the formation of stable sulfate phases (e.g., ettringite) [9]. These phases fill the pores, thereby enhancing flexural strength. Under exposure to 5 wt% Na_2SO_4 solution, the flexural strength of all groups decreased. This degradation is due to the conversion of portlandite to low-strength gypsum ($CaSO_4{\cdot}2H_2O$) [10], which compromises internal cohesion. Mortars incorporating 10% CPA-B or CPA-D showed further reduced sulfate resistance compared to the control, although the reduction was less pronounced for CPA-D. Exposure to 5 wt% $MgSO_4$ solution also led to a decrease in flexural strength for all groups. However, the mortar with 10% CPA-D demonstrated a significantly higher K value, exceeding the control by 24%.

These results indicate that CPA can enhance the sulfate resistance of FA-BFS-PC mortars to some extent. The mechanism involves the high content of portlandite and anhydrite ($CaSO_4$) in CPA, which promotes the dissolution of FA and BFS, elevating the hydration degree and refining the pore structure (reducing porosity), consequently improving sulfate resistance. Notably, CPA-D outperformed CPA-B in enhancing sulfate resistance.

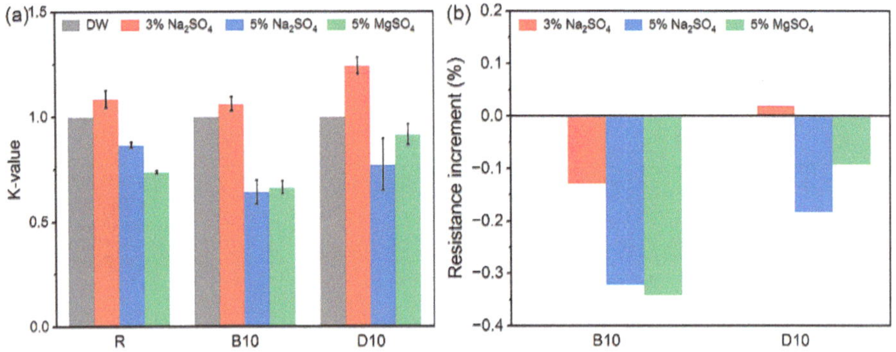

Fig. 7. The test results of sulfate corrosion resistance of the mortar: (a) the K value of the sample and (b) the proportion of the increase in K value relative to sample R1.

4 Conclusions

This study demonstrates that a Composite Functional Admixture (CPA), derived from calcium hydroxide-containing industrial solid wastes, significantly enhances the performance of FA-BFS-Portland cement systems. By synergistically supplying reactive $Ca(OH)_2$ and anhydrite, CPA accelerates early FA/BFS dissolution and hydration, boosting early-age mechanical strength and refining the paste pore structure. Long-term curing confirms CPA maintains later-age strength and hydration products, achieving equivalent pore refinement to controls. Crucially, sulfate resistance exhibits particle-size dependency: CPA-D outperforms CPA-B, particularly under $MgSO_4$ attack, due to its unique composition/size-modulated formation of stable sulfate phases. This waste-derived CPA offers multifunctional merits—waste valorization, performance enhancement, and cost efficiency—providing an innovative pathway for sustainable cementitious materials.

References

1. Giergiczny, Z.: Fly ash and slag. Cem. Concr. Res. **124**, 105826 (2019)
2. Kolani, B., Buffo-Lacarrière, L., Sellier, A., Escadeillas, G., Boutillon, L., Linger, L.: Hydration of slag-blended cements. Cem. Concr. Compos. **34**, 1009–1018 (2012)
3. De Weerdt, K., Ben, H.M., Le Saout, G., Kjellsen, K.O., Justnes, H.: Lothenbach B.: hydration mechanisms of ternary Portland cements containing limestone powder and fly ash. Cem. Concr. Res. **41**, 279–291 (2011)

4. He, P., Yu, J., Xue, L., Han, X.: Influence of ion chelator on pore structure, water transport and crack-healing properties of cement pastes incorporating high-volume fly ash and blast-furnace slag. J. Build. Eng. **55**, 104696 (2022)
5. Nedunuri, S.S.S.A., Sertse, S.G., Muhammad, S.: Microstructural study of Portland cement partially replaced with fly ash, ground granulated blast furnace slag and silica fume as determined by pozzolanic activity. Constr. Build. Mater. **238**, 117561 (2020)
6. Jeong, Y., Oh, J.E., Jun, Y., Park, J., Ha, J., Sohn, S.G.: Influence of four additional activators on hydrated-lime [ca(OH)2] activated ground granulated blast-furnace slag. Cem. Concr. Compos. **65**, 1–10 (2016)
7. Yong, C.L., Mo, K.H., Koting, S.: Phosphorus slag in supplementary cementitious and alkali activated materials: a review on activation methods. Constr. Build. Mater. **352**, 129028 (2022)
8. Li, X., Wanner, A., Hesse, C., Friesen, S., Dengler, J.: Clinker-free cement based on calcined clay, slag, portlandite, anhydrite, and C-S-H seeding: an SCM-based low-carbon cementitious binder approach. Constr. Build. Mater. **442**, 137546 (2024)
9. Hou, W. et al.: Calcium transfer process of cement paste for ettringite formation under different sulfate concentrations. Constr. Build. Mater. **348**, 128706 (2022)
10. Ragoug, R. et al.: Durability of cement pastes exposed to external sulfate attack and leaching: physical and chemical aspects. Cem. Concr. Res. **116**, 134–145 (2019)

Waste-to-MOF: Sustainable Microwave Fabrication of MIL-53(Al) MOF Using Reclaimed PET and Industrial Aluminum Dross Waste for CO_2 Sorption Application

Mehboob Ur Rahman[1], Teera Butburee[2], Pakorn Opaprakasit[1], Rafaqat Ali Khan[3], Chariya Kaewsaneha[1], Chalita Ratanatawanate[2], Wanida Chooaksorn[4], Nurak Grisdanurak[5], and Paiboon Sreearunothai[1](✉)

[1] Functional Advanced Materials Engineering Research Center (FAME), School of Integrated Science and Innovation (ISI), Sirindhorn International Institute of Technology (SIIT), Thammasat University, Pathum Thani, Thailand
paiboon@siit.tu.ac.th

[2] National Nanotechnology Center, National Science and Technology Development Agency, 111 Thailand Science Park, Pathum Thani 12120, Thailand

[3] Department of Chemistry, COMSATS University Islamabad, Abbottabad Campus, Abbottabad 22060, Khyber Pakhtunkhwa, Pakistan

[4] Department of Environmental Science, Faculty of Science and Technology, Thammasat University, Pathum Thani, Thailand

[5] Department of Chemical Engineering, Faculty of Engineering, Thammasat University, Pathum Thani 12121, Thailand

Abstract. Grey aluminum dross, a byproduct of aluminum smelting and recycling, finds applications in several sectors; however, its applications are still limited. Metal-organic frameworks (MOFs) are porous materials comprised of metal nodes and organic linkers. MOFs have pore size and shape control, large surface area, and structural stability, among other qualities that make them versatile for a variety of uses. MIL-53(Al) MOF was successfully synthesized by microwave synthesis utilizing polyethylene terephthalate (PET)-derived BDC and acid leachate from grey Al dross waste. Under microwave radiation, the complete coordination of Al metal clusters and BDC (benzene dicarboxylic acid) linker generated from PET took place in 60 min. The structures and characteristics of the grey dross solid waste and the synthesized materials were investigated using X-ray diffraction (XRD), scanning electron microscopy (SEM), X-ray fluorescence spectrometry (XRF), inductively coupled plasma optical emission spectroscopy (ICP-OES), Fourier-transform infrared spectrometry (FTIR), nuclear magnetic resonance (NMR), thermogravimetric analysis (TGA), and Brunauer-Emmett-Teller (BET). The waste-derived MOF was also assessed as an adsorbent for CO_2, a gas that is widely produced by industrial processes and a significant cause of environmental issues. The waste-derived MIL-53(Al) showed the CO_2 uptake capacity of 1.39 mmol/g (1 atm, 25 °C). The material demonstrated encouraging CO_2 adsorption capabilities, highlighting its potential as a sustainable sorbent. This strategy upcycles industrial and plastic waste into useful materials with added value, addressing environmental issues and the circular economy concepts.

Y. Zhu (Ed.): ISESIE 2025, LNCE 000, pp. 30–50, 2026.
https://doi.org/10.1007/978-981-92-2389-3_4

Keywords: Grey dross upcycling · PET upcycling · MIL-53 (Al) · MOF · CO_2 sorption

1 Introduction

An important environmental issue nowadays is the rise in toxic waste production brought on by the world's fast industrialization. It is imperative to create more sustainable procedures to mitigate these environmental effects. The world's cities generated more than two billion tons of solid trash in 2016, which is equivalent to 0.74 kg per person each day [1]. The amount of waste generated annually is predicted to increase by 70% between 2016 and 2050 due to fast urbanization and population growth, reaching 3.40 billion tons. Nonbiodegradable plastic and metal particles are two of the main elements of solid waste. Because of its poor biodegradability and inadequate recovery, the large amount of polyethylene terephthalate (PET) waste produced more than twenty-four million tons, resulting in serious environmental issues. Because the "one-time-use" approach generates so much solid trash globally, garbage disposal has turned into a persistent issue. With effective scientific and technical intervention as well as public awareness, the "Reduce, Reuse, and Recycle" concepts could offer beneficial answers to the global problem of solid waste management and usage [1].

Among the metals that are most used globally is aluminum. Al dross is a byproduct of the melting of aluminum (Al). Two distinct types of this byproduct entrapment are created during the melting process: primary aluminum dross, which is created when virgin or primary Al is melted [2, 3], while the melting of Al scrap results in the production of secondary Al dross. The normal composition of primary aluminum dross is 15–75% metallic Al with less than 5% salts, but secondary aluminum dross has an Al concentration of 5–20% of the secondary dross's total weight percentage. The dross composition significantly changes depending on its processing conditions. Most of the metallic Al, Al_2O_3, fluorides, and nitrides are present in white dross [4]. The secondary Al dross (black dross) is made up of several metal salts, nitrides, fluorides, and Al_2O_3. Due to its high concentration of salt, secondary Al dross poses environmental problems and is dangerous to dispose of in landfill. Recycling is becoming a more sustainable way to deal with different kinds of waste. Aluminum dross has been used as a precursor material in investigations to date to synthesize layered double hydroxides and zeolites [5, 6].

Metal–organic frameworks, or MOFs, are a new class of porous materials made up of both metals and organic components. MOFs may regulate pore structure features, including pore size and shape, as well as physicochemical traits like stability, hydrophilicity, and hydrophobicity by utilizing the qualities of their constituents [7]. These characteristics make MOFs appropriate for a range of uses, such as drug delivery [8], gas storage [9], catalysis [10], and molecule separation [11]. Thus, in recent years, MOFs have been the focus of research [12]. One noteworthy area of research in MOF is in synthesis of MOF from waste and low-grade materials, including PET and grey Al dross waste as precursors. Dross normally comprises contaminants including metal nitrides and oxides, as well as 8–80% aluminum [13]. Therefore, the development of optimum synthesis techniques is necessary to make it possible to produce MOFs from waste. Concurrently, the severity of air pollution has increased recently, prompting research into efficient treatment techniques for absorbing gaseous pollutants such sulfur dioxide (SO_2) [14], nitrogen oxides (NO_n) [15], carbon dioxide (CO_2) [16], and volatile organic compounds (VOCs) [17]. The ease and affordability of adsorption systems make them one of the best ways for removing pollutants.

One important greenhouse gas that contributes significantly to global warming is carbon dioxide (CO_2), which is mostly released by the burning of fossil fuels, industrial operations, and transportation. Removal of CO_2 has garnered a lot of interest since it has the potential to lessen the effects of climate change. Through its acceleration of the greenhouse effect, which causes ocean acidification, extreme weather patterns, and rising global temperatures, CO_2 poses a serious threat to the environment. CO_2 is also very difficult to remove due to its chemical stability and long atmospheric lifetime, which makes the creation of effective adsorption materials for its capture and sequestration necessary [16].

The focus of this study was MIL-53(Al), a MOF composed of grey aluminum dross-derived Al leachate solution and PET-derived BDC. The primary objective was to create a rapid synthesis method for MIL-53(Al) using aluminum leachate and PET-derived BDC as the starting material. The potential use of the synthesized waste-derived MIL-53(Al) as an effective CO_2 capture agent for gas separation and climate mitigation was further assessed by evaluating its CO_2 sorption capabilities.

2 Experimental Section

2.1 Chemicals

The solid grey (Al) dross waste was obtained from a local Al scrap smelting facility. The elemental contents of the grey solid dross waste, as determined by XRF analysis, are shown in Table 1 and Fig. 1. After use, PET drinking water bottles were gathered from local plastic recycling bins. $Al(NO_3)_3{\cdot}9H_2O$ (> 98%, ACS reagent), 1,4-Benzenedicarboxylic acid (H_2BDC, 98%), DMF (99.5%), HCl, NaOH (98%), HNO_3, and ethanol of analytical-reagent grade were purchased from Sigma-Aldrich. Double-distilled water was used to make aqueous solutions.

Table 1. The metallic concentration (mg/l) of grey Al dross waste measured by the ICP-OES analysis.

Metallic elements	Concentration (mg/l)
Aluminum (Al)	1862.5
Iron (Fe)	87

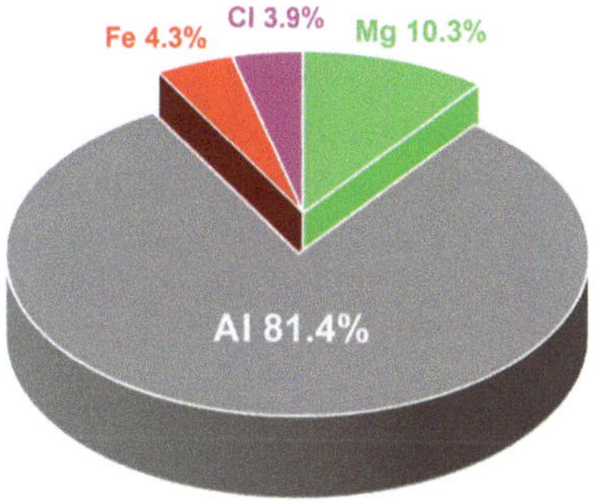

Fig. 1. Pie chart of the elemental composition of the grey aluminum dross waste obtained from the XRF analysis.

2.2 BDC Recovery from Post-consumer Bottles

The process for producing BDC was developed based on the findings published by Song et al. [18]. After the labels and lids were removed, the clear mineral water bottles were broken into small pieces. Using soapy water, the small flakes were cleaned, then baked at 60° Celsius for the entire night to dry. The PET hydrolysis experiment was carried out in a 100 mL Teflon-lined stainless-steel autoclave, and the product was identified as BDC. 6.0 g of PET flakes (0.01 M), 60 ml of water, and 2.496 g of NaOH (0.03 M) were added to the autoclave. It was then heated to 200 °C and left there for 6 h. 2 M NaOH was added to the liquid above until pH 13 was reached after a time of natural cooling. This made it possible to extract any PET that wasn't hydrolyzed using centrifugation. After the supernatant was brought to pH 1 with 6 M HCl, the white precipitated BDC was separated using centrifugation. The BDC was then cleaned twice using ethanol and H_2O, respectively, and vacuum dried for 12 h at 90 °C.

2.3 Synthesis of MIL-53(Al) MOF from Grey Aluminum Dross Leachate and PET-Derived BDC Via the Microwave Method

Firstly, 5 g of aluminum-containing grey dross waste was mixed into a 100 mL nitric acid solution prepared from 20 ml of concentrated HNO_3 and 80 ml of H_2O. A simple centrifugal separation method was used to obtain the leachate solution after stirring at room temperature for 24 h.

Secondly, 30 ml of leachate solution and 0.83 g (5 mmol) PET-derived BDC were added to the 30 ml of water. Then, the reaction mixture was sonicated and stirred (at 550 rpm) for 10 min. After that, the solution was moved to a microwave pressure vessel with a 100 ml Teflon lining. The reaction was then run for one hour at 200 °C in a microwave reactor using 500 W of power. Using a FlexiWAVE Advanced Flexibility Microwave Synthesis Platform (Milestone Srl, Italy), microwave-assisted synthesis was performed. To obtain the desired temperature of 200 °C, the reaction mixture was heated using a 20-min ramp period. An isothermal hold was then maintained for an hour. The temperature of the reaction was monitored with a non-contact infrared (IR) sensor.

The final product was allowed to cool naturally to room temperature and washed repeatedly with water, DMF and ethanol. To eliminate any unreacted BDC that may have remained in the cavities or tunnels, the obtained MOF was calcined at 320 °C in a box furnace for 3 days.

MOF was also synthesized using directly solid dross waste (500 mg) and PET-derived BDC (5 mmol, 0.83 g). Pristine MIL-53(Al) was also produced using the same process (mentioned above), using commercial benzene-1,4-dicarboxylic acid (BDC, 5 mmol, 0.83 g) as the linker and $Al(NO_3)_3·9H_2O$ (10 mmol) as the metal precursor.

2.4 Material Characterization

A Bruker D8 Advance was utilized to perform powder X-ray diffraction patterns (PXRD) for manufactured materials at 40 kV and 40 mA with a step size of 0.02° utilizing Cu-Kα (1.5418 Å) radiation. FTIR spectra were captured using a diamond crystal ATR on a Thermo-Scientific Nicolet iS5. Surface morphologies were analyzed using a Hitachi SU8030 field emission scanning electron microscope (SEM) set to an acceleration voltage of 3–10 kV. The zeta potentials were measured with a Malvern NanoZS90 zeta-sizer. An additional 1 mM NaCl was added to ensure that the solution had sufficient conductivity. Thermal stability was examined by thermal gravimetric analysis (Metler-Toledo simultaneous thermal analysis TGA/DSC 3 +). 1H NMR (Bruker AVANCE III HD 600 MHz NMR spectrometer) was utilized to analyze the chemical depolymerization of PET to BDC following the product's immersion in DMSO-d6 (Dimethyl Sulfoxide-99.9%, Cambridge Isotope Laboratories, Inc.). The samples were degassed at 200 °C and 10–5 mmHg for at least six hours, and then their N_2 adsorption-desorption isotherms were determined at 77 K using a Quanta Chrome NOVA2000e. The micropore volume and specific surface area were measured using the BET and the BJH techniques. The medium refractive index was 1.330, the relative dielectric constant was 80.4, and the medium viscosity was 1.0 mPa s.

2.5 Utilizing Waste-Derived MIL-53 (Al) MOF as an Adsorber for CO_2 Adsorption Studies

Waste-derived MIL-53 (Al) MOFs were tested for CO_2 adsorption capability at 25 °C and low pressures between 0 and 1 bar using a Quanta-chrome model Autosorb iQ-CMP, USA, gas adsorption analyzer. Before the experiment, a sample holder with about 100 mg of synthetic MIL-53 (Al) MOF adsorbents was degassed for two hours at 200 °C to get rid of any pre-adsorbed gases.

3 Results and Discussion

3.1 Characterizations of Al Dross Waste and its Leachate

The XRF data showed that the grey (Al) dross waste consists mostly of Al (81.4%), Mg (10.3%), Cl (3.9%), Fe (4.3%), and a small quantity of other elements (< 0.1%) by weight (Fig. 1). XRD (Fig. 2) revealed that the primary phases of the solid grey Al dross waste were metallic aluminum, aluminum nitride, Al_2MgO_4, and aluminum oxide (Al_2O_3) [19]. The grey dross was leached using a nitric acid solution prepared by mixing 20 mL of HNO_3 with 80 mL of water. Five grams of grey dross waste were stirred in the solution at room temperature for 24 h. The supernatant was extracted using a straightforward centrifugal separation technique. The amount of Al in the leachate solution was then evaluated by ICP-OES (PerkinElmer Optima-8000) to be about 1863 mg/L, or 69 mM, which will be used as the Al source for the synthesis of MIL-53(Al).

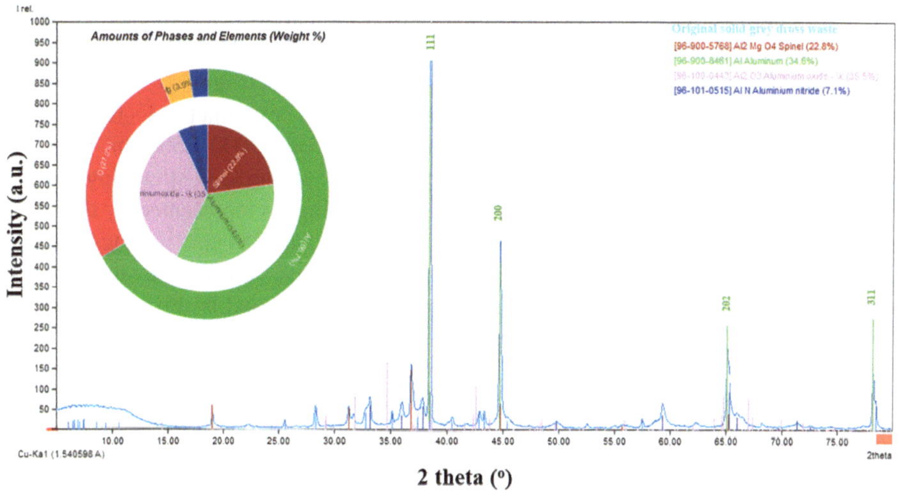

Fig. 2. XRD profile of the original solid grey aluminum dross waste

3.2 Alkaline Depolymerization of PET Plastic into BDC

A chemical recycling process has been developed to transform PET bottles from final consumers into items that can be used. Following its production by alkaline hydrolysis and acidification, BDC was utilized as an upcycling linker source in the development of MIL-53 (Al) MOF.

A PET transformation rate of over 97% was achieved, and the linker yield was 88% (5.28 g/6 g). The 1H NMR spectra were used to calculate the percentage of byproducts generated during the chemical depolymerization of PET using Eq. (1). It is calculated by dividing the total integrated peak areas of all products by the total integrated peak areas of the protons in the byproducts.

$$Byproduct\ (\%) = \sum Area\ of\ byproduct\ peaks / \sum Area\ of\ all\ product\ peaks \quad (1)$$

The purity of the PET-derived BDC linker was assessed using ^{1}H NMR spectroscopy, with DMSO-D6 acting as a solvent. The distinctive peaks of the recycled PET products are seen in Fig. 1(c) and are the same as those of commercial BDC. A broad singlet signal from the carboxylic group protons at 13.2 ppm and a particular singlet signal from the aromatic ring protons at 8.0 ppm are the two main signals shown in the ^{1}H NMR spectrum in Fig. 3(a). According to the NMR data, the compounds are mainly BDC with high purity (88% based on ratio of integration areas) was obtained [25].

To verify the chemical structures of the resulting BDC, FTIR spectroscopy was employed. The absorption bands of the recycled PET products are like those of commercial BDC, as illustrated in Fig. 3(b). The benzene rings' asymmetric C–H stretching band first appeared around 3064 cm^{-1} [20]. At 1678 cm^{-1} and 3000–2500 cm^{-1}, respectively, are the broad C=O stretching mode peak and the strong O-H stretching band of carboxylic acid [21]. The C=C stretching of aromatic rings is represented by the bands at 1137, 1511, and 1575 cm^{-1}. While the absorption band at 1284 cm^{-1} is caused by the carboxyl dimer's C–O stretching mode, the one at 1425 cm^{-1} is caused by the carboxyl dimer's O–H in-plane deformation [22]. The O-H bending band is situated in BDC at 932 cm^{-1} [23]. MIL-53 (Al) can be effectively synthesized as an organic linker using the resulting BDC.

UV-vis spectrophotometer (GENESYS 10S) was also used to examine the produced BDC (Fig. 3c). Peak maxima at 240 nm are visible in the UV-visible spectra of both commercial and developed BDC, and these values are consistent with those earlier published for BDC [1, 24]. The produced BDC's FTIR and UV-vis spectra closely matched those found in the literature, indicating that the product's synthesis was successful.

3.3 XRD of the Synthesized Waste-Derived MIL-53 (Al)

The PXRD patterns reveal the successful production of the MIL-53(Al) MOFs from waste and commercial materials. Figure 4 displays the PXRD pattern of the original solid dross waste, and the MIL-53 (Al) MOF synthesized from the dross leachate. Notable peaks at 2θ locations 9.36°, 12.41°, 17.85°, 18.66°, 23.26°, 25.16°, 27.15°, 28.28°, and 38.02° which correspond to the typical XRD pattern of two common forms of MIL-53(Al): the lt form (large pore, CCDC#220476) and the ht. form (narrow pore, CCDC#220477) [26–28].

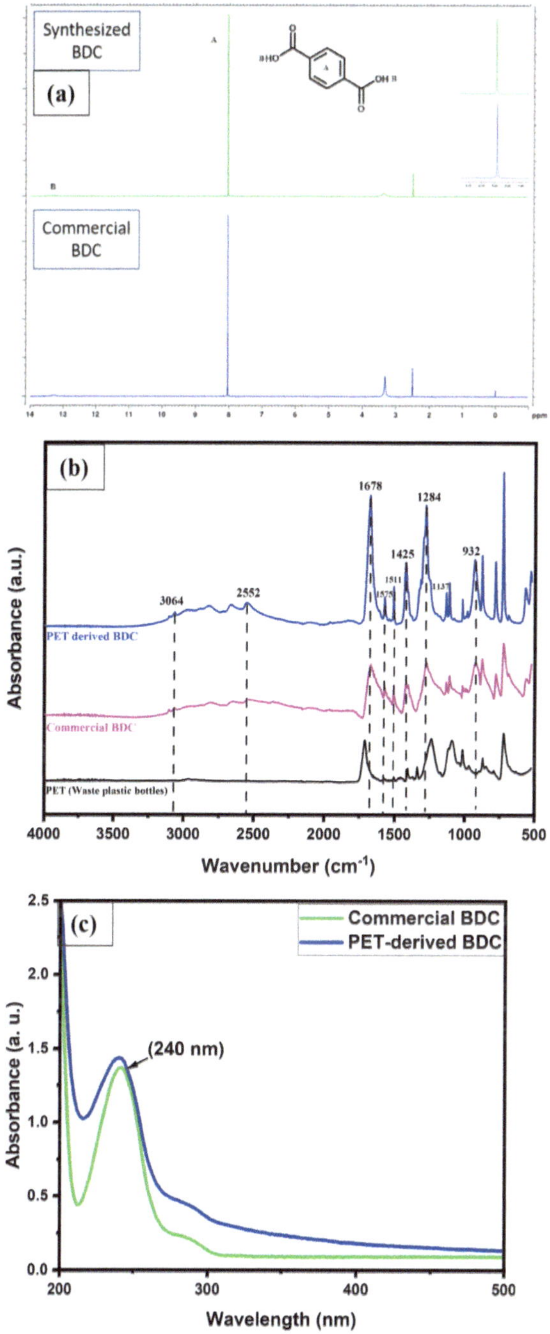

Fig. 3 (a-c). (a) ^{1}H NMR spectra of commercial BDC and PET-derived BDC, (b) FTIR spectra of PET, commercial BDC and PET-derived BDC, (c) UV-vis. spectra of commercial BDC and PET-derived BDC.

MIL-53 (Al) can exist in two phases, according to earlier reports: the large-pore (lp) phase, which has larger holes because of water molecules adhering to the micropores, and the narrow-pore (np) phase, that has closed and constricted pores.

The PXRD pattern's intensity and sharpness can be used to indicate the degree of crystallinity of the synthesized MOF. To ascertain the crystallite size (D) of the resultant samples, Debye-Scherrer's method was applied. The crystallite sizes of the synthesized samples ranged from 25 to 33 nm. PXRD results of the MIL-53 (Al) (derived from leachate/PET-derived BDC) and MIL-53 (Al) (commercial chemicals) samples match the identified MIL-53 (Al) lt (CCDC#220476), large pore and MIL-53 (Al) ht. (CCDC#220477), narrow pore phases after additional activation at 320 C for three days. The selected aluminum source was effectively transformed into MIL-53(Al) within one hour, as seen in Fig. 4. The highest levels of crystal formation and the most noticeable diffraction patterns are found in the resulting products manufactured from aluminum waste.

Waste-derived MIL-53 (Al) MOFs were gradually converted from waste to the finished product (Fig. 4). The gradual elimination of the reflected pattern corresponding to the crude grey dross resources and the formation of new peaks attributed to the typical MIL-53(Al) patterns served as evidence for this.

Regarding the transformation of grey dross into aluminum-based MOF after one hour at 200 °C, the typical MIL-53 (Al) patterns were generated. These results suggest that a reaction temperature of 200 °C for one hour is optimal for the successful and efficient formation of MOF from waste. The leaching was quite important as when employing PET-derived BDC and the solid grey dross waste without any additional purification or leaching procedures, no MOF structure was formed. This result shows the effectiveness of the transformation process.

3.4 Morphological Analysis

SEM was used to investigate the morphologies of the synthesized samples to assess the development of the synthetic technique. SEM micrographs of original solid grey dross waste and synthesized MIL-53 (Al) MOFs are displayed in Fig. 5 (a-c). The size and shape of the crystals remarkably illustrated the conversion processes of the leachate obtained from grey (Al) dross waste to MIL-53(Al) MOFs. Agglomerated rod-shaped and μm-sized MIL-53(Al) crystals (0.2–0.9 μm) were formed from agglomerated spherical and μm-sized raw material (dross waste/0.1 μm) under microwave treatment [29, 30].

The resulting MIL-53 (Al) MOF particles are typically in the μm range in size. Temperature and time are two environmental factors that have been demonstrated to affect MOF crystal sizes. Throughout the reaction, notable changes occurred in the size and shape of the product. The original grey (Al) dross waste was spherical, rough-surfaced, and larger than 0.1 μm in cluster size. After a one-hour reaction between PET-derived BDC and dross-derived leachate, rod-shaped particles were obtained with an average length of around 1 μm were obtained.

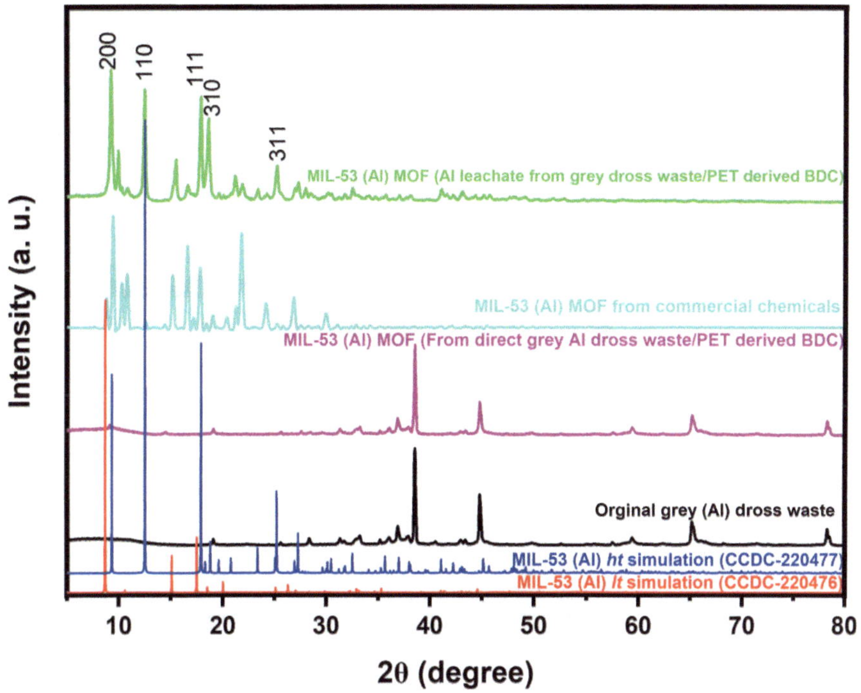

Fig. 4. PXRD patterns of original grey (Al) dross waste, MIL-53(Al) synthesized from commercial chemicals, and MIL-53(Al) synthesized from dross leachate + PET-derived BDC.

3.5 EDX Analysis

The main components of the grey (Al) dross waste are Al, O, and C, while the minor elements are Mg, Cl, Fe, Si, Mn, K, Ca, and Ar, as shown by the elemental mapping EDX spectrum (Fig. 6 a and b).

The efficacy of MOF synthesis and aluminum extraction may be impacted by the complex mixture of metallic and non-metallic elements found in grey dross, as indicated by these minuscule levels. According to the EDX spectrum of MIL-53 (Al) MOFs (Fig. 6 b), the primary elements for the resultant MIL-53 (Al) MOFs are Al, O, and C.

3.6 FT-IR Analysis

The FTIR spectra of the natural grey dross waste and the synthetic MOF made from waste materials and commercial chemicals are displayed in Fig. 7. The outcomes were contrasted with MIL-53(Al) MOF made from commercial chemicals. Markers for the-COOH group in the 1400–1700 cm^{-1} range show that MOF was formed effectively using PET-derived BDC as linkers as well as leachate from grey dross waste as a source of (Al) metal [1]. Two bands at 1593 and 1506 cm^{-1} for the pure and waste-derived MIL-53 (Al) MOFs samples indicate the-COO−asymmetric stretching, whereas the bands at 1442 and 1406 cm^{-1} are responsible for the-COO−symmetric stretching [31]. The short, sharp peak that can be seen at 1696 cm^{-1} in both pure and waste-derived

Fig. 5 (a-c). (a) SEM micrograph of original grey (Al) dross waste, (b) MIL-53(Al) MOFs from commercial chemicals and (c) MIL-53(Al) MOFs from grey dross waste-derived leachate + PET-derived BDC.

MIL-53(Al) is attributed to the unreacted carboxylic acid that is trapped in the porous structure of the material [32]. The low-range FTIR peaks, such as 584 cm^{-1} (Al-O) and 730-1100 (stretched bonds of C-H), are associated with the functionalities present in MIL-53 (Al) MOFs [1, 33, 34]. In general, the synthesized MIL-53 (Al) MOFs FTIR spectra match the results of the literature quite well.

3.7 Thermogravimetric Analysis

Due to its predominantly inorganic (metallic oxide or alloy) makeup, solid grey (Al) dross waste has a high thermal stability. The conversion process was further investigated using TGA (Fig. 8), and the results are shown in Table 2. First, weight loss in MIL-53 (Al) MOF occurs when the temperature drops from room temperature to less than 120 °C, which is associated with the removal of adsorbed water. Second, weight loss in MIL-53 (Al) MOF occurs when the temperature changes from 360 °C to 450 °C, which is associated with the removal of unreacted carboxylic acid trapped in the material's porous structure. Above 500 °C, the activated MIL-53(Al) MOF TGA curve typically only showed a significant weight-loss phase, during which the inorganic component changed into amorphous alumina and the organic ligand BDC was eliminated from the structures. In the range of 360–450 °C and 450–630 °C, it seems to exhibit weight loss in two stages.

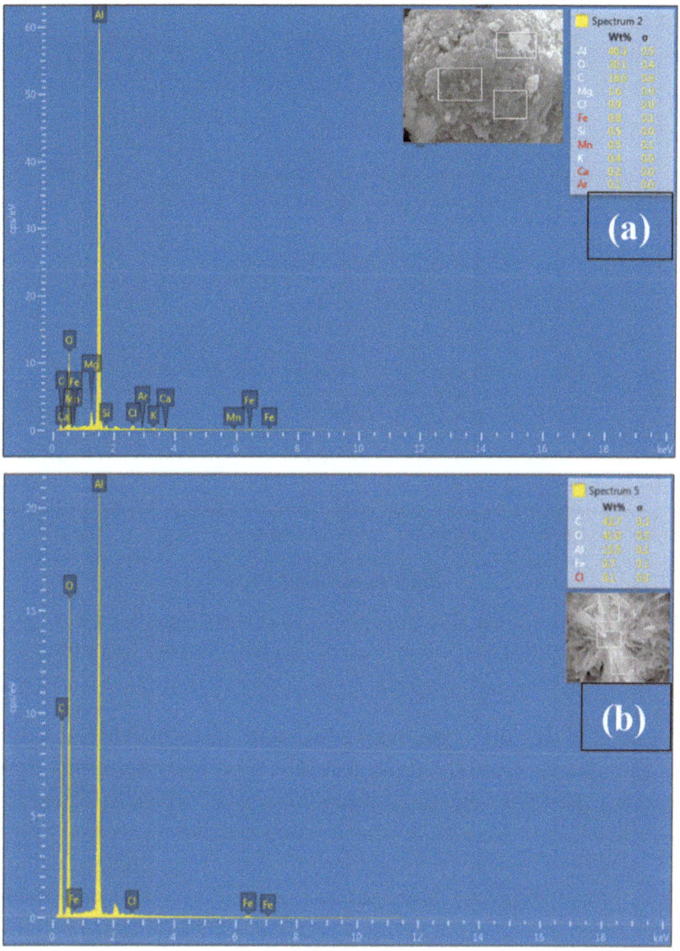

Fig. 6 (a-b). Energy-dispersive X-ray spectroscopy (EDX) (a) elemental analysis of original solid grey (Al) dross waste, (b) MIL-53 (Al) MOF from grey (Al) dross waste derived leachate & PET derived BDC.

This could be due to two types of terephthalic acids: the loosely bound BDC linker in the crystalline MIL-53(Al) structure and the firmly bound BDC trapped in the structure. Further evidence for the presence of these two terephthalic acid forms was supplied by the previously published FTIR results. In the MIL-53(Al) structure, strongly attached terephthalate linkers to the Al^{3+} break down at significantly higher temperatures, around 450–620 °C [35], while weakly bound terephthalic acids seem to break down at lower temperatures, approximately 360–450 °C, that is typical for the breakdown of carboxylic acids. With a total mass loss of roughly 45%, the waste-derived MOF (made with PET-derived BDC and grey dross-derived leachate) demonstrated a higher residual mass at 1000 °C than MOFs made from commercial chemicals, which showed a larger mass loss of roughly 67% during heating. An analysis of the primary decomposition phase (about

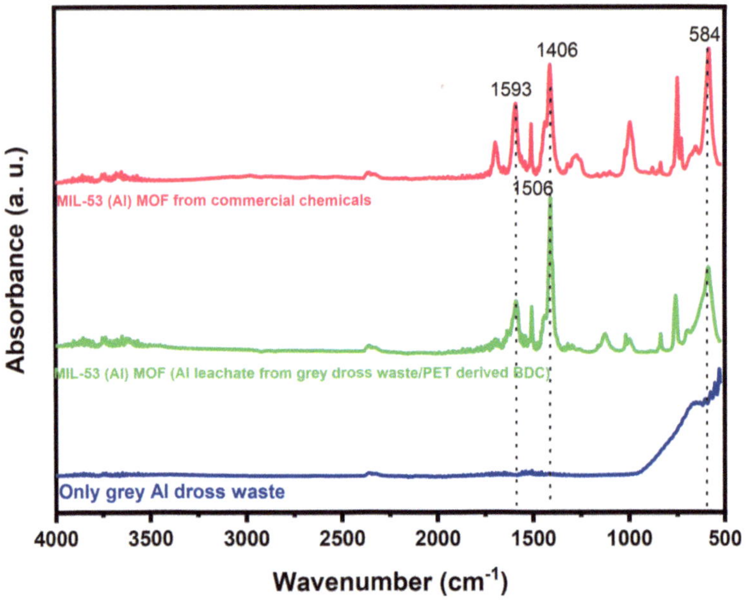

Fig. 7. FTIR spectra of original grey (Al) dross waste, MIL-53(Al) MOFs from commercial chemicals and MIL-53(Al) MOFs from grey dross waste-derived leachate/PET-derived BDC.

400–600 °C) reveals that the waste-derived sample has a higher residual weight due to inorganic components from the dross leachate. This does not imply higher thermal stability regarding the MOF structure, where the waste-derived MOF exhibited weight loss at a temperature lower than MOF from commercial chemicals, and hence the waste-derived MOF exhibited less thermal stability [36–38]. These findings demonstrate how combining plastic and industrial waste streams in MOF synthesis can promote circular economy strategies.

Table 2. The thermal behavior of MIL-53 (Al) MOFs

Samples names	Weight loss (%)		
	50–120	360–450	450–700
(a) Original (Al) grey dross waste	0	0	6
(b) MIL-53 (Al) MOF from dross waste derived leachate and PET-derived BDC	5	3	45
(c) MIL-53 (Al) MOF from commercial precursors	5	6	67

3.8 Brunauer-Emmett-Teller (BET)

The nitrogen sorption isotherm was examined to understand the architectural characteristics of the MIL-53 (Al) MOF materials. All samples displayed similar isotherm

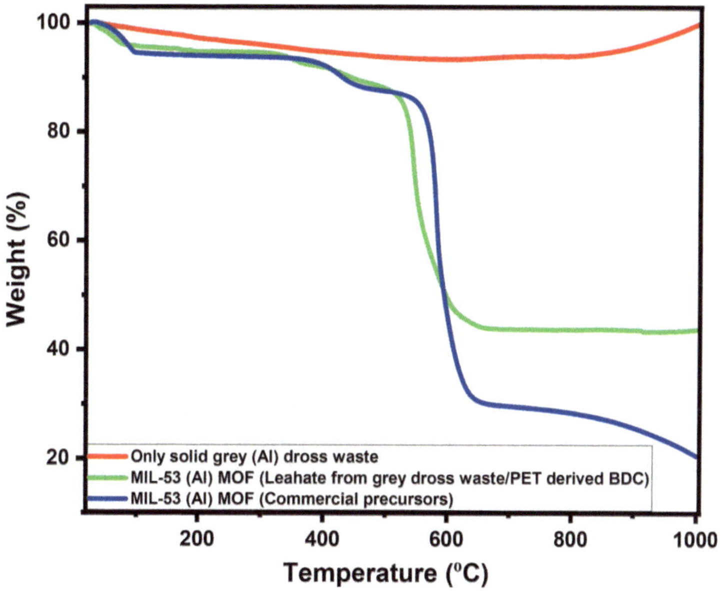

Fig. 8. Thermal behavior of solid grey dross waste, MIL-53 (Al) MOFs synthesized from commercial and grey dross (Al) waste.

patterns, consistent with type IV isotherms at 77 K and 1 bar, indicating adsorption behavior. Table 3 compiles the surface area, pore volume, and pore diameter of the MIL-53 (Al) MOFs. Using nitrogen-sorption measurements, time-lapse porosity analysis of the final products offers a thorough understanding of the synthetic process. A mesoporous structure (pore diameters between 2 and 50 nm) is indicated by the nitrogen-sorption isotherms for grey dross waste, which have a characteristic type IV shape. However, the MIL-53 (Al) MOFs isotherm showed a type IV hysteresis loop in the relative pressure range $P/P^0 = 0$–1. This may be explained by the micropores (pore widths < 2 nm) that resulted from the buildup of the particles under study and the starting material, which was waste mesoporous grey aluminum dross. The surface area of the produced product increased dramatically after one hour, as shown in Fig. 9. In the one-hour process, almost all the grey (Al) dross waste was transformed into MIL-53(Al) MOF, according to the PXRD data.

The initial product was nonporous at this reaction time due to its small pore volume and surface area. The changes in the product porosity throughout the experiment supported the X-ray, FT-IR, and SEM investigations by showing that the transformation happened in a typically progressive way while staying different from the crude precursor. Table 2 lists the nitrogen-sorption measurements of the finished products made from aluminum nitrate and grey (Al) dross waste. Accessible surface and void volume of MIL-53 (Al) MOF (produced from waste materials) at 200 °C/1 h were higher than those of MIL-53 (Al) MOF prepared from commercial chemicals following the microwave reaction with the ligand BDC. The synthesis from waste materials may have introduced structural defects and disorder, which would have increased the surface area and pore volume and provided more sites for N_2 adsorption.

When comparing MIL-53 MOF from waste materials to commercial compounds, the former was greater. Based on the characterization data above, MIL-53(Al) can be produced using leachate from grey dross (an aluminum source) as raw materials without the requirement for additives. Raw materials are continuously transformed into new products through a methodical process called conversion. The relevant products are porous, crystalline, and have a high conversion efficiency [25, 38].

For MOF building, we suggest an atom pre-alignment reaction method that is informed by the alternative manufacture of MIL-53(Al) using grey aluminum dross waste as the source, in contrast to standard synthesis from soluble precursors like $Al(NO_3)_3{\cdot}9H_2O$. This procedure is not the same as the commercial liquid-state MOF manufacturing process. It is believed that BDC functions as an organic linker, organizing aluminum atoms to form MOF. In the conventional microwave production of MIL-53(Al), the aqueous solution contains the aluminum ion, $Al(H_2O)_6{}^{3+}$. It can be explained by the fact that it creates $AlO_4(OH)_2$ centres by substituting a H_2O molecule from the $[Al(H_2O)_6]^{3+}$ ion. From BDC and the remaining nitrate ion, H^+ deprotonation yields nitric acid.

According to our proposed atom pre-alignment reaction mechanism, an atom of oxygen or OH-group-bonded polyhedral units are already present in the insoluble aluminum species aluminum oxide, aluminum hydroxide, and aluminum oxyhydroxide. By substituting oxygen sites, deprotonated BDC easily attaches to pre-aligned Al_2O units in MIL-53(Al), creating the $AlO_4(OH)_2$ center. This coordination environment is further encouraged to develop when hydroxyl groups are added to the aluminum precursor. In grey dross waste, the aluminum octahedral center, which is bonded by two OH groups and four O_2 atoms, changes from sharing edges or faces to sharing corners. These octahedral sites then work with dicarboxylates to create MIL-53(Al) [38, 39]. Therefore, it is possible to produce MOFs on a wide scale with higher safety standards, affordability, and environmental sustainability by using reaction systems with a pH environment that is close to neutral, noncorrosive raw materials, and preparation conditions devoid of organic solvents. Finally, the new process will encourage MOFs to be used in more useful applications [38].

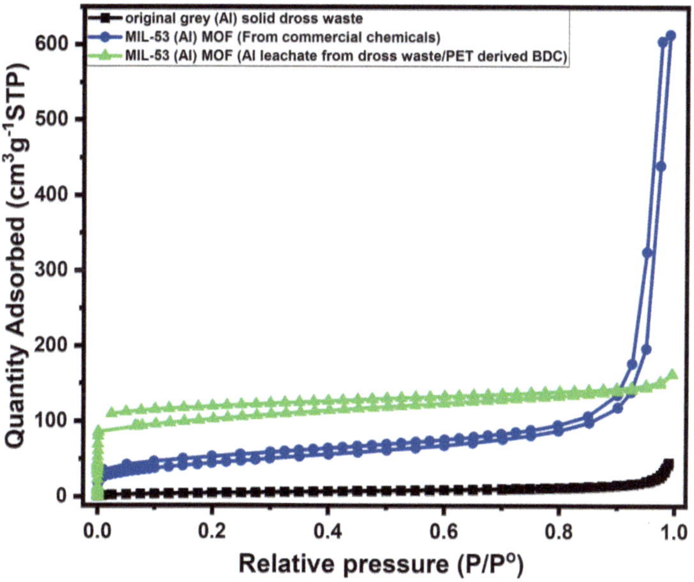

Fig. 9. N_2 adsorption isotherm of grey dross waste, MIL-53(Al) frameworks from commercial chemicals, MIL-53(Al) frameworks from direct solid grey dross waste-derived leachate/PET-derived BDC.

Table 3. Surface area, pore volume, and pore diameter of grey (Al) dross waste, MIL-53 (Al) MOFs from commercial chemicals and wastes using N_2 adsorption-desorption isotherm.

Sample	Surface area (BET) (m^2/g)	Pore volume (BJH) (cm^3/g)	Pore diameter (BJH) (nm)
1) Grey (Al) dross waste	19	0.067	3.138
2) MIL-53 (Al) MOF (from commercial chemicals)	201	0.054	1.504
3) Al leachate from grey dross waste/ PET derived BDC	636	0.123	1.607

3.9 CO_2 Sorption Studies Using Grey (Al) Dross Waste and MIL-53(Al) MOF as an Adsorber

CO_2 is one of the primary contributors to the greenhouse effect, and significant efforts are being undertaken to absorb it and prevent it from entering the atmosphere. Due to solvent temperature instability, physical or chemical absorption using aqueous basic solvents, such as amine solutions, has been used to accomplish this type of CO_2 sequestering with variable degrees of success. Solid sorbents with higher stability, such as zeolites, functionalized silica, and carbon-based compounds, were then used. Thus, MOFs are an important class of high-surface-area materials that have a high capacity to collect CO_2 and remarkable thermal stability. Our investigation into the as-synthesized (Al) MOFs for CO_2 sorption was motivated by these features. When using a volumetric methodology, Fig. 10 and Table 4 demonstrate the CO_2 sorption capabilities of grey dross waste and waste-derived MIL-53(Al) metal-organic frameworks (MOFs) at room temperature (25 °C) and low pressure. All these materials have a type I isotherm that suggests microporous architectures [1, 40, 41].

This type of isotherm shows strong adsorption at low relative pressures, indicating important interactions between CO_2 molecules and the framework pores. In accordance with the material's typical physisorption activity, MOF CO_2 capture capability increased as pressure increased. MIL-53(Al) MOF (Al Leachate from Grey dross Waste/PET Derived BDC) has an adsorption capacity of 1.38 mmol/g at a relative pressure (p/p_0) of 1 (Table 3). These materials' microporous structure, which promotes the effective CO_2 adsorption, offers a large surface area and numerous active sites for physisorption. The benzene-1,4-dicarboxylate linkers and Al centers in MIL-53(Al) MOF facilitate the capture of CO_2 through the interaction of van der Waals forces, allowing CO_2 molecules to be effectively retained inside the porous network. The fact that CO_2 adsorption had not reached equilibrium at 1 bar of pressure demonstrated that these MIL-53(Al) MOFs are efficient at absorbing CO_2.

This implies that increasing pressure may result in more adsorption. Higher CO_2 sorption is achieved in the waste-derived MIL-53 MOF (Al Leachate from grey dross waste/PET generated BDC) because of structural flaws, enhanced framework flexibility, and greater CO_2 interactions [42, 43]. Despite maintaining its fundamental structure, the MOF produced from MIL-53(Al)-waste (Al Leachate from grey dross waste/PET derived BDC) has a higher capacity for CO_2 adsorption. This is mostly due to structural flaws produced during the recycled material production process. The affinity for CO_2 is increased by these defects, which also increase the number of adsorption sites. These findings raise the prospect of developing efficient MOFs for CO_2 capture through waste valorization [44, 45].

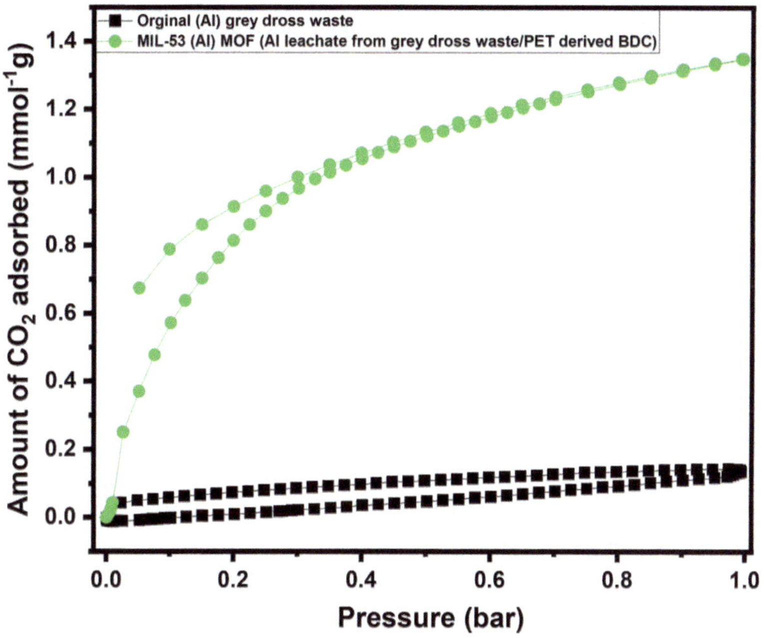

Fig. 10. CO_2 adsorption isotherm of grey dross waste, MIL-53(Al) frameworks, and MIL-53(Al) frameworks from direct solid grey dross waste-derived leachate/PET-derived BDC.

Table 4. CO_2 gas adsorption properties of grey (Al) dross waste, MIL-53 (Al) MOFs from wastes

Samples names	q_{CO2} (mmol/g) 25 °C
1) Grey (Al) dross waste	0.139
2) MIL-53 (Al) MOF (Al Leachate from grey dross waste/PET-derived BDC)	1.388

4 Conclusion

Waste-derived MIL-53(Al) samples were developed for this investigation using a rapid microwave-assisted technique. Nitrogen adsorption isotherm analysis, TGA, SEM, Zeta-sizer, FTIR spectroscopy, and XRD were used to analyze the acquired samples. According to the characterization results, waste-derived MIL-53(Al) showed crystallinity that was comparable to that of regular MIL-53(Al). The dross-derived MIL-53(Al) demonstrated outstanding absorption ability in CO_2 adsorption experiments, mainly because of its microporous structure and advantageous interactions between the CO_2 molecules and the MOF framework. These results demonstrate that dross impurities had no discernible effect on adsorption behavior in aqueous solutions or crystal formation during synthesis. In summary, this work demonstrated that a sustainable strategy for MIL-53

(Al) could be employed as a gas purification agent and successfully manufactured it in a short amount of time using dross-derived leachate and PET-derived BDC as a starting material for the first time. Using a microwave, the quick and energy-efficient process produced a MOF with promising CO_2 sorption capabilities. This method advances circular economy-based materials for environmental applications in addition to valuing industrial and plastic wastes.

Acknowledgments. Mehboob Ur Rahman gratefully acknowledges the EFS Scholarship from Sirindhorn International Institute of Technology, Thammasat University. Research support from Thailand Science Research and Innovation Fundamental Fund (2568) through Thammasat University to P. Sreearunothai and M. Rahman is also gratefully acknowledged.

References

1. Panda, D., et al.: Lab-cooked MOF for CO_2 capture a sustainable solution to waste management. J. Chem. Educ. **97**(4), 1101–1108 (2020)
2. Lin, W.C., et al.: Recycling of aluminum dross for producing calcinated alumina by microwave plasma. Sustain. Environ. Research. **32**(1), 50 (2022)
3. Mahinroosta, M., A.: Allahverdi hazardous aluminum dross characterization and recycling strategies: A critical review. J. Environ. Manag. **223**, 452–468 (2018)
4. Wang, Y., He, T., Cao, J., et al.: Waste-to-microporous framework: Al/Si fractionation of coal fly ash coupled with waste PET conversion for separated synthesis of high-porosity MIL-53 (Al) and ZSM-5 toward efficient dyes and CO_2 capture. J. Mol. Struct. **1335**, 141984 (2025)
5. Kang, Y., et al.: Synthesis of zeolite using aluminum dross and waste LCD glass powder a waste-to-waste integration valorization process. Metals. **9**(12), 1240 (2019)
6. Murayama, N., et al.: Effect of structure-directing agent on $AlPO_4$-n synthesis from aluminum dross. Int. J. Miner. Process. **93**(2), 110–114 (2009)
7. Kitagawa, S., et al.: Functional porous coordination polymers. Angew. Chem. Int. Ed. **43**(18), 2334–2375 (2004)
8. Wang, L., Zheng, M., Xie, Z.: Nanoscale metal–organic frameworks for drug delivery: a conventional platform with new promise. J. Mater. Chem. B. **6**(5), 707–717 (2018)
9. Liu, Y., et al.: Assembly of metal-organic frameworks (MOFs) based on indium-trimer building blocks A porous MOF with soc topology and high hydrogen storage. Angew. Chem.-Int. Edition. **46**(5), 3278–3283 (2007)
10. Wang, Q., Astruc, D.: State of the art and prospects in metal–organic framework (MOF)-based and MOF-derived nano-catalysis. Chem. Rev. **120**(2), 1438–1511 (2019)
11. Gucuyener, C., et al.: Ethane/ethene separation turned on its head: selective ethane adsorption on the metal-organic framework ZIF-7 through a gate-opening mechanism. J. Am. Chem. Soc. **132**(50), 17704–17706 (2010)
12. Someya, M., Iwaya, S., Konno, H.: Synthesis of PBT-derived metal–organic frameworks as adsorbents for water treatment. Colloids Surf. A Physicochem. Eng. Asp. **703**, 135418 (2024)
13. David, E., Kopac, J.: Hydrolysis of aluminum dross material to achieve zero hazardous waste. J. Hazard. Mater. **209**, 501–509 (2012)
14. Martínez-Ahumada, E., et al.: MOF materials for the capture of highly toxic H_2S and SO_2. Organometallics. **39**(7), 883–915 (2020)
15. Han, S., et al.: MOF stability and gas adsorption as a function of exposure to water, humid air, SO_2, and NO_2. Microporous Mesoporous Mater. **173**, 86–91 (2013)

16. Ghanbari, T., Abnisa, F., Daud, W.M.A.W.: A review on production of metal-organic frameworks (MOF) for CO_2 adsorption. Sci. Total Environ. **707**, 135090 (2020)
17. Siu, B., et al.: Selective adsorption of volatile organic compounds in metal-organic frameworks (MOFs). Coord. Chem. Rev. **485**, 215119 (2023)
18. Song, K., et al.: Efficient upcycling electroplating sludge and waste PET into Ni-MOF nanocrystals for the effective photoreduction of CO_2. Environ. Sci. Nano. **8**(2), 390–398 (2021)
19. López-Delgado, A., et al.: Zero-waste process for the transformation of a hazardous aluminum waste into a raw material to obtain zeolites. J. Clean. Prod. **255**, 120178 (2020)
20. Hollauer, E., Mondragon, M., Castaño, V.M.: Fourier transform infrared and Raman spectra, vibrational assignment, and ab initio calculations of terephthalic acid and related compounds. Spectrochim. Acta A Mol. Biomol. Spectrosc. **57**(5), 993–1007 (2001)
21. Kim, S.-H., et al.: Hydrogen production by electrochemical reaction using ethylene glycol with terephthalic acid. RSC Adv. **11**(4), 2088–2095 (2021)
22. Valh, J.V., et al.: Conversion of polyethylene terephthalate to high-quality terephthalic acid by hydrothermal hydrolysis: the study of process parameters. Text. Res. J. **90**(13–14), 1446–1461 (2020)
23. Čolnik, M., Knez, Ž., Škerget, M.: Sub-and supercritical water for chemical recycling of polyethylene terephthalate waste. Chem. Eng. Sci. **233**, 116389 (2021)
24. Liang, F.-Y., et al.: Waste-to-resource strategy through green synthesis of PET-derived metal-organic frameworks for efficient photocatalytic dye degradation. Microporous Mesoporous Mater. **384**, 113431 (2025)
25. Jindakaew, J., et al.: Upcycling of post-consumer polyethylene terephthalate bottles into aluminum-based metal-organic framework adsorbents for efficient orthophosphate removal. Sci. Total Environ. **935**, 173394 (2024)
26. Yamane, R., et al.: One-step synthesis of aluminum dross-derived MIL-53 (Al) as an aniline adsorbent. Microporous Mesoporous Mater., 113549 (2025)
27. Férey, G., et al.: Hydrogen adsorption in the nanoporous metal-benzenedicarboxylate m (oh)(o 2 c–c 6 h 4–co_2)(m= al^{3+}, cr^{3+}), mil-53. Chem. Commun. **24**, 2976–2977 (2003)
28. Serre, C., et al.: Very large breathing effect in the first nanoporous chromium (III)-based solids: MIL-53 or CrIII (OH)⊙$\{O_2C-C_6H_4-CO_2\}$⊙$\{HO_2C-C_6H_4-CO_2H\}$x⊙ H_2Oy. J. Am. Chem. Soc. **124**(45), 13519–13526 (2002)
29. Li, C., et al.: The strengthening role of the amino group in metal–organic framework MIL-53 (Al) for methylene blue and malachite green dye adsorption. J. Chem. Eng. Data. **60**(11), 3414–3422 (2015)
30. Alaerts, L., et al.: Selective adsorption and separation of ortho-substituted alkylaromatics with the microporous aluminum terephthalate MIL-53. **145**(33), 223–334 (2008)
31. Wang, C., et al.: Superior removal of arsenic from water with zirconium metal-organic framework UiO-66. Sci. Rep. **5**(1), 16613 (2015)
32. Katugampalage, T.R., et al.: A smart magnetically separable MIL-53 (Al) MOF-coated nano-adsorbent for antibiotic pollutant removal with rapid and non-contact inductive heat regeneration. Chem. Eng. J. Adv. **8**, 100160 (2021)
33. Loiseau, T., et al.: A rationale for the large breathing of the porous aluminum terephthalate (MIL-53) upon hydration. Chem.–A European J. **10**(6), 1373–1382 (2004)
34. Rahmani, E. and M. Rahmani.: Al-based MIL-53 metal organic framework (MOF) as the new catalyst for friedel–crafts alkylation of benzene. Ind. Eng. Chem. Res. 57(1), 169–178 (2018).
35. Sánchez-Sánchez, M., et al.: Synthesis of metal–organic frameworks in water at room temperature: salts as linker sources. Green Chem. **17**(3), 1500–1509 (2015)

36. Ashling, C.W., et al.: Synthesis and properties of a compositional series of MIL-53 (Al) metal–organic framework crystal-glass composites. J. Am. Chem. Soc. **141**(39), 15641–15648 (2019)
37. Li, L., et al.: A synthetic route to ultralight hierarchically micro/mesoporous Al (III)-carboxylate metal-organic aerogels. Nat. Commun. **4**(1), 1774 (2013)
38. Li, Z., et al.: The metal–organic framework MIL-53 (Al) constructed from multiple metal sources alumina, aluminum hydroxide, and boehmite. Chem.–A European J. **21**(18), 6913–6920 (2015)
39. Zhou, X., et al.: Adsorption of phthalic acid esters (PAEs) by amphiphilic polypropylene nonwoven from aqueous solution: the study of hydrophilic and hydrophobic microdomain. J. Hazard. Mater. **273**, 61–69 (2014)
40. Rallapalli, P., et al.: An alternative activation method for the enhancement of methane storage capacity of nanoporous aluminum terephthalate MIL-53 (Al). J. Porous. Mater. **17**, 523–528 (2010)
41. Sing, K.S.: Reporting physisorption data for gas/solid systems with special reference to the determination of surface area and porosity (provisional). Pure Appl. Chem. **54**(11), 2201–2218 (1982)
42. Millange, F., Walton, R.I.: MIL-53 and its isoreticular analogues: a review of the chemistry and structure of a prototypical flexible metal-organic framework. Israel J. Chem. **58**(9–10), 1019–1035 (2018)
43. Knebel, A., et al.: Smart metal-organic frameworks (MOFs): switching gas permeation through MOF membranes by external stimuli. Chem. Eng. Technol. **41**(2), 224–234 (2018)
44. Finsy, V., et al.: Separation of CO_2/CH_4 mixtures with the MIL-53 (Al) metal–organic framework. Microporous Mesoporous Mater. **120**(3), 221–227 (2009)
45. Andonova, S., et al.: Adsorption forms of CO_2 on MIL-53 (Al) and MIL-53 (Al)–OH x as revealed by FTIR spectroscopy. J. Phys. Chem. C. **121**(34), 18665–18673 (2017)

Construction and Optimization of a Green Infrastructure Network in an Island City Based on the MSPA-MCR Model: A Case Study of Lingkun Island

Mingzhe Yang[1], Shaoyi Wang[2], Chenyu Chen[1], Huabin Shentu[2], Yang Ping[1], Hailan Yu[2], Zhenzhou Zhang[1], Shuai Zhang[1], and Haitian Wu[2](✉)

[1] PowerChina Eco-environmental Group Co., Ltd., Shenzhen 518102, China
[2] Department of Ecology and Environment, PowerChina Huadong Engineering Co., Ltd., Hangzhou, Zhejiang 311122, China
wu_ht1@hdec.com

Abstract. Island ecosystems face severe threats to their ecological integrity from rapid urbanization due to their inherent fragility. To address this, this study presents a systematic framework for constructing and optimizing a Green Infrastructure (GI) network on Lingkun Island, a rapidly developing island city, to enhance its ecological resilience. The framework integrates Morphological Spatial Pattern Analysis (MSPA) with the Minimum Cumulative Resistance (MCR) model. MSPA was first employed to identify 31 core ecological source habitats, totaling 87.74 hm^2, which are predominantly concentrated along the island's vital coastal and intertidal zones. Subsequently, the MCR model was used to delineate an optimized network of ecological corridors connecting these sources. A comparative analysis demonstrates that the planned GI network exhibits substantial structural improvements over the current landscape: the number of corridors increases from 39 to 68, while the average corridor length decreases significantly from 1584 m to 1078 m. These structural enhancements translate to greater functional connectivity and system efficiency, mitigating habitat fragmentation. This research provides a scientifically robust and replicable methodology for ecological planning in highly developed island and coastal regions, offering a crucial tool to help balance urban expansion with environmental sustainability.

Keywords: Green Infrastructure · Morphological Spatial Pattern Analysis · Minimum Cumulative Resistance · Ecological Network · Island City Planning · Ecological Resilience

1 Introduction

Island ecosystems are uniquely vulnerable to the pressures of urbanization due to their spatial constraints and inherent ecological fragility [1]. The expansion of built environments often leads to habitat fragmentation, biodiversity loss, and the degradation of essential ecosystem services [2]. Lingkun Island, situated at a critical marine-terrestrial

Y. Zhu (Ed.): ISESIE 2025, LNCE 000, pp. 51–59, 2026.
https://doi.org/10.1007/978-981-92-2389-3_5

interface, exemplifies the challenges faced by such territories undergoing intense development (Source Material, p. 4). To mitigate these adverse impacts, Green Infrastructure Strategic Planning (GI) is imperative [3]. GI, Natural and semi natural regional networks, is a cornerstone of Urban sustainable development, enhancing ecological resilience and human well-being [4].

Traditional land-use planning often lacks the quantitative ecological basis needed to design effective GI networks [5]. In response, this study employs a scientifically robust framework based on the "ecological source-resistance surface-corridor" paradigm [6]. Specifically, this research integrates Morphological Spatial Pattern Analysis (MSPA), a tool for identifying structurally significant habitat patches [7], with the Minimum Cumulative Resistance (MCR) model, which simulates ecological flows to identify optimal pathways [8]. The objective of this study is to apply this integrated model to construct and optimize a coherent and functional GI network for Lingkun Island [9], offering scientific strategies for its ecological sustainability.

2 Material and Methods

2.1 Study Area

Lingkun Island is located in Wenzhou, China, and functions as a critical ecological node connecting the Oujiang River's terrestrial ecosystem with the East China Sea's marine environment (see Fig. 1). Its transitional geography makes it a vital area for regional ecological stability, yet it faces significant development pressure.

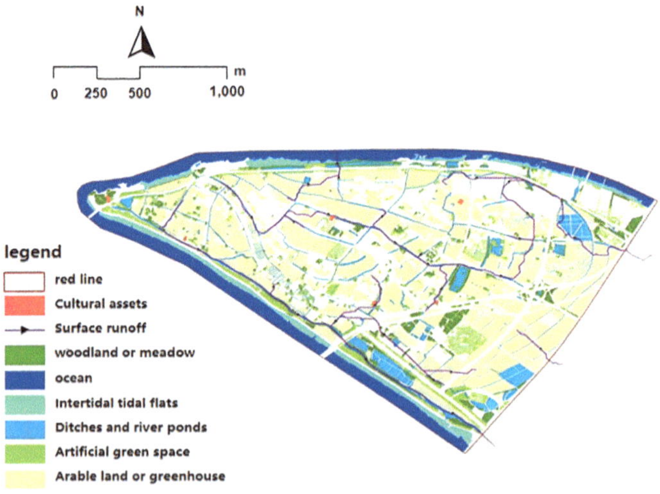

Fig. 1. Ecosystem space distribution of the study area, Lingkun isalnd.

2.2 Methodological Framework

The methodology was implemented using Geographic Information System (GIS) software and comprised three sequential steps [10]:

Ecological Source Identification via MSPA. To identify core habitats, MSPA was employed. Green spaces (e.g., forests, grasslands, water bodies) were classified as the "foreground" in a binary landscape raster [11]. MSPA categorizes the foreground into seven types based on spatial morphology, as shown in Table 1. For this study, the "Core" areas, which represent large, contiguous habitat patches vital for interior species, were selected. A minimum area threshold of 4000 m^2 was applied to ensure the ecological viability of these patches. This process identified 31 distinct patches as the ecological sources for the GI network, as shown in Fig. 2.

Development of Ecologically Resilient Surfaces. An ecological resistance surface was generated to model the cost of movement for organisms across the landscape. Different land-use types were assigned resistance coefficients based on their permeability to ecological flow [12], ranging from low resistance (e.g., woodland) to high resistance (e.g., impervious built-up areas), as detailed in Table 2. These coefficients were used to create a continuous raster surface representing the cumulative resistance across the study area, as shown in Fig. 2.

Ecological Corridor Extraction via MCR Model. the modle of MCR was utilized to identify the most efficient ways between the ecological sources [13]. This model calculates the path of least cumulative resistance from each source to all other sources across the resistance surface. The resulting least-cost paths represent the optimal potential ecological corridors, forming the structural links of the GI network.

3 Results and Discussion

3.1 Identification of Ecologically Strategic Sources

The MSPA analysis identified 31 core habitat patches, totaling 87.74 hm^2, which constitute the foundational ecological sources of the GI network, as shown in Fig. 3. The spatial distribution of these sources is predominantly concentrated along the island's northern and southern coastlines and within intertidal zones. This finding underscores the strategic importance of the marine-terrestrial interface to Lingkun Island's ecological integrity. These coastal and intertidal ecosystems are not merely peripheral but are the primary reservoirs of biodiversity and must be prioritized for conservation [14]. Their role as the foundational "anchors" of the network makes their protection essential for maintaining regional ecological functions.

3.2 Structure of the Optimized Green Infrastructure Network

Figure 4(a) and 4(b) shows a network of 68 ecological corridors connecting the 31 identified sources generated by MCR model. A key characteristic of this network is the strong spatial correlation between the delineated corridors and existing hydrographic

Table 1. MSPA landscape categories and their ecological significance.

Category	Definition	Ecological significance
Core	A set of foreground pixels whose distance from the background is greater than a specified edge-width parameter	Represents large, contiguous habitat patches (e.g., extensive green spaces, wildlife habitats) that are crucial for interior species and overall ecosystem stability
Islet	A patch of foreground area that is not connected to any other foreground component and is smaller than the minimum area threshold for a core	Small, isolated, and fragmented habitat patches, such as small parks within built-up areas, that are disconnected from the main ecological network. They may serve as stepping-stones but are not stable habitats
Perforation	An area of background pixels completely surrounded by a core area	Represents non-habitat land use (e.g., developed land) within a Core area, indicating internal fragmentation and possessing negative ecological value
Edge	The outer boundary of a foreground patch, adjacent to background pixels	The transitional zone between a core habitat and the surrounding non-habitat matrix. It is subject to edge effects, which can alter microclimates and species composition
Bridge	A strip of foreground pixels that connects two or more separate core areas	A structural corridor connecting separate Core areas. Functionally, it is a key component of the Green Infrastructure that facilitates species migration, gene flow, and energy transfer, thus maintaining network connectivity
Loop	A strip of foreground pixels that connects back to the same core area from which it originates	A pathway that increases internal connectivity within a large habitat patch but does not connect to external patches, thus having limited value for inter-patch dispersal
Branch	A strip of foreground pixels connected at only one end to another foreground component (e.g., core, edge, bridge)	A dead-end corridor or appendage extending from another component. It represents a pathway with limited connectivity, functioning more as a habitat extension than a true connector
Category	Definition	Ecological significance

features, such as river channels and drainage ditches. This indicates a natural synergy between the island's green (vegetated) and blue (water-based) infrastructure. This finding suggests a highly efficient planning strategy: the protection and restoration of existing riverine systems can simultaneously enhance both aquatic and terrestrial connectivity

Table 2. Landscape ecological resistance values for different land-use types

Land-use type	Sub-category	Resistance value
Nature reserves, wetland parks	National level	1
	Provincial level	3
	Municipal level & other	5
Forest Parks, Geoparks	National level	3
	Provincial level	5
	Municipal level & other	7
Wetland, woodland	Area > 100 ha	3
	Area > 50 ha	5
	Area < 50 ha	9
Scenic areas	–	15
Mountainous land	Area > 100 ha	10
	Area < 100 ha	15
Farmland/cropland	–	30
Garden land/orchard	–	30
Grassland	–	30
Water bodies	River/reservoir surface	20
Ocean	–	400
Developed land	Residential land	600
	Public management & service land	800
	Commercial & business facility land	800
	Industrial land	1000
	Logistics & warehousing land	1000
	Public utility land	1000
	Rural settlement	600
Urban green space	Park green space	20
	Protective green space	30
	Roadside green space	60
	Other green space	20

[15], maximizing conservation returns on investment. The resulting network forms a cohesive ecological matrix that facilitates critical ecological flows across a landscape otherwise dominated by anthropogenic features.

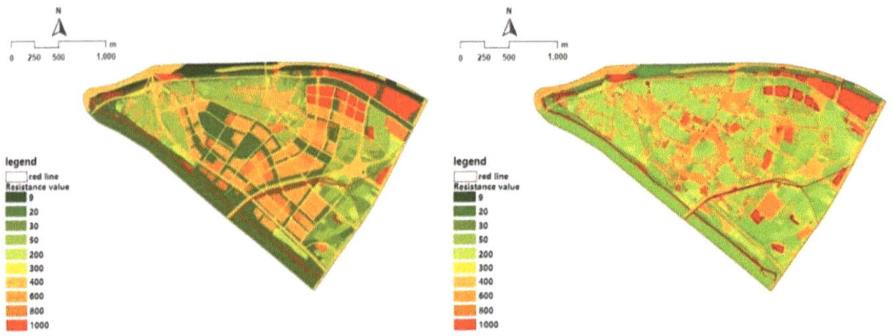

(a)Planned ecological resistance surface (b) Current ecological resistance surface

Fig. 2. Ecological resistance surface of Lingkun island.

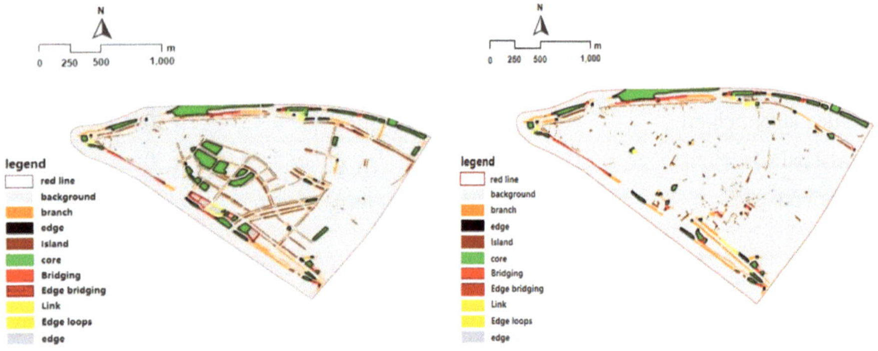

(a) Planned ecological spatial pattern (b) Current ecological spatial pattern

Fig. 3. MSPA analysis results of Lingkun island.

3.3 Network Optimization and Enhanced Ecological Resilience

Figure 5 shows the improved ecological space network according to MCR analysis. A quantitative comparison between the planned network and the current landscape reveals significant structural improvements. The number of corridors increases from 39 to 68, while the average corridor length decreases from 1584 m to 1078 m. This structural enhancement has direct implications for ecological function. The increased corridor density directly mitigates habitat fragmentation by linking previously isolated patches [16], thereby improving the overall structural connectivity.

Furthermore, the reduction in average corridor length is a critical indicator of enhanced functional connectivity. Shorter travel paths reduce the energy expenditure and predation risk for mobile species, increasing the probability of successful dispersal and gene flow between populations [17]. This creates a more resilient meta population structure. A denser, more interconnected network with greater pathway redundancy is inherently more robust to future disturbances. The loss of a single corridor is less likely

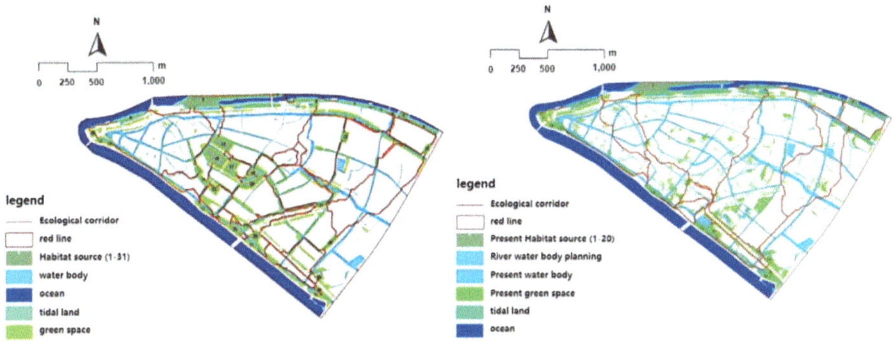

(a)Planned ecological space network (b) Current ecological space network

Fig. 4. MCR Analysied ecological space network of Lingkun island.

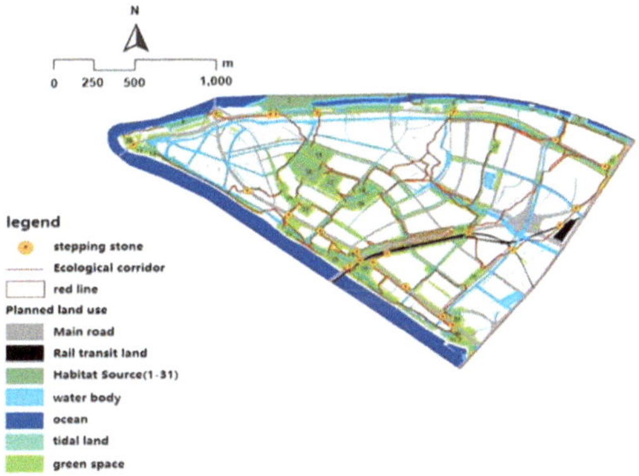

Fig. 5. Optimized ecological space network of Lingkun island.

to cause a systemic failure, thereby enhancing the long-term ecological resilience of the island [18].

3.4 Implications for Spatial Planning and Policy

The results of this study offer direct and actionable insights for urban planning. The delineated GI network provides a scientific basis for spatial conservation planning. The identified ecological sources and corridors should be integrated into Lingkun Island's official land-use master plan, potentially as designated ecological protection zones with stringent development controls. This framework enables a proactive, rather than reactive, approach to development, allowing planners to guide high-impact projects away from ecologically sensitive areas [19]. Where infrastructure development must intersect with corridors [20], this analysis provides the basis for mandating mitigation measures, such as wildlife culverts or green bridges. This study thus demonstrates the successful translation of quantitative ecological modeling into evidence-based policy for sustainable urban development [21].

4 Conclusion

This research successfully designed and optimized a Green Infrastructure network for Lingkun Island by integrating the MSPA and MCR models. The resulting network, comprising 31 ecological sources and 68 corridors, exhibits demonstrably superior structural and functional connectivity compared to the existing landscape. The primary contribution of this study is the application of a robust, data-driven methodology for GI planning in an ecologically sensitive and rapidly urbanizing island context. This approach provides a scientifically defensible and replicable tool for planners and policymakers. It is recommended that future urban planning initiatives adopt such integrated ecological models to proactively embed ecological resilience into urban design, ensuring that development proceeds in harmony with environmental conservation.

References

1. Vogt, P., et al.: Mapping functional connectivity. Ecol. Indic. **9**(1), 64–71 (2009)
2. Banks-Leite, C., et al.: Countering the effects of habitat loss, fragmentation, and degradation through habitat restoration. One Earth. **3**(6), 672–676 (2020)
3. Pricope, N.G., Dalton, E.G.: Mapping coastal resilience: precision insights for green infrastructure suitability. J. Environ. Manag. **376**, 124511 (2025)
4. Liu, L., Palaiologou, F., Schmidt-III, R.: Green infrastructure morphology: integrating urban morphological perspectives for sustainable urban development. J. Urban Manag (2025)
5. Zhang, J., et al.: A strategy for green infrastructure network based on multifunctional synergy. Ecol. Front. **45**(2), 483–496 (2025)
6. Li, C., et al.: Construction of ecological security pattern of national ecological barriers for ecosystem health maintenance. Ecol. Indic. **146**, 109801 (2023)
7. Chen, M., et al.: MSPA-based green space morphological pattern and its spatiotemporal influence on land surface temperature. Heliyon. **10**(11), e31363 (2024)
8. Huang, C., Hou, X., Li, H.: An improved minimum cumulative resistance model for risk assessment of agricultural non-point source pollution in the coastal zone. Environ. Pollut. **312**, 120036 (2022)
9. Honeck, E., et al.: Methods for identifying green infrastructure. SN Appl. Sci. **2**(11), 1916 (2020)
10. Lange, S., Müller, F.: Geographic information systems (GIS) and terrestrial ecosystems. In: Reference Module in Earth Systems and Environmental Sciences. Elsevier (2024)
11. Lin, J., et al.: An assessment framework for improving protected areas based on morphological spatial pattern analysis and graph-based indicators. Ecol. Indic. **130**, 108138 (2021)
12. Li, Y., Wang, X., Dong, X.: Delineating an integrated ecological and cultural corridor network: a case study in Beijing, China. Sustainability. **13**, 412 (2021)
13. Guo, Z., et al.: Analysis of ecological network evolution in an ecological restoration area with the MSPA-MCR model: a case study from Ningwu County, China. Ecol. Indic. **170**, 113067 (2025)
14. Spalding, M.D., et al.: The role of ecosystems in coastal protection: adapting to climate change and coastal hazards. Ocean. Coast. Manag. **90**, 50–57 (2014)

15. Portela, A.P., et al.: Regional planning of river protection and restoration to promote ecosystem services and nature conservation. Landsc. Urban Plan. **211**, 104101 (2021)
16. Wang, Y., et al.: Optimal configuration of a wildlife corridor system. Glob. Ecol. Conserv. **46**, e02560 (2023)
17. Silva, C.M., et al.: Birds' gap-crossing in open matrices depends on landscape structure, tree size, and predation risk. Perspect. Ecol. Conserv. **18**(2), 73–82 (2020)
18. Travers, E., Härdtle, W., Matthies, D.: Corridors as a tool for linking habitats – shortcomings and perspectives for plant conservation. J. Nat. Conserv. **60**, 125974 (2021)
19. Huang, X., et al.: Constructing and optimizing urban ecological network in the context of rapid urbanization for improving landscape connectivity. Ecol. Indic. **132**, 108319 (2021)
20. Wang, C., Wu, J., Wu, S.: Urban agglomeration policy and coordinated road infrastructure development. Transp. Res. A Policy Pract. **195**, 104433 (2025)
21. Deng, Y., et al.: Reassessing the ecological effectiveness of ecological restoration programs: evidence from a quasi-natural experiment in China. Ecol. Eng. **212**, 107506 (2025)

Hydro-Physical Behavior of Bermuda Grass Roots in Fly Ash–Calcined Clay Porous Concrete for Green Infrastructure

John Bosco Niyomukiza(✉) and Amin Eisazadeh

School of Civil Engineering and Technology, Sirindhorn International Institute of Technology, Thammasat University, Bangkok, Thailand
niyojayb1992@gmail.com

Abstract. Urban green infrastructure increasingly depends on engineered substrates that support vegetation while enhancing stormwater infiltration, erosion control, and material sustainability. This study investigates the hydro-physical adaptation of *Cynodon dactylon* (Bermuda grass) roots in porous concrete substrates modified with fly ash and calcined clay for green infrastructure applications. While root biomechanics in natural soils are well understood, little is known about how roots adapt to rigid substrates such as porous concrete, especially in terms of their hydro-physical behavior. To address this gap, key hydrological traits, including root suction and gravimetric water content (GWC) were measured to assess root responses to the water retention and structural conditions of the substrate. The porous concrete substrate where Bermuda grass was planted exhibited a porosity of 24.14% and an initial pH 11.37, which was reduced through SCM integration, creating favorable conditions for vegetation establishment. Root diameters ranged from 0.27 to 0.59 mm, suction values from − 115 MPa to − 140 MPa, and GWC from 0.099 to 0.892 gg^{-1}, indicating diverse water retention responses at the individual root level. These findings highlight the viability of engineered bio-compatible substrates for green infrastructure applications. By integrating sustainable materials and plant-compatible chemistry, this study supports the design of multifunctional systems that enhance vegetation performance while promoting circular use of industrial by-products.

Keywords: Bermudagrass · green infrastructure · porous concrete · root suction · slope stabilization · sustainable construction

1 Introduction

Urbanization has led to an increasing prevalence of impermeable surfaces, resulting in severe environmental challenges such as increased stormwater runoff, urban heat islands, and progressive land degradation. In response, green infrastructure has emerged as a multifunctional solution that integrates ecological processes into urban systems, offering co-benefits in stormwater management, microclimate regulation, and biodiversity enhancement [1]. At the core of many green infrastructure applications is the strategic incorporation of vegetation within engineered substrates to achieve synergistic ecological and mechanical performance [2–5].

Y. Zhu (Ed.): ISESIE 2025, LNCE 000, pp. 60–67, 2026.
https://doi.org/10.1007/978-981-92-2389-3_6

Among the suite of green-compatible materials, porous concrete has gained attention due to its unique ability to support vegetation while maintaining structural integrity and hydrological permeability [2]. This material's inherent porosity facilitates both root penetration and water infiltration, making it suitable for applications in bio-retentive pavements, vegetated retaining walls, and slope stabilization [4, 6–8]. However, the success of such systems depends not only on the physical properties of the substrate but also on its compatibility with biological growth and mechanical reinforcement by root systems. Parallel to advances in material performance, sustainability in construction has become paramount, prompting widespread substitution of traditional Portland cement with supplementary cementitious materials (SCMs) such as fly ash and calcined clay. These pozzolanic materials, derived from industrial by-products and thermally activated clays, respectively, have demonstrated significant environmental benefits by lowering CO_2 emissions, reducing clinker usage, and enhancing durability through chemical resistance [2, 9, 10]. Specifically, blends of calcined kaolinite clay and fly ash have shown to refine microstructure, and improve both early-and later-age strength, which are parameters critical for sustaining vegetation and resisting degradation under variable environmental loads [9]. Despite these material advancements, the biological response of root systems to cementitious substrates remains underexplored. While natural soils provide an optimal balance of mechanical support and nutrient availability, concrete environments present unique challenges due to high alkalinity, limited nutrient content, and mechanical impedance. These factors can drastically alter root morphology, tensile strength, and spatial architecture, with implications for both plant health and soil-structure interactions. Fine roots (< 2 mm), known for their high tensile strength and dense distribution, are particularly important in reinforcing soil matrices and stabilizing slopes [11–13].

Biomechanical studies have established that root tensile strength and modulus are critical parameters influencing their capacity to resist pullout and shear, especially in engineered substrates [14]. However, the interplay between substrate composition, root adaptation, and hydro-mechanical function remains insufficiently characterized. Limited research has been conducted to evaluate how roots adjust their physiological and mechanical traits within cement-based ecological substrates, but these studies are scarce and often confined to empirical observation without mechanistic insight [15]. Furthermore, engineered porous concretes designed for vegetation referred to as porous vegetation concrete (PVC) or planting concrete, must balance trade-offs between porosity, pH, strength, and plant survivability [2]. For instance, PVC has demonstrated potential in urban thermal regulation and moisture balance, functioning as both structural infrastructure and microclimatic moderator [6]. The integration of plant-compatible materials like fly ash and calcined clay into these systems may therefore yield dual benefits: environmental circularity and improved bio-functionality.

This study addresses these interdisciplinary challenges by investigating the hydro-behavior of *Cynodon dactylon* (Bermuda grass) roots in porous concrete matrices modified with fly ash and calcined clay. It evaluates matric potential (suction), gravimetric water content (GWC), and diameter, which are key traits linked to mechanical performance and ecological viability. By systematically examining how engineered substrates influence root development and functionality, the research aims to provide empirical foundations for eco-compatible, performance-optimized green infrastructure systems.

2 Materials and Methods

2.1 Porous Concrete Preparation and Vegetation Establishment

The porous concrete substrate was formulated using a blend of Portland-limestone cement (PCC), Class F fly ash (FA), and calcined clay (CC). Both FA and CC were incorporated as partial cement replacements, each at 15%, to reduce substrate alkalinity and improve compatibility with plant growth. The coarse aggregate used was crushed limestone (9.5–19 mm), with an aggregate-to-binder ratio of 5 and a water-to-binder ratio of 0.35. These proportions were selected to ensure sufficient porosity for root penetration. Fine aggregates were deliberately excluded to preserve interconnected void spaces that are critical for water retention and root development. The final mix composition of the porous concrete substrate where vegetation was grown consisted of 70% PCC, 15% FA, and 15% CC. Detailed material properties are available in the recent publication by Niyomukiza et al. [2].

After batching the constituents, the concrete was mixed thoroughly and cast into rectangular molds measuring 40 cm × 40 cm × 5 cm. The specimens were compacted and then cured in water for 28 days. The porosity and pH of the fabricated porous concrete substrate were 24.13%, and 11.37, respectively. Following the curing period, the cast concrete was transferred into a transparent container. A slurry mixture comprising topsoil, Bermuda grass seeds, and a root growth accelerator was applied evenly across the concrete surface to initiate vegetation growth (Fig. 1).

Fig. 1. Vegetation growth atop porous concrete (a) prior to vegetation establishment, (b) after full vegetation cover.

Water was sprayed on the surface at least twice daily to ensure consistent moisture levels during the establishment phase. Within five days, seed germination and sprouting were observed. The specimens were monitored under natural conditions for 201 days, after which, the roots were carefully extracted for laboratory analysis.

2.2 Determining the Suction-Water Content Characteristics of Bermuda Grass Roots

Following the uprooting of Bermuda grass roots from porous concrete substrate, they were carefully washed to remove residual soil, and stored in a refrigerator at 4 °C to maintain their freshness until the time of testing. This careful handling ensured that the roots remained in a stable state, minimizing any potential damage prior to the biomechanical testing process. Before refrigeration, the root architecture was examined and documented. Observations revealed that Bermuda grass possesses a fibrous root system, as shown in Fig. 2(a). Based on the classification by Styczen and Morgan [16], this root system corresponds to the M-type structure, shown in Fig. 2(b). According to their classification, M-type roots exhibit maximum development at greater depths, although approximately 80% of the root matrix is concentrated within the top 30 cm of soil. Fibrous roots, such as those of Bermuda grass, are known for their capacity to enhance soil cohesion more effectively than taproot systems of equivalent root area ratio [11, 17]. This attribute makes them particularly valuable for soil reinforcement applications.

(a) (b)

Fig. 2. Bermuda grass root architecture (a) photo (b) schematic view.

Following specimen preparation, root suction and GWC were immediately measured for all root samples. Each root specimen was placed in a dewpoint potentiometer (Model WP4C) to measure its water potential (in MPa), as shown in Fig. 3 (a). Afterward, the sample was weighed on a scale with a precision of 0.0001 g, as shown in Fig. 3 (b). The sample was then placed in an oven at 105 °C for 24 h to determine the corresponding GWC.

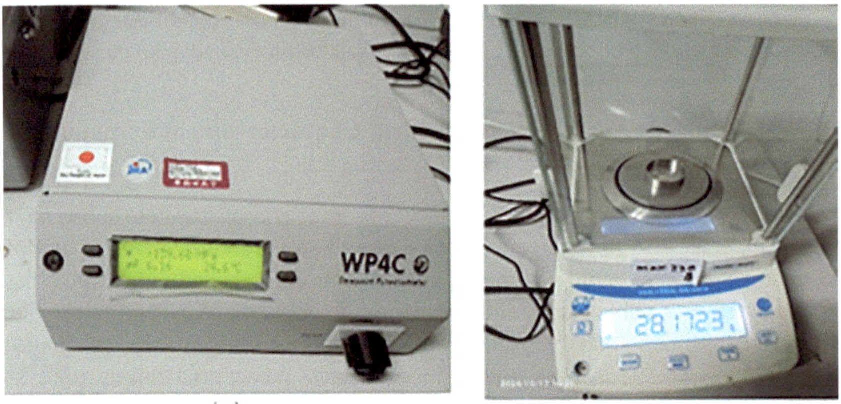

(a) (b)

Fig. 3. Determination of root suction (a) and measuring weight of the root (b).

3 Results and Discussions

3.1 Root Suction-Gravimetric Water Content Characteristics

The relationships between root diameter, GWC, and suction observed in the laboratory experiments are presented in this section. The results indicate that root suction values range between − 115 MPa and − 140 MPa. These high values are typically associated with dense fibrous root systems, which enhance matric forces and reduce pore water pressure. Meanwhile, the GWC varies from 0.099 gg^{-1} to 0.892 gg^{-1}, which falls within the range observed for various species under differing hydration conditions [18, 19].

3.2 Relationship Between Root Diameter and Root Hydro-Properties

The relationships between root diameter and root hydro-properties are illustrated in Fig. 4, where Fig. 4(a) shows the relationship between average root diameter and suction, Fig. 4(b) depicts the relationship between average root diameter and GWC, and Fig. 4(c) highlights the relationship between suction and GWC. From these relationships, the following empirical equations were derived:

$$D = 2E - 05S^{2.0166} \tag{1}$$

$$D = 0.5079GWC^{0.1755} \tag{2}$$

$$\mathrm{GWC} = 25.9\mathrm{S}^{7.3909} \tag{3}$$

The results revealed a positive correlation between root diameter, GWC, and suction. Specifically, both GWC and suction increase with increasing root diameter. These findings have significant implications for soil reinforcement and slope stability. The positive correlation observed in Fig. 4(a) and Fig. 4(b) suggests that thicker roots are more

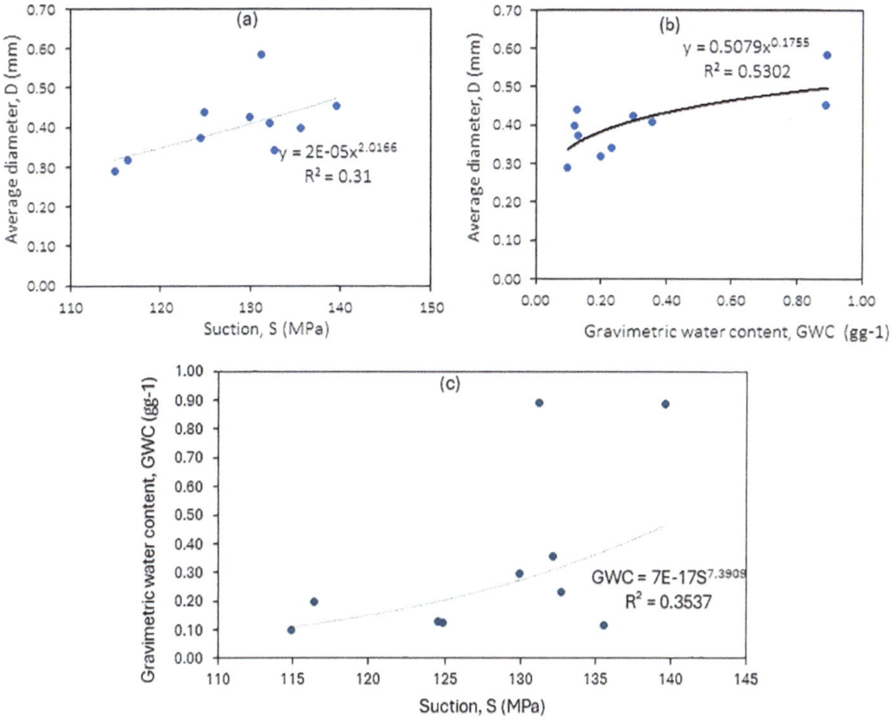

Fig. 4. The relationship between root average diameter, suction and GWC (a) average root diameter and suction (b) average root diameter and GWC (c) GWC and suction.

effective at influencing the soil's hydrological behavior. Higher suction values indicate that roots can create stronger matric forces within the soil, reducing pore water pressure and enhancing the shear strength of the soil. This is particularly beneficial for stabilizing slopes, as the reduced pore water pressure minimizes the risk of soil failure. As shown in Fig. 4(c), the relationship between GWC and suction highlights the role of roots in regulating soil moisture dynamics. Root systems, particularly those with dense fibrous structures, can enhance water retention and matric suction, contributing to soil cohesion. These traits help mitigate desiccation cracking, which is critical for maintaining slope stability.

These findings highlight that higher suction enhances overall soil stability by strengthening soil–root interactions, even if the intrinsic mechanical properties of the roots slightly decline. Such interactions are particularly important in arid and semi-arid environments, where maintaining adequate matric suction is essential for slope stability and erosion control. The results emphasize the dual role of suction: it not only contributes directly to increasing soil shear strength, but also facilitates the effective mobilization of tensile forces within the root network, ensuring that the vegetative reinforcement system remains robust [20].

4 Conclusions

This study examined the hydro-physical behavior of *Cynodon dactylon* (Bermuda grass) roots established within porous concrete containing fly ash and calcined clay. The investigation focused on suction, and gravimetric water content to evaluate the root system's adaptation and functionality in a non-traditional, engineered substrate. The major conclusions are as follows:

- Root diameters in this study ranged from 0.27 to 0.59 mm, indicating a predominance of fine roots. This morphological characteristic may influence root distribution within the substrate and suggests adaptability to pore structure conditions in the porous concrete matrix.
- A positive correlation was identified between root diameter, gravimetric water content (0.099 to 0.892 gg^{-1}), and root suction (− 115 to − 140 MPa), indicating that Bermuda grass roots exhibit greater moisture retention and hydraulic adaptability. These findings suggest that root morphology plays a critical role in water uptake dynamics within porous concrete substrates, contributing to more resilient green infrastructure systems.
- The fly ash and calcined clay–based porous concrete provided a suitable environment for root growth. It supported both mechanical anchorage and hydrological viability, thereby fulfilling the structural and ecological requirements for integrated green infrastructure systems.

These findings demonstrate the feasibility of combining biological and industrial systems to create functionally integrated substrates for urban sustainability. Future research should address long-term biological–material interactions, assess root behavior under cyclic environmental stresses, and investigate other plant species to generalize the applicability of such composite systems in diverse urban contexts.

Acknowledgments. The authors acknowledge Sirindhorn International Institute of Technology (SIIT) and Thammasat University (TU) for awarding the Excellent Foreign Students (EFS) and TU scholarships to the first author to support his Ph.D. studies.

References

1. Liu, C., Xia, Y., Chen, J., Huang, K., Wang, J., Wang, C., et al.: Research and application progress of vegetation porous concrete. Materials. **16**(21), 1–20 (2023)
2. Niyomukiza, J.B., Eisazadeh, A., Tangtermsirikul, S.: Synergistic effect of calcined clay and fly ash on the performance of porous vegetation concrete. Constr. Build. Mater. **458**(Aug), 139749 (2025)
3. Liu, Y., Wang, B., Qian, Z., Yu, J., Shi, T., Fan, Y., et al.: State-of-the-art on preparation, performance, and ecological applications of planting concrete. Case. Stud. Constr. Mater. **20**(Mar.), e03131 (2024)
4. Zheng, C., Zhang, Z., Huang, Z., Wang, D., Zhang, W., Zhou, Z., et al.: Review of porous vegetation eco-concrete (PVEC) technology: from engineering requirements to material design. Compos. Part B Eng. **279**(Mar.), 111442 (2024)

5. Shi, M., Fang, Y., Zhang, F., Chu, H., Jin, W., Li, W., et al.: A bionic mineral precipitation method to improve the mechanical and vegetative properties of ecological concrete. Constr. Build. Mater. **476**(Apr.), 141294 (2025)
6. Rahman, M.S., MacPherson, S., Akbarzadeh, A., Guerini, A., Chapelat, J., Lefsrud, M.: A study on heat and mass transfer through vegetated porous concrete for environmental control. J. Clean. Prod. **366**(Jul.), 132984 (2022)
7. Cheng, J., Luo, X., Shen, Z., Guo, X.: Study on vegetation concrete technology for slope protection and greening engineering. Pol. J. Environ. Stud. **29**(6), 4017–4028 (2020)
8. Huang, F., Zhou, Y., Yang, J., Zeng, H., Tao, S., Chu, H., et al.: A novel struvite-modified planting concrete for sustainable slope protection and environmental restoration. Environ. Res. **283**(Feb.), 122194 (2025)
9. Chen, W., Zhu, H., Li, Y., Yang, L., Cheng, S., Yu, H.: Engineered cementitious composites using blended limestone calcined clay and fly ash: mechanical properties and drying shrinkage modeling. Case. Stud.Constr. Mater. **20**(Feb.), e02960 (2024)
10. Niyomukiza, J.B., Eisazadeh, A., Tangtermsirikul, S.: Recent advances in slope stabilization using porous vegetation concrete in landslide-prone regions: a review. J. Build. Eng. **76**(Mar.), 107129 (2023)
11. De Baets, S., Poesen, J., Reubens, B., Wemans, K., De Baerdemaeker, J., Muys, B.: Root tensile strength and root distribution of typical Mediterranean plant species and their contribution to soil shear strength. Plant Soil. **305**(1–2), 207–226 (2008)
12. Mattia, C., Bischetti, G.B., Gentile, F.: Biotechnical characteristics of root systems of typical Mediterranean species. Plant Soil. **278**(1–2), 23–32 (2005)
13. Su, X., Zhou, Z., Liu, J., Cao, L., Liu, J., Wang, P.: Estimating slope stability by the root reinforcement mechanism of Artemisia sacrorum on the loess plateau of China. Ecol. Model. **444**(Feb.), 109473 (2021)
14. Schwarz, M., Preti, F., Giadrossich, F., Lehmann, P., Or, D.: Quantifying the role of vegetation in slope stability: a case study in Tuscany (Italy). Ecol. Eng. **36**(3), 285–291 (2010)
15. Likitlersuang, S., Phan, T.N., Boldrin, D., Leung, A.K.: Influence of growth media on the biomechanical properties of the fibrous roots of two contrasting vetiver grass species. Ecol. Eng. **178**(Jan.), 106574 (2022)
16. Styczen, M.E., Morgan, R.P.C.: Engineering properties of vegetation. In: Morgan, R.P.C., Rickson, R.J. (eds.) Slope Stabilization and Erosion Control: a Bioengineering Approach, pp. 4–58. E & F N Spon, London (1995)
17. Ekeoma, E.C., Boldrin, D., Loades, K.W., Bengough, A.G.: Drying of fibrous roots strengthens the negative power relation between biomechanical properties and diameter. Plant Soil. **469**(1–2), 321–334 (2021)
18. Jotisankasa, A., Hossain, M., Leksungnoen, N., Dubois, L., Nishimura, S.: Root tensile strength and suction-characteristic as indicators for screening species for slope stabilization in extreme climates. Rhizosphere. **31**(Jun.), 100949 (2024)
19. Zhang, C., Zhou, X., Jiang, J., Wei, Y., Ma, J., Hallett, P.D.: Root moisture content influence on root tensile tests of herbaceous plants. Catena. **172**(Oct.), 140–147 (2019)
20. Fatahi, B., Khabbaz, H., Indraratna, B.: Parametric studies on bioengineering effects of tree root-based suction on ground behaviour. Ecol. Eng. **35**(10), 1415–1426 (2009)

Energy Saving Potential and Economic Analysis of a Waste-to-Energy and Solar Thermal Coupling System Based on Cascade Utilization Principle

Xiaobo Qi[1], Peng Yi[2], Liaokun Mo[3], Xin Luo[4], Xiangru Li[4], Jie Wang[2], and Junhao Wang[5](✉)

[1] Shaoxing Environmental Industry Co., Ltd., Shaoxing, Zhejiang 312000, P. R. China
[2] Qingneng Environmental Protection Co., Ltd., Shaoxing, Zhejiang 312000, P. R. China
[3] Shaoxing Public Utilities Group Co., Ltd., Shaoxing, Zhejiang 312000, P. R. China
[4] PowerChina Huadong Engineering Co. Ltd., Hanghzou, Zhejiang 311122, P. R. China
[5] Hohai University, Nanjing, Jiangsu 211100, P.R. China
241609010109@hhu.edu.cn

Abstract. Conventional Waste-to-Energy (WtE) cogeneration plants suffer from low energy efficiency, typically around 21%. This study proposes and evaluates an integrated system to enhance performance by applying the principle of cascade energy utilization. Using the Shaoxing WtE plant as a case study, a thermodynamic analysis reveals significant exergy destruction from using high-grade extraction steam for low-temperature auxiliary heating. The proposed scheme substitutes this steam with on-site solar thermal energy and recovered flue gas waste heat. Results indicate this thermal substitution can increase net power generation by 1.76 MW, boosting overall energy efficiency from 21.0 to 22.1%. The economic assessment is highly compelling, yielding a static payback period of only 0.6 years, which is substantially superior to the 10 year payback for an alternative photovoltaic system. This research validates direct thermal integration as a highly profitable and effective strategy for upgrading WtE plants, offering a practical solution to improve sustainability in the waste management sector.

Keywords: Waste-to-Energy · Solar Thermal Integration · Cascade Energy Utilization · Waste Heat Recovery · Multi-energy System

1 Introduction

Waste-to-Energy (WtE) cogeneration plants are pivotal for sustainable urban waste management, yet they are consistently challenged by low thermal efficiency, with net electrical efficiencies often falling in the 20–25% range [1]. A primary cause of this underperformance is systemic exergy destruction, where high-grade energy is used for low-grade thermal tasks [2]. In a typical WtE plant, high-pressure steam extracted from the turbine is used for auxiliary heating processes, such as deaeration, representing a significant loss of potential work that could otherwise be converted to electricity.

Y. Zhu (Ed.): ISESIE 2025, LNCE 000, pp. 68–75, 2026.
https://doi.org/10.1007/978-981-92-2389-3_7

To mitigate these inefficiencies, various enhancement strategies have been explored. Techno-economic analyses of waste heat recovery (WHR) from flue gas using technologies like absorption heat pumps have shown promise in improving overall plant performance [3]. Concurrently, the integration of solar thermal energy into industrial processes has been identified as a viable pathway to decarbonize industrial heat demand [4]. While these technologies have been investigated independently, a significant research gap persists in the holistic analysis of a deeply coupled system where both solar thermal and flue gas heat are strategically used to displace high-grade turbine extraction steam within a WtE thermal cycle.

This paper addresses this gap by proposing and evaluating a multi-energy complementary system based on the principle of cascade energy utilization, using the detailed engineering data of the Shaoxing WtE plant as a case study. The objective is to quantify the increase in net power generation and overall efficiency achieved by substituting extraction steam with on-site renewable and recovered heat sources. Furthermore, this study conducts a rigorous economic viability analysis, calculating the static payback period and critically comparing the proposed thermal integration scheme against a conventional solar photovoltaic (PV) installation. The findings provide a data-driven, practical framework for the energy-saving retrofitting of WtE facilities, presenting a financially robust pathway to enhance their sustainability and resource recovery performance.

2 System Description and Methods

2.1 Baseline System: The Shaoxing WtE Cogeneration Plant

This analysis is founded on the technical specifications and feasibility data provided for the Shaoxing Circular Economy Industrial Park (Phase II) project. The reference system is a modern WtE (Waste-to-Energy) plant designed to process 1,500 t per day of municipal solid waste via two 750 t/d moving-grate incinerators [5]. The plant operates as a cogeneration (CHP) facility, utilizing waste heat recovery boilers to generate superheated steam to drive a 45 MW turbine-generator unit. The key design and operational parameters of the baseline system are summarized in Table 1.

Table 1. Key operating parameters of the baseline WtE plant.

Parameter	Value	Unit
Waste Processing Capacity	2 × 750	t/d
Main Steam Pressure	6.4	MPa
Main Steam Temperature	450	°C
Deaerator Steam Consumption	12	t/h
Primary Air Heating Steam	10.1	t/h
Reported Energy Efficiency	21	%

The thermal architecture of the baseline system is characterized by the use of high-grade extraction steam for low-temperature auxiliary processes [6]. This represents a significant thermodynamic inefficiency ("exergy destruction"), as high-quality steam extracted from the turbine is used for tasks that could be fulfilled by lower-grade heat sources. This "high-grade for low-grade use" scenario presents the primary opportunity for system optimization [7].

2.2 Design of the Multi-Energy Integrated Scheme

The fundamental principle guiding the design of the new schemes is cascade utilization of energy. This approach aims to substitute high-grade turbine extraction steam with appropriately matched medium-and low-grade heat sourced from on-site solar and waste heat recovery systems. This substitution allows the preserved high-grade steam to expand further through the low-pressure stages of the turbine, thereby increasing power generation. The characteristics of the two primary alternative heat sources, as identified in the project materials, are detailed in Table 2.

Table 2. Characteristics of available on-site energy sources for integration.

Energy source	Key parameter	Value	Unit
Solar Thermal	Available Rooftop Area	5,000	m^2
	Annual Direct Normal Irradiance (DNI)	804.3	kWh/m^2
	Proposed Steam Generation	1.2	t/h
	Steam Temperature Range	120–180	°C
	Estimated Investment	2.87	Million CNY
Flue Gas Waste Heat	Flue Gas Temperature (Post-Boiler)	150	°C
	Potential Recoverable Heat	15	MW
	Latent Heat Portion	52	%

The integrated system re-routes thermal energy by coupling these alternative sources at thermodynamically appropriate points [8]. The conceptual framework for this integration aligns with the multi-energy hub model presented in the project overview (see Fig. 1). The specific modifications are as follows:

Solar Thermal to Deaerator Coupling: A parabolic trough solar thermal system, as depicted in the project's technical proposal (see Presentation, Slide 17), is integrated to supply heat directly to the deaerator. The deaerator's operating temperature (~130 °C) is well-matched with the solar system's output temperature range (120–180 °C). This coupling directly substitutes the high-grade extraction steam originally designated for this purpose.

Flue Gas Heat to Low-Temperature Processes: Heat recovered from the 150 °C flue gas stream is allocated to lower-temperature auxiliary processes. Technologies such as absorption heat pumps can upgrade this heat for use in water and leachate treatment [9], which require heating to only 30–35 °C, thus replacing another source of low-pressure steam or electrical heating.

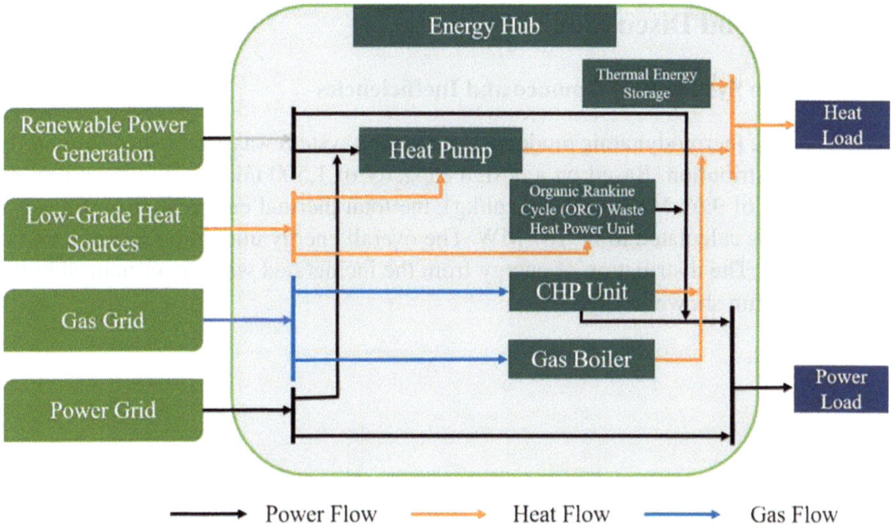

Fig. 1. Conceptual framework with energy flow of the integrated system

2.3 Design of the Multi-Energy Integrated Scheme

To quantify the benefits of the proposed coupling scheme, a comparative analysis was conducted based on thermodynamic calculations and a static economic assessment [10].

Thermodynamic Performance Analysis: The primary metric for energy-saving performance is the additional electricity generated from the conserved extraction steam. The incremental power output (*ΔP*, in kW) is calculated based on the enthalpy drop of the saved steam through the subsequent turbine stages. The enhanced overall energy utilization efficiency (η_{new}) is then calculated as:

$$\eta_{new} = \left(P_{original} + \Delta P + Q_{heat}\right)/E_{waste} \tag{1}$$

where $P_{original}$ is the original power output, Q_{heat} is the cogeneration heat output, and E_{waste} is the total energy input from waste incineration [11].

Economic Viability Analysis: The economic feasibility is assessed using the Static Payback Period (*SPP*) [12]. The *SPP* is thus determined by:

$$SPP = C_{inv}/R_{annual} \tag{2}$$

where R_{annual} is the annual revenue increase calculated from the additional electricity generation, based on the local industrial tariff of 0.6413 CNY/kWh. C_{inv} is the initial investment cost sourced from the project data in Table 2.

This analytical framework provides a direct and robust method for evaluating the proposed energy integration scheme, leveraging the specific and reliable data available from the project documentation.

3 Results and Discussion

3.1 Baseline System Performance and Inefficiencies

A steady-state thermodynamic model of the baseline system was established to quantify its energy distribution. Based on a design capacity of 1,500 t/day and a lower heating value (LHV) of 9.62 MJ/kg (2300 kcal/kg), the total thermal energy input from waste incineration is calculated to be 167 MW. The overall energy utilization rate of approximately 21%. The distribution of energy from the incinerated waste is visualized in the Sankey diagram shown in Fig. 2.

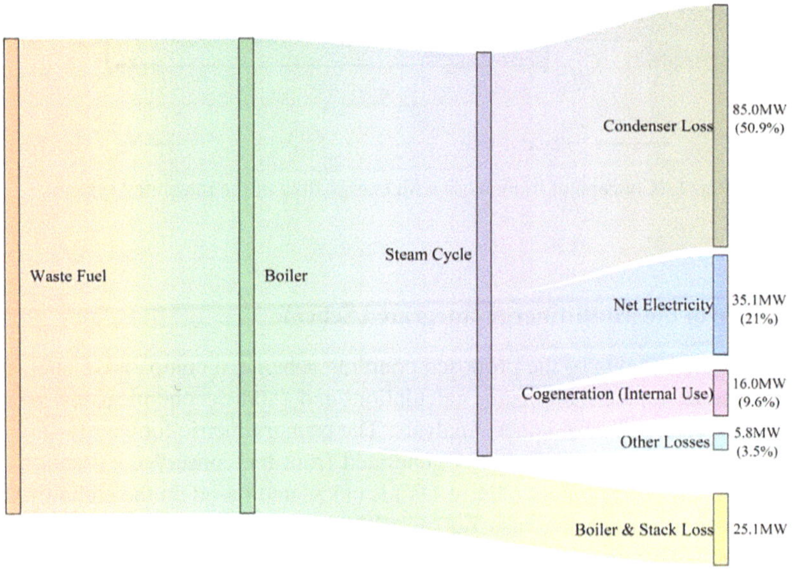

Fig. 2. Distribution of energy from the incinerated waste

3.2 Energy and Environmental Benefits of the Integrated Scheme

The integrated scheme substitutes 2.4 t/h of turbine extraction steam with heat from the solar thermal system (1.2 t/h) and flue gas recovery unit (1.2 t/h). This conserved steam expands through the low-pressure turbine stages, yielding a calculated additional power output of 1.76 MW. This enhancement significantly boosts the plant's net output and efficiency. Over an operational year of 8,000 h, this amounts to an additional 14,080 MWh of electricity generation. The corresponding reduction in CO_2 emissions is estimated at 11,968 t per year, based on the regional grid's average emission factor. The key performance improvements are summarized in Table 3.

Table 3. Comparative Energy Performance of Baseline vs. Integrated Scheme.

Performance metric	Baseline system	Integrated scheme	Improvement
Net Electrical Output (MW)	35.1	36.86	+ 1.76 MW (+ 5.0%)
Annual Net Electricity (MWh)	280,800	294,880	+ 14,080 MWh
Overall Energy Utilization Efficiency	21.00%	22.10%	+ 1.1 percentage points
Annual CO_2 Emission Reduction (tonnes)	0	11,968	–

3.3 Economic Viability and Technology Comparison

The economic analysis demonstrates the strong financial case for the integrated system. The total investment for the solar thermal and heat recovery systems is estimated at 5.37 million CNY. The annual revenue from the additional electricity generation, calculated using the local industrial tariff of 0.6413 CNY/kWh, is 9.03 million CNY. This yields an exceptionally short Static Payback Period (SPP) of just 0.6 years, or approximately 7 months.

This result stands in stark contrast to the alternative solar photovoltaic (PV) proposal detailed in the project materials. As visualized in Fig. 3, the solar thermal component of the integrated scheme, despite a higher investment cost, offers a dramatically faster return than the PV system.

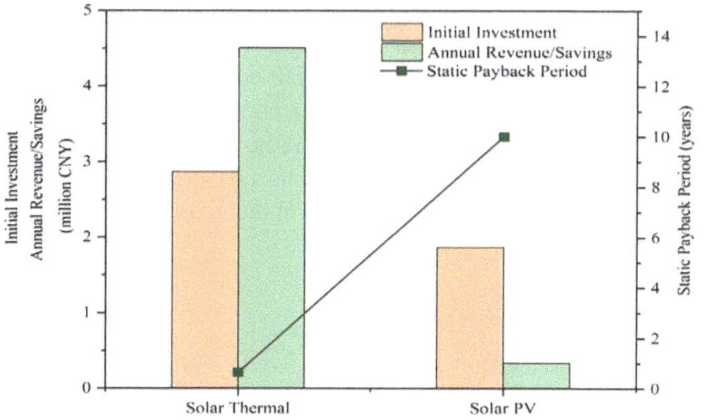

Fig. 3. Techno-economic comparison of solar thermal (for heat) vs. solar

The comparison highlights the superior economic efficiency of using solar thermal energy to directly displace a thermal load within a CHP plant. The PV route (Light → Electricity) is inherently less efficient for this purpose due to multiple energy conversion steps if its output is valued against displacing heat. The solar thermal route (Light → Heat) is a more direct and economically compelling solution for this specific industrial application, maximizing the value derived from the initial solar energy [13].

3.4 System Adaptability and Challenges

The proposed integrated system enhances the plant's operational resilience by diversifying its internal heat sources. This can help buffer the steam cycle against fluctuations in the calorific value of the incoming municipal waste [14].

However, practical implementation must address key challenges identified in the project documentation. The primary challenge is managing the intermittency of solar energy, which necessitates a robust control system to seamlessly manage the dynamic switching between solar heat, turbine extraction, and thermal storage [15]. Secondly, the risk of acid dew point corrosion from deep flue gas cooling is a critical material science and operational issue. The final design must incorporate advanced corrosion-resistant materials for the heat exchangers and a control strategy to bypass deep cooling when flue gas conditions approach the calculated dew point [16], ensuring the long-term integrity and reliability of the equipment.

3.5 Conclusion

This study demonstrates that the integration of solar thermal and flue gas waste heat recovery into a Waste-to-Energy (WtE) cogeneration plant, guided by the principle of cascade energy utilization, offers a technologically sound and economically compelling pathway for significant efficiency enhancement. Based on the case study of the Shaoxing facility, the proposed scheme successfully substitutes 2.4 t/h of high-grade turbine extraction steam, resulting in an additional net power generation of 1.76 MW and increasing the overall system energy utilization efficiency from a baseline of 21.0 to 22.1%. Furthermore, the economic analysis reveals a remarkably short static payback period of just 0.6 years, confirming that in a CHP context, direct thermal substitution provides superior techno-economic returns over photovoltaic integration [17]. Therefore, this research provides a robust and financially viable framework for the energy-saving retrofitting of existing and future WtE plants, contributing a practical solution for enhanced sustainability and resource utilization in the solid waste management sector.

References

1. Chen, H., et al.: Design and evaluation of a municipal solid waste incineration power plant integrating with absorption heat pump. Energy. **294**, 131007 (2024)
2. Chen, L., et al.: Performance evaluation and optimization design of integrated energy system based on thermodynamic, exergoeconomic, and exergoenvironmental analyses. Appl. Energy. **326**, 119987 (2022)

3. Ling, J., Chen, X.: Energy and economic analysis of a new combination Cascade waste heat recovery system of a waste-to-energy plant. Energies. **17**, 5196 (2024)
4. Gil, J.D., et al.: A review from design to control of solar systems for supplying heat in industrial process applications. Renew. Sust. Energ. Rev. **163**, 112461 (2022)
5. Kitto, J.: Solid Waste Management & Modern Waste-to-Energy Technology (2025)
6. Xu, G., et al.: A novel flue gas waste heat recovery system for coal-fired ultra-supercritical power plants. Appl. Therm. Eng. **67**(1), 240–249 (2014)
7. Zhang, L., et al.: Performance analysis of turbine extraction superheat utilization schemes of built power plant. Energy Rep. **6**, 2200–2210 (2020)
8. Yang, D., Wang, M.: Optimal operation of an integrated energy system by considering the multi energy coupling, AC-DC topology and demand responses. Int. J. Electr. Power Energy Syst. **129**, 106826 (2021)
9. Ayou, D.S., Wu, W., Coronas, A.: Absorption-based heat pumps for decarbonization of industrial process heating: performance, current status, and new developments. Therm. Sci. Eng. Prog. **62**, 103679 (2025)
10. Skorek-Osikowska, A., et al.: Thermodynamic, economic and environmental assessment of energy systems including the use of gas from manure fermentation in the context of the Spanish potential. Energy. **200**, 117452 (2020)
11. Li, F., et al.: Measuring the energy production and utilization efficiency of Chinese thermal power industry with the fixed-sum carbon emission constraint. Int. J. Prod. Econ. **252**, 108571 (2022)
12. Rafiq, A., et al.: Life cycle environmental and economic viability analysis of CO2 utilization for chemical production in the cement sector. Sustainable Prod. Consumption. **58**, 364–384 (2025)
13. Pandey, A.K., R, R.K., Samykano, M.: Chapter 1—solar energy: direct and indirect methods to harvest usable energy, in dye-sensitized solar cells. Pandey, A.K., Shahabuddin, S., Ahmad, M.S., Pandey, A.K., Shahabuddin, S. and Ahmad, M.S. (eds.) 2022, Academic. p. 1–24
14. Ren, H., et al.: Integrated optimization of a regional integrated energy system with thermal energy storage considering both resilience and reliability. Energy. **261**, 125333 (2022)
15. Poddar, S., et al.: Changes in solar resource intermittency and reliability under Australia's future warmer climate. Sol. Energy. **266**, 112039 (2023)
16. Zuo, W., Zhang, X., Li, Y.: Review of flue gas acid dew-point and related low temperature corrosion. J. Energy Inst. **93**(4), 1666–1677 (2020)
17. Mahmud, M.A.P., et al.: Environmental life-cycle assessment and techno-economic analysis of photovoltaic (PV) and photovoltaic/thermal (PV/T) systems. In: 2018 IEEE International Conference on Environment and Electrical Engineering and 2018 IEEE Industrial and Commercial Power Systems Europe (EEEIC/I&CPS Europe) (2018)

The Impact of Recyclability on Environmental Performance of Coated Paper Packaging for Food Products

Tomasz Nitkiewicz[1(✉)] and Marcin Rychwalski[2]

[1] Faculty of Management, Czestochowa University of Technology, ul. Dabrowskiego 96, 42201 Czestochowa, Poland
tomasz.nitkiewicz@pcz.pl

[2] Institute of Quality Sciences and Product Management, Department of Products Technology and Ecology, Cracow University of Economics, ul. Rakowicka 27, 31510 Cracow, Poland

Abstract. The strive towards circularity within the packaging sector has been an ever-growing challenge. Regular paper seems to be a good solution due to its natural and renewable origins and relative ease in its end-of-life processing. But food products often require barrier and seal functions or mechanical properties that cannot be provided by regular paper or cardboard. Possibly, the implementation of innovative, highly functional and recyclable coated paper/cardboard materials for food packaging applications would allow for the replacement of traditional multi-layered plastics. The objective of the paper is to verify the environmental performance of selected coated paper materials over their counterparts from multi-layered plastics with the use of LCA methodology. LCA screening is used to compare the environmental impacts of selected food products' life cycles with the use of different packaging materials. Special focus is on the recyclability issue of the packaging materials and their contribution to overall environmental performance. The underlying objective of the paper is to indicate the best environmental practices and scenarios within the development of the innovative coated paper materials and to recommend possible updates to its recyclability scenarios.

Keywords: packaging recyclability · coated paper packaging · packaging for food products · LCA

1 Introduction

The food sector is heavily dependent on packaging as far as its performance is concerned. Packaging in the food supply chain plays a crucial role as it protects and preserves the quality and safety of food products and extends the shelf-life. Extensive use of packaging materials also has side effects in high material and energy consumption within the food life cycle, as well as contribution to its emissions, waste generation and related environmental impacts. Packaging contributes to the impacts that occur in most of the life cycle phases of the food value chain, including also end-of-life phase when the product

Y. Zhu (Ed.): ISESIE 2025, LNCE 000, pp. 76–88, 2026.
https://doi.org/10.1007/978-981-92-2389-3_8

has already been consumed. Therefore, the sustainability, circularity, or simply environmental performance of the food product will be highly affected by the performance of its packaging, with regard not only to its manufacturing but also to its availability for widespread recycling and the effectiveness of related processes.

The circular economy (CE) concept has a long history of conceptual development and application, but about a decade ago, it was introduced as a mainstream policy in EU (EC 2015). Despite many differences in defining circularity, the researchers agree on its dual aim: economic prosperity and environmental quality (Kirchherr et al. 2017). Geissdoerfer et al. (2017) The proposal to define circularity states that it is "a regenerative system in which resource input and waste, emission, and energy leakage are minimized by slowing, closing and narrowing material and energy loops". Additionally, the CE concept has been adopted as a key approach in pursuing sustainability. From the perspective of our paper, we are shifting the focus from the economy as a whole towards only one narrowed sector, which is the packaging in the food sector. The ambition of the European Commission is to achieve 100% reusable and/or recyclable packages in 2030 in EU countries (EC, 2019). Moreover, the packaging companies are under pressure of consumers, supply chain partners, and the recycling sector, making them one of the most important parts in progressing CE forward (Niero and Hauschild 2017).

Recycling plays a crucial role in reducing the environmental impact of paper packaging throughout its life cycle. Studies show that paper and paperboard contribute significantly to packaging waste, but also have high recycling rates (Pilli et al. 2021; Ewijk et al. 2018). Life cycle assessments (LCAs) reveal that recycling can lead to substantial reductions in energy use, greenhouse gas emissions, and other environmental impacts compared to virgin paper production (Manda et al. 2012; Huang and Matthews 2008). However, the environmental benefits of recycling vary considerably across regions due to differences in energy mix, process efficiency, and waste treatment methods (Provost-Savard et al. 2022). While packaging production and end-of-life management contribute to environmental impacts, they can also prevent food waste and associated impacts (Almeida et al. 2021). Innovations in coating materials, such as biodegradable polymers, may improve recyclability without significantly affecting recycled pulp quality (Vukoje and Bota 2025; Nitkiewicz et al. 2024a, 2024b). Overall, increasing recycling rates and improving recycling processes can significantly reduce the life cycle impacts of paper packaging (Ewijk et al. 2018; Nitkiewicz et al. 2023).

The sustainability and circularity challenges have been widely supported by diversified assessment and decision support methodologies and frameworks. Life cycle assessment seems to be one of the most used and perhaps the most trusted methods for complex support of sustainability and circularity-driven decisions (Nitkiewicz and Cappelletti 2022; Rychwalski and Nitkiewicz 2024).

The paper is a follow-up of REPAC2 project, which has proposed a coated paper packaging for food products as a potential alternative to currently used packaging materials, but with a higher recyclability rate and lower environmental impact. The task is to verify whether the CE-driven development of packaging brings in significant improvements in environmental impact and to verify what is the role of recycling in these improvements. The paper presents the partial results of REPAC2 project. The approach is focused on measuring the impact of coated paper packaging with a simple LCA approach.

2 Research Methodology

The research is based on LCA screening of six different food products within current and potential packaging. Since the objectives are focused on checking overall improvement, Carbon Footprint is selected as the impact assessment methodology while preserving all the recommendations in LCA screening procedure. The following section of the paper presents the assessment methodology within a typical LCA structure.

2.1 Goal and Scope

The goal of the assessment is to compare the environmental impacts of innovative packaging for food products with their currently used alternatives. Table 1 presents the list of products included with their current and potential packaging. In each case, the packaging innovation is based on developing packaging material from coated paper and achieving the properties that meet the requirements of different food products. The detailed aim of this study is to estimate the recycling and end-of-life processing with regard to the environmental impact.

The subjects of the assessment are 6 research cases from REPAC2 project. The functional unit is always defined as life cycle of:

1. 1 kg of food product in
2. Its packaging (either coated-paper packaging or its currently used packaging)

The assessment includes the phases of supply, manufacturing, transportation, and end-of-life processing, while the phases of distribution and use are excluded. The coating materials and its composition are currently being developed by the partners of REPAC2 project and remain confidential. Therefore, only basic information on coatings is disclosed, as presented in Table 1.

The reference flow covers all flows related to the 1 kg of functional unit. The reference flow is modelled with primary data from packaging and food product manufacturers and secondary data. The physical allocation approach is used. Any allocation issues within the assessment are solved by estimating the volume of specific flow that is prescribed to the specific functional unit(Nitkiewicz et al. 2024a, 2024b).

Table 1. Products and their current and potential packaging

No.	Type of food product	Proposed coated-paper packaging	Currently used packaging
CS1	diced salami	vacuum metallized CP	foil PET/PE/EVOH/ EVOH/PE
CS2	fruit rolls	acrylic-and vinyl copolymer CP	metalized PP or paper-plastic laminate
CS3	crackers	extrusion polyolefin CP	paper-PP laminate
CS4	truffles	PVOH CP	metalized PP
CS5	tea in tea bags	extrusion EVOH CP	laminated paper PE-EVOH-PE
CS6	chocolate tablet	CP with cold seal	flowpack OPP transparent/ OPP pearl + cold seal

2.2 Life Cycle Inventory

The informational needs of life cycle inventory for the purposes of CF/PEF screening are not so extensive, and therefore, the data providers and sources are limited to:

1. Direct stakeholders

- Food product manufacturers
- Coated paper packaging manufacturers
- Research units from REPAC2 consortium

2. Indirect data sources

- LCI databases (ecoinvent 3, agrifootprint databases)
- Scientific papers and reports and patents

In case of lack of data, the processes or life cycle phases have been estimated based on primary and secondary data. If it was not possible, the processes or life cycle phases were excluded from the assessment.

Recent research shows that the recycling rate of paper and cardboard packaging waste for the EU was 81.5%, 10% of waste was landfilled and 8.5% was incinerated (Eurostat 2021). Research also shows that the level of recycling of post-consumer plastic packaging waste for Europe (EU27 + 3) has reached 46%, 17%—landfill, 37% energy recovery—here we have thermal treatment of waste, also by pyrolysis (Report Plastics-Facts-2022). The recycling processes, as well as remaining end-of-life processing, are modelled according to the data presented above.

2.3 Life Cycle Impact Assessment

The Carbon Footprint assessment is made with the Global Warming Potential (GWP) method, denoted as IPCC 2021 GWP100 v. 1.01. The method takes the time horizon of 100 years as a point of reference and its impacts are expressed in a single unit of emitted kg of CO_2-eq. The GWP100 method includes three categories of impact factors, which are differentiated on the basis of their source: fossil, biogenic and land transformation. Additionally, the category of CO_2 uptake is included in order to evaluate the potential of mitigating greenhouse gas emissions in a manner (PRé Sustainability 2022). The assessment is made in the form of CF screening (Field & Simmons, 2014; ISO 2018). Table 2 presents the basic assumptions on the allocation method, system boundaries and shelf-life of food products. It is important to notice that the shelf-life parameters are only simulated for food products within coated-paper, while original packaging materials are already tested.

Table 2. Summary of basic assumptions for LCA screening

Variable / Parameter	Assumptions
Allocation	Any allocation issues within the assessment are solved by estimating the volume of specific flow that is prescribed to the specific functional unit. Physical allocation is used.
System borders	Cradle-to-grave with exclusion of use phase
Shelf-life	Shelf-life is modelled in the same manner for CP and alternative packaging

2.4 Interpretation and Sensitivity Analysis

Sensitivity analysis is devoted to the investigation of the possibility of improvement of the overall impact by applying recycling scenarios for new packaging. Therefore, the scenario of recycling improvement is considered in order to check how significant it could be while overall CF is concerned. The basic scenario assumes the recycling rate on the basis of EU average for paper packaging (81,5%). Improved recycling scenario assumes an increase in the recycling rate by 10%.

On the basis on recycling of coated paper material experiments, the following parameters have been introduced into the assessment:

1. Yield—a numerical yield parameter expressed in % with regard to the completely recyclable non-coated paper. It is applied to calculate the quantity of pulp obtained after the recycling.
2. Recyclability score—a qualitative parameter calculated on the basis of yield, visual impurities, and sheet adhesion properties, expressed in %. It is applied as the weighting indicator for the quality of the obtained pulp.

Recyclability score can have values within the range of −100 to 100, and its score below 0 indicates that no recycling activities are assumed.

Additionally, sheet adhesion properties are assessed as coherent (could be recycled) and non-coherent (could not be recycled), and influence the recyclability score accordingly, despite other factors. Table 3 presents the recyclability scores for CP packaging included in the assessment.

Table 3. Recyclability scores for investigated CP packaging

Parameter	CS1	CS2	CS3	CS4	CS5	CS6
Recyclability score [pts]	69	70	−10	89	−2	−21

As mentioned before, CP packaging with a score below 0 should not be recyclable with regular paper due to the lack of susceptibility for recycling within the existing technological and organizational framework. In that way, cases 3, 5, and 6 are excluded from the recycling sensitivity analysis.

3 Results

Firstly, the general results of the CF assessment for whole life cycles of products within its packaging variants are presented. When looking at various cases (Fig. 1), it is typically found that packaging has an average impact of about 10% on the entire food product. The results presented here are normalized and show only the share of packaging within the overall impact of food products. The whole life cycle typically includes production processes, transportation with collective packaging and recycling processes. However, in CS1 the contribution of packaging is lower compared to the other cases, but it has the highest impact among the packages examined, giving a carbon dioxide equivalent of 5.0374 [kg CO_2-eq]. This is due to the use of metallized paper in this case, which may increase the emissions associated with packaging production. For comparison, in CS5, where the packaging share is also low and amounts to 2.6%, the carbon dioxide emission equivalent is approximately 0.87 [kg CO_2-eq]. For the remaining cases, the values are below 1.6 [kg CO_2-eq].

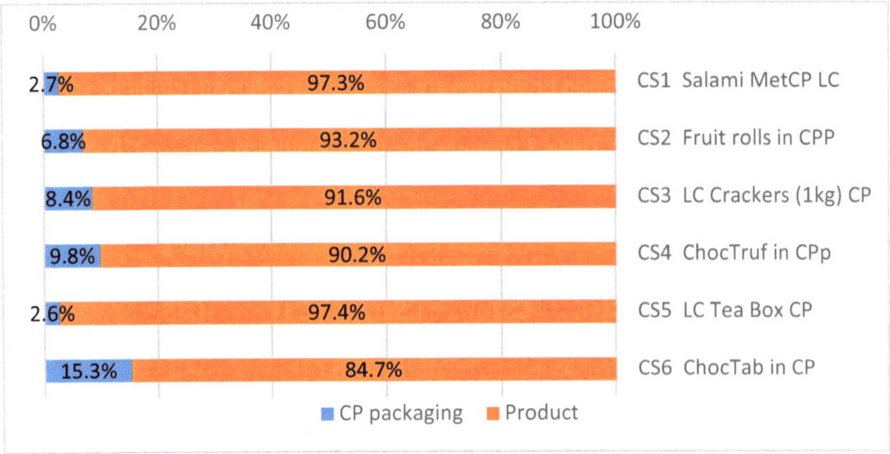

Fig. 1. The contribution of the CP packaging to the overall impact

The overall impact, as measured with CF method, with CO_2 uptake, shows the slight advantage of coated-paper packaging. This is the case in CS1 (0.007 kg CO_2-eq decrease), CS2 (0,2 and 0,7 kg CO_2-eq decrease in comparison to metPP and paper-plastic laminate packaging, respectively), CS3 (0,03 kg CO_2-eq) and CS6 (2,5 kg CO_2-eq). For CS4 and CS5 coated paper packaging, have overall higher impact that amounts to 0,004 and 0,0009 kg CO_2-eq, respectively.

Fig. 2 presents the overall structure of the process contribution to overall CF. In general, four types of processes are considered: manufacturing, transportation, waste disposal (fraction of waste that is not recovered), and recycling (recovered fraction). For clarity reasons, the CS5 has the division of positive ("-" score on the graph) and negative ("+") impacts introduced for selected categories. The high share of positive impacts is due to the use of CF method with CO_2 uptake indicator. The cultivation of herbs for tea, using paper-based packaging, and the biodegradability of the product contribute to CO_2 uptake observed along the life cycle. The recycling capability could supposedly be one of the key advantages of CP packaging materials, especially while sustainability and circularity issues are at stake. This is compared with different paper laminates or plastic alternatives. One of the key objectives of REPAC2 project is to investigate the recyclability of CP packaging and also to check it with regard to environmental consequences.

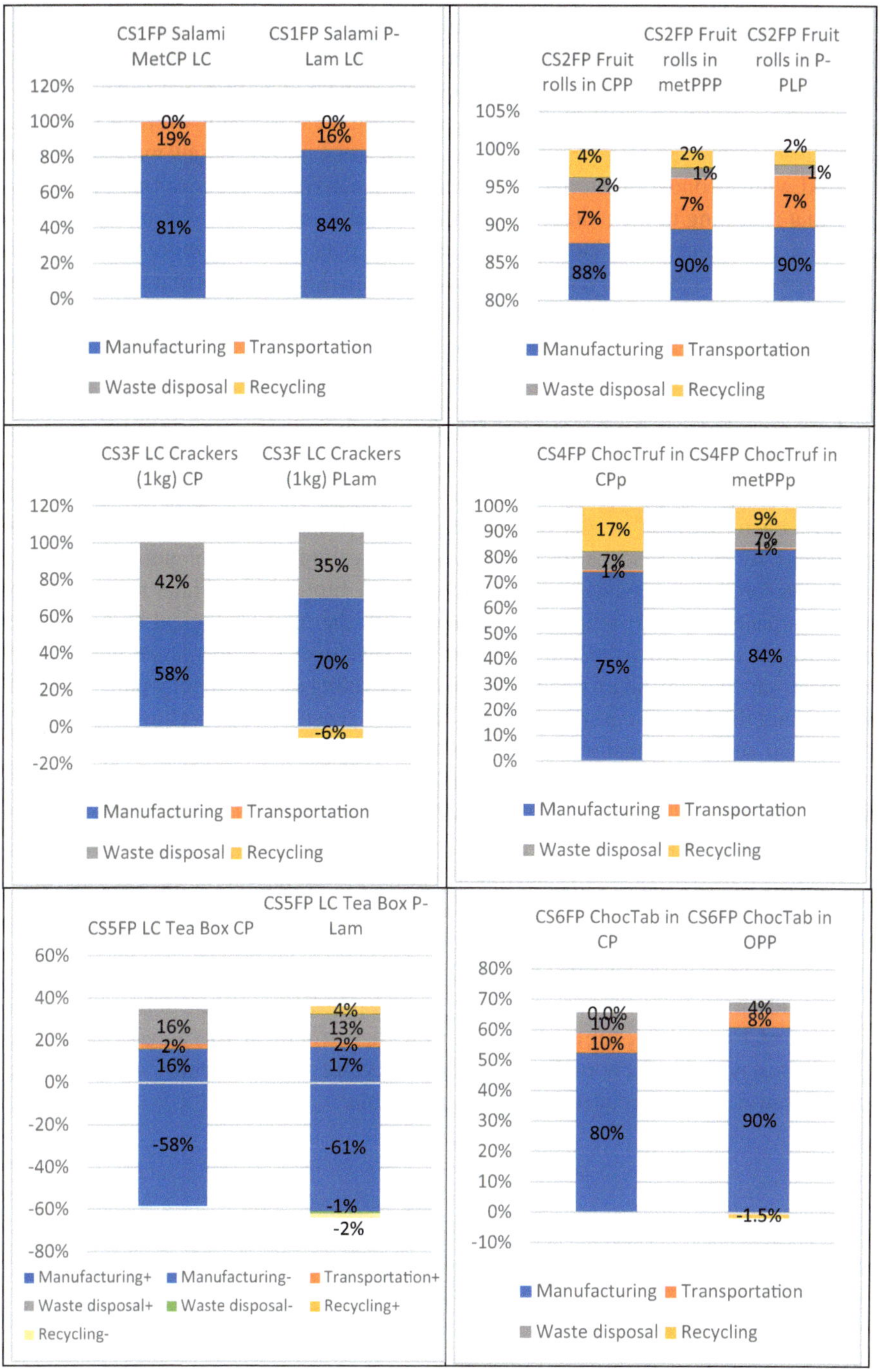

Fig. 2. The structure of processes contribution to the overall impact

According to the results presented on Fig. 2 the recycling has a relatively low impact on overall CF, while the manufacturing phase shows the most significant one. In fact, the end-of-life processes have a two-directional impact. The first one is related to the end-of-life processing and incinerating or landfilling the waste. However, the capacity to decrease the impacts is related to the avoidance of some portion of raw materials that are replaced with secondary material due to recycling activities. The resulting calculations show that returning to the market a material that has a much more harmful potential in relation to paper and plastic laminates or metallized plastic is much more valued in terms of avoided impacts. Secondly, bringing back to the market materials that have significant CO_2 uptake during its vegetation process (like wood), just removes the proportional value of this uptake.

4 Interpretation and Sensitivity Analysis

Fig. 3 sums up the results of the sensitivity analysis on the improvement recycling scenario. Each case is shown for variants with actual packaging, coated-paper packaging, and sensitivity analysis scenarios with assumed recycling improvements. The CF scores are shown for its impact categories, namely: fossil, biogenic, and land transformation-related impacts and CO_2 uptake. As mentioned in the methodological sect. CS3, CS5 and CS6 are excluded from the recycling improvement scenario since the recycling parameters are not satisfactory and do not allow recycling them as regular paper. We could observe that the biogenic impact category is more sensitive to recycling improvement, while fossil and land transformation impacts are rather resistant.

The sensitivity analysis recycling contribution is based on a simple assumption that the recycling scenario would improve by 10% quantity of materials finally recycled. This assumption concerns only the packaging and its elements that are parts of individual packaging, collective packaging and transportation packaging for food products. It means that the effects are calculated for all the materials used for any type of packaging within the whole life cycle.

On average, the biggest influence on CF impact with recycling scenario improvement is observed for the biogenic category and for some part CO_2 uptake category, while food products in CP packaging life cycles are concerned. As for the fossil category, on average, the influence is higher for non-CP packaging life cycles are concerned. Finally, the effect on land transformation-related impacts is negligible.

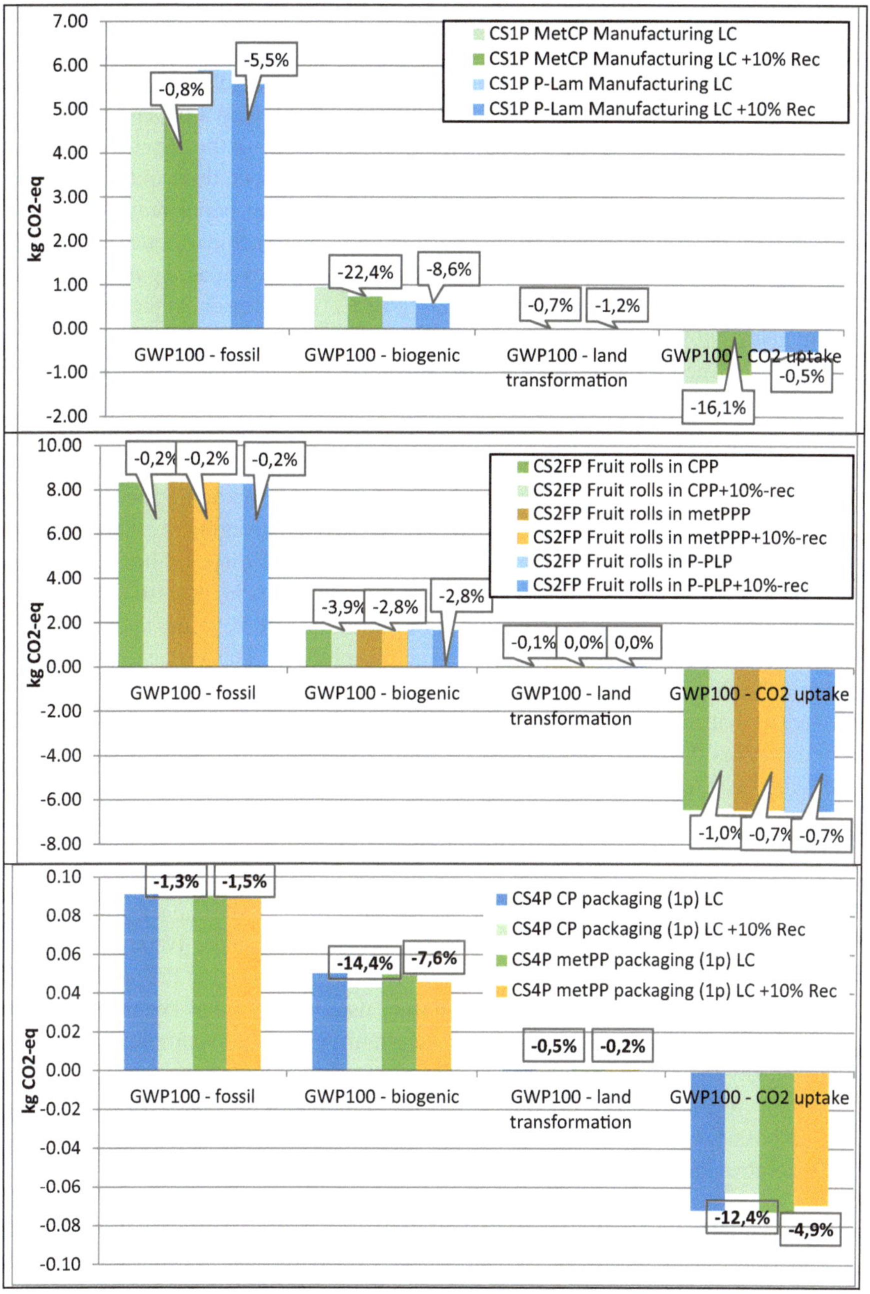

Fig. 3. Sensitivity analysis for recycling improvement within CS1, CS2 and CS4

5 Conclusion

The overall results of the sensitivity analysis with regard to recycling show the potential for a significant overall CF decrease. The challenge is to develop a paper-like recycling scenario for CP packaging and to improve the overall capacity of the paper recycling sector and recycled paper applications. As documented in the paper, three out of six coated-paper materials perform reasonably well within the existing paper recycling system. The remaining three should be developed further in order to achieve better recyclability scores, or just abandoned as sustainable and circular alternative packaging.

The problem is not related to the recycling of CP as paper or its technical aspects only, but also, and perhaps more significantly, to the market and societal solutions that could influence the efficiency of selective waste collection or the capacity of its processing within different recycling processes. Strict and rigorous policy on packaging labelling, a clear and accessible selective waste collection system, continuous development of recycling facilities and their high level symbiosis with industry are only necessary prerequisites while efficiency of introducing CP packaging is concerned.

The contribution of key processes to overall CF from packaging is quite diversified, but some observations could be made. It seems that waste disposal has a significant share, and it usually has less impact for CP solutions than it is in the case of counterpart packaging (4 out of 6 cases). Surprisingly, the contribution of recycling to overall CF for the packaging life cycle is not always better for CP solutions. This is due to the relatively high impact reduction to recovery of plastics.

It seems that the challenge of coated paper use is to find the right technical solution first that fits the recycling capacity. Without it, the potential gains are heavily burdened with waste treatment. Secondly, the requirements of the food sector are very often too strict for any coated paper materials, and the development of packaging is striving towards single-material but plastic construction or biobased variants of laminated plastics or even paper. However, as some cases prove it, coated paper packaging is still an attractive perspective for the food sector, and bearing in mind the capacity and development of the paper recycling system, especially in a longer perspective and with regard to economic terms. The future research should focus on optimizing the performance of the packaging within the strict limits of coating use. It seems that keeping the lowest possible share of coating, avoiding metallized coatings, and a too complex composition would be a key to decreasing the carbon footprint and match the capacity existing recycling system. We could also rely on the fast development of the paper recycling system itself, but possibly due to its current functioning on high workload, it will not be easy to quickly achieve instant progress on an industrial scale.

Acknowledgement. The paper presents partial results from the project titled *Functional and recyclable coated paper packaging for food products* (*REPAC*[2]) that is co-financed by Polish National Centre for Research and Development (NCBR) within CORNET initiative.

References

Almeida, C. et al.: Packaging environmental impact on seafood supply chains: a review of life cycle assessment studies. J. Ind. Ecol. **26**(6), 1961–1978 (2021). https://doi.org/10.1111/jiec.13189

EC: Closing the Loop—an EU Action Plan for the Circular Economy. https://eur-lex.europa.eu/resource.html?uri=cella%0Ar:8a8ef5e8-99a0-11e5-b3b7-01aa75ed71a1.0012.02/DOC_1&format=PDF (2015)

EC: The green deal: sustainable industry. https://ec.europa.eu/commission/presscorner/api/files/attachment/859389/sustainable_industry_nl.pdf.pdf (2019)

EUROSTAT: Recycling rates of packaging waste for monitoring compliance with policy targets, by type of packaging. https://doi.org/10.2908/ENV_WASPACR (2021)

Ewijk, S., Stegemann, J., Ekins, P.: Global life cycle paper flows, recycling metrics, and material efficiency. J. Ind. Ecol. **22**(4), 686–693 (2018). https://doi.org/10.1111/jiec.12613

Field, B., Simmons, C.: Product carbon footprinting for beginners guidance for smaller businesses on tackling the carbon footprinting challenge. https://www.bsigroup.com/LocalFiles/en-GB/standards/BSI-sustainability-guide-product-carbon-footprinting-for-beginners-UK-EN.pdf (2014). Accessed 13 Feb 2023

Geissdoerfer, M., Savaget, P., Bocken, N.M.P., Hultink, E.J.: The circular economy—a new sustainability paradigm? J. Clean. Prod. **143**, 757–768 (2017). https://doi.org/10.1016/j.jclepro.2016.12.048

Huang, Y.A., Matthews, H.S.: Seeking opportunities to reduce life cycle impacts of consumer goods-an economy-wide assessment, pp. 1–6 (2008). doi: https://doi.org/10.1109/isee.2008.4562890

ISO: ISO 14067:2018—Gazy cieplarniane—Ślad węglowy wyrobów—Wymagania i wytyczne dotyczące kwantyfikacji (2018)

Kirchherr, J., Reike, D., Hekkert, M.: Conceptualizing the circular economy: an analysis of 114 definitions. In: Resources, Conservation and Recycling, vol. 127, pp. 221–232. Elsevier B.V (2017). https://doi.org/10.1016/j.resconrec.2017.09.005

Manda, B.M.K., Blok, K., Patel, M.K.: Innovations in papermaking: an LCA of printing and writing paper from conventional and high yield pulp. Sci. Total Environ. **439**, 307–320 (2012). https://doi.org/10.1016/j.scitotenv.2012.09.022

Niero, M., Hauschild, M.Z.: Closing the loop for packaging: finding a framework to operationalize circular economy strategies. Procedia CIRP. **61**, 685–690 (2017). https://doi.org/10.1016/j.procir.2016.11.209

Nitkiewicz, T., Cappelletti, G., Russo, C.: New packaging solutions in food sector for circular business models, Scientific Papers of Silesian University of Technology Organization and Management Series, (211), (2024a). doi: https://doi.org/10.29119/1641-3466.2024.211.29

Nitkiewicz, T., Cappelletti, G.: Verification of circular economy solutions and sustainability of products with life cycle assessment. In: Wojnarowska, M., Ćwiklicki, M., Ingrao, C. (eds.) Sustainable Products in the Circular Economy (1st Edition). Routledge (2022)

Nitkiewicz, T., Wiszumirska, K., Jachimowski, A., Rychwalski, M.: Managing circularity of packaging for food products—carbon footprint assessment of innovative coated paper packaging. In: Current Trends in Quality Science Innovative and Sustainable Products, Materials, and Technologies, pp. 168–178. Poznańskie Towarzystwo Przyjaciół Nauk (2024b). https://doi.org/10.56091/CTQS.Innov-14

Nitkiewicz, T., Wiszumirska, K., Rychwalski, M.: Circular solutions for food packaging. Innovative coated paper packaging and its carbon footprint, Scientific Papers of Silesian University of Technology Organization and Management Series 2023, (186), 513–524, (2023). doi: https://doi.org/10.29119/1641-3466.2023.186.36.

Pilli, T., Baiano, A., Lopriore, G., Russo, C., Cappelletti, G.: An overview on the environmental impact of food packaging. In: Pilli, T., Baiano, A., Lopriore, G., Russo, C., Cappelletti, G. (eds.) Sustainable Innovations in Food Packaging SpringerBriefs in Molecular Science, pp. 1–14 (2021). https://doi.org/10.1007/978-3-030-80936-2_1

PRé Sustainability: SimaPro database manual—Methods library. Https://Support.Simapro.Com/s/Article/SimaPro-Methods-Manual (2022)

Provost-Savard, A., Legros, R., Majeau-Bettez, G.: Parametrized regionalization of paper recycling life-cycle assessment. Waste Manag. **156**, 84–96 (2022). https://doi.org/10.1016/j.wasman.2022.11.018

Rychwalski, M., Nitkiewicz, T.: The role of life cycle assessment in the design and development of sustainable products, pp. 53–69. Routledge eBooks, (2024) doi:https://doi.org/10.4324/9781032710693-5

Vukoje, M., Bota, J.: Effect of PCL nanocomposite coatings on the recyclability of paperboard packaging. Recycling. **10**(2), 53 (2025). https://doi.org/10.3390/recycling10020053

Exploring the Role of Bio-Inspired Designs for Ecosystem Services in Built Environments

Ruaa M. Ismail, Merhan M. Shahda, Sara Eltarabily, and Naglaa A. Megahed(✉)

Architectural Engineering and Urban Planning Department, Faculty of Engineering, Port-Said University, Port Said, Egypt
naglaaali257@eng.psu.edu.eg

Abstract. Through pollution, biodiversity loss, climate change, and other effects, urbanization has a major detrimental influence on ecosystems and the services they provide. This necessitates finding solutions that lessen the harm that the built environment causes to the natural environment. This study investigates how bio-inspired designs contribute to or improve ecosystem services, such as water management, energy provision, climate regulation, and purification. This study analyzes different real-world case studies at various urban scales concerning climate and bio-inspired design, along with the potential ecosystem services they can provide. The findings indicate that while energy and water provisioning are the most prevalent provisioning services, climate regulation and purification are the most common regulating services, while other services like food and raw materials are fairly rare. The findings demonstrate the effectiveness of bio-inspired designs and their potential to increase ecosystem services on a range of scales, contributing to more resilient and sustainable built environments.

Keywords: Bio-inspired design (BID) · ecosystem services (ES) · urbanization · built environment · nature · sustainability

1 Introduction

Ecosystems provide essential goods and services for human well-being, such as providing fresh water and food, regulating climate, forming soil, and providing aesthetic and recreational opportunities. Over the past 50 years, human activity has caused about 60% of ecosystem services (ES) to be degraded or exploited in an unsustainable manner. Through various land-use strategies, human activities impact ecosystems, changing ecological processes and, consequently, ES [1]. Moreover, a number of environmental issues, such as habitat degradation, biodiversity loss, heat island effect, water bloom effect, and congestion effect, have emerged with the fast growth of urbanization. These issues pose a threat to human life and do not support sustainable development [2]. As a result, the structure and function of ecosystems are seriously threatened by the transformation of natural ecosystems into semi-natural, semi-artificial, and artificial ecosystems that occur during the urbanization process. However, as urban civilization develops, there are increasing ecological needs for ES to be provided sustainably. In other words, urbanization's effects on the environment have an effect on the city itself [3].

Y. Zhu (Ed.): ISESIE 2025, LNCE 000, pp. 89–99, 2026.
https://doi.org/10.1007/978-981-92-2389-3_9

Some studies have explored to use green spaces to supply services [4, 5]. Climate regulation, biodiversity, nature experience, recreation, and health are among the urban ES that were evaluated. A study by Yusuf et al. [6] shows how open spaces help mitigate heat, sequester carbon, and regulate water. For example, by promoting energy balance and biogeochemical cycling, areas with high levels of natural vegetation and biodiversity—such as botanical gardens and woodlands—improve urban metabolism. Other research has focused on land use change and how it affects the shortage of services, as well as the need to model and map the spatial flow of ES during the planning process [7, 8]. Furthermore, the infrastructure network offers ES and sustainable urban systems in spite of shifting climate and landscapes, according to the integrated framework for "urban ecological infrastructure" that Li et al. [9] provided. At an ecosystem scale, this infrastructure is an organic fusion of grey (non-living), green (vegetated), and blue (water-based) landscapes, together with arteries (corridors) and exits (outflows, treatment, or recycling).

Having a solid understanding of how ES might be incorporated into built environments can help rethink the design and functioning of cities. Regarding this, the following outlines the concept of ecosystem services and their classifications.

1.1 Ecosystem Services (ES) Classification

The primary definition of ecosystem services (ES) is "the benefits that humans gain from ecosystems, either directly or indirectly." [10]. The 2005 Millennium Ecosystem Assessment (MEA) emphasized the connection between science and policy by distinguishing between providing, regulating, supporting, and cultural ES, while the Final Ecosystem Goods and Services (FEGS) framework in 2007 defined ecosystem services (ES) as natural components used for human welfare [11]. The 2010 The Economics of Ecosystems and Biodiversity (TEEB) project expanded ecological economics, and the 2011 UK National Ecosystem Assessment (UK-NEA) classified ecosystems as processes, intermediate, final, and beneficial products [12]. The Common International Classification of Ecosystem Services (CICES) prioritizes provisioning, regulating, and cultural services [13, 14]. Fig. 1 displays the common ES from various classifications, according to Ismail et al. [14]. These services will be used for analysis in this study.

Fig. 1. Common ecosystem services (ES) from different classifications. *Source* Authors after [14].

ES offer as a reminder that nature is a vital source of life-supporting processes rather than merely a physical location. Thus, the challenge is how we might use nature as inspiration to create bio-inspired designs (BIDs) across urban scales that offer these services in the built environments.

1.2 Bio-Inspired Design (BID)

Bio-inspired design (BID) is a field of study that combines biology/nature, engineering design, and computer science to address design challenges by systematically applying biological knowledge. BID research is ongoing and produces noteworthy findings in a variety of fields. Even while biological systems provide sophisticated solutions, fully utilizing the biological domain's understanding presents difficulties. Because engineers struggle to exploit the vast amount of biological data that comes from the natural world, there have been challenges in the transition from engineering to biological study [15].

Although the concept of ES has helped us better understand the benefits that nature provides to urban environments, its integration into actual design and planning processes is still fragmented and often superficial. Traditional urban planning strategies usually ignore or treat these services independently, falling short of their interrelated potential. This generates an urgent need for innovative techniques that recognize ES and integrate them into practical design solutions. This creates an urgent need for innovative approaches that recognize ES and turn them into practical design solutions. As a result, the current study aims to provide an analysis of real-world bio-inspired case studies and the ES they can provide in built environments at various urban scales.

2 Methodology

This study examines the potential of real-world nature-inspired case studies to provide or improve ES in built environments. To investigate this, we established selection and overview criteria to guide our case study analysis as shown in Fig. 2. The selection criteria centered on finding real-world applications that are widely dispersed across various urban scales and climatic situations, recognized as bio-inspired, and provided by adequate documentation. Following their selection, the instances were examined using a set of overview criteria, including each intervention's design inspiration, implementation scale, and the particular ES it sought to address. These case studies were then thoroughly analyzed in Table 1 to determine their scale: micro (building scale), meso (street and neighbourhood), or macro (city scale), climate according to Koppen classification which divides climates into five groups based on seasonal precipitation and temperature patterns: A (tropical), B (arid), C (temperate), D (continental), and E (polar) with seasonal precipitation subgroups assigned to climates. The inspired ideas and their associated applications were also recognized, highlighting the most significant outcomes and outlining the services that may be offered or improved by these ideas.

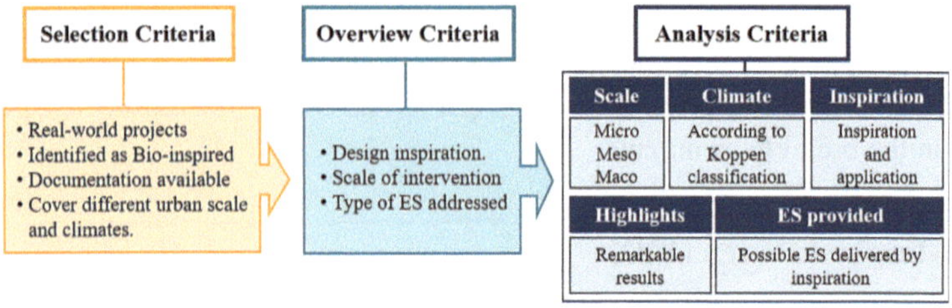

Fig. 2. Methodology chart for case studies selection. *Source* Authors.

Table 1. Real-world case studies of bio-inspired designs.

Scale	Case study	Climate	Inspiration	Highlights	ES provided	References
Micro	Manuel Gea González Hospital, Mexico 	Cwb	Self-cleaning façade inspired by lotus leaf	A 2500 m^2 facade uses Prosolve 370e sheets and titanium dioxide to neutralize pollutants, reducing daily pollution from 1000 cars and minimizing air conditioning emissions	R2, C2	[16]
	Expo Milano—Italy Pavilion, Italy 	Cfa		Over 700 active BIODYNAMIC panels capture air pollutants and convert them into innocuous salts, while solar glass on the roof emits no pollutants and consumes 40% less energy		[17]
	BIQ House, Germany 	Cfb	Bio-reactive façade inspired by Photosynthesis	Uses 120 facade-mounted boards and 200 m^2 closed photobioreactors to produce heat and algal biomass, enabling thermal control, dynamic shading, and solar thermal heat energy	P3, R1	[18]

(*continued*)

Table 1. (*continued*)

Scale	Case study	Climate	Inspiration	Highlights	ES provided	References
	Media-TIC, Barcelona 	Csa	Adaptive breathing envelope inspired by plant stomata	Adaptive envelope with smart sensors and an ETFE (ethylene tetrafluoro ethylene) sun filter to cut CO_2 by 65%. Up to 90% of solar heat gain is reduced by the envelope	R1, R2	[19, 20]
	Al Bahar Towers, Abu Dhabi 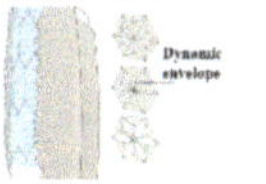	BWh	Dynamic façade inspired by plant adaptive mechanisms	It is anticipated that this dynamic facade will cut CO_2 emissions by 40%, solar heat gain by 50%, and solar radiation entering it by 20%	R1, R2	[21]
	California academy of sciences, USA 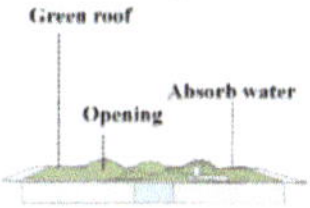	Csb	Green roof inspired by green spaces, natural photosynthesis.	The Living Roof converts carbon dioxide into oxygen, collects 98–100% of stormwater, and offers insulation, reducing energy demand. Its seven hills are covered in biodegradable vegetation trays, filled with 1.7 million plants, creating an oasis for various species	P2, R1, R3, C1, C2, C3	[22]
	Austrian Post distribution Centre, Wernberg 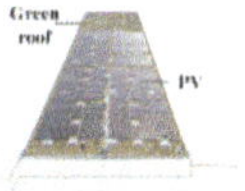	Cfb	PV-green roof inspired by plants photosynthesis	A massive farm of 3,528 solar panels was converted into a 35,000 m^2 green roof. This bio-solar roof provides habitats. 89,000 kWh of energy are produced on average each year	P3, R1	[23]
	Bosco Verticale, Milan, Italy 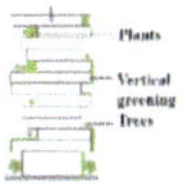	Cfa	Green spaces inspired vertical greening	The building, with over 40 trees and plants, absorbs 30 tons of CO_2 annually, maintains a 2–3 °C lower temperature, and uses solar panels for irrigation	P1, R1, R2, C2	[24]

(*continued*)

Table 1. *(continued)*

Scale	Case study	Climate	Inspiration	Highlights	ES provided	References
	Eastgate Centre—Harare, Zimbabwe 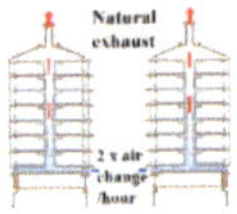	Cwa	Passive cooling inspired by termite mound	uses only 10% of the energy required by a typically cooled structure through passive cooling, and 35% less energy to maintain the same temperature when actively cooled	R1	[25]
	Gando Primary School, Burkina Faso 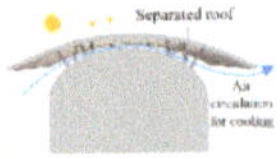	Aw	Separated roof inspired by desert snail	Curved, separated surfaces reflect more heat than flat ones, preserving lower temperatures and minimizing heat transfer to the interior from dispersed solar radiation	R1	[26]
Meso	The Supertrees, Singapore 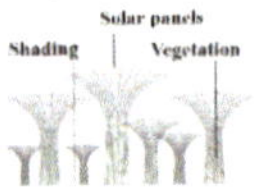	Af	Trees inspired tree-like structure	Super-trees, resembling trees, are large metal structures covered in vegetation, serving habitat, cooling, and shade. Six of them generate electricity through solar panels	P2, P3, R1, R2, C1, C2, C3	[27]
	CityTree, Germany 	Cfb	Benches inspired by street trees.	Uses an automated nutrient and water supply system with 72 sensors, solar system, wooden seat, and sustainable rainwater management, reducing tension and cooling the environment by up to 17 K through evaporation	P2, R1, R2, C2	[28]
	Smog-Free Tower, Beijing 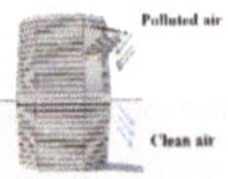	Dwa	Towers inspired by the lungs and leaves remove pollutants	The device filters 30,000 m^3 of dirty air per day, using solar power, and creates dark gemstones from collected soot, which includes 42–45% carbon	P4, R2	[29]

(continued)

Table 1. *(continued)*

Scale	Case study	Climate	Inspiration	Highlights	ES provided	References
	Metropol Parasol, Seville 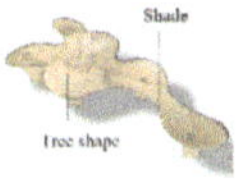	Csa	Outdoor shading inspired by Ficus trees.	It's climatic strategies significantly reduced temperatures by over 3 °C compared to nearby open spaces	R1, C2, C3	[30]
	the Terra Pavilion, Dubai's Expo City 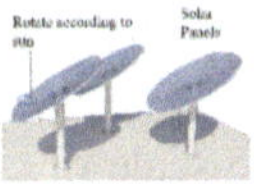	BWh	Energy trees Inspired by Ghaf Tree	The canopy, which consists of eight rotating Energy Trees, is intended to provide tourists with much-needed shade while producing 4 gigawatt hours of electricity annually	P3, R1, C3	[31]
Macro	Chicago Green Alleys Program 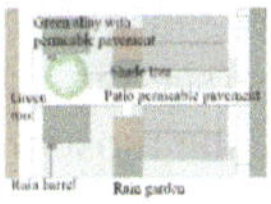	Dfa	Inspired by natural reflective surfaces and water filtration systems	Permeable pavement lowers runoff, relieves pressure on sewer systems, recharges ground water, and filters pollutants. High albedo pavement lowers UHI effects and enhances air quality	P2, R1, R3	[32]
	Vauban, Freiburg, Germany 	Cfb	Developing self-sufficient urban district that resembles natural ecosystems	The project utilized green roofs, gardens, permeable surfaces, and drainage sloughs to conserve water, reduce rainfall, and produce biogas, with 93% of the neighborhood's electricity coming from renewable sources	P2, P3, R1, R3, C1, C2, C3	[33, 34]
	Floating Houses, IJburg 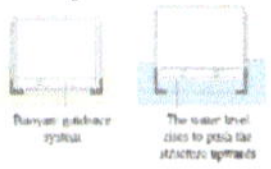	Cfb	Floating houses inspired by seaweeds such as kelp and Venus's flower	The building's flood exposure is adjusted by rising and lowering with water levels, ensuring stability through diagonal mooring poles and sliding connections to absorb vertical movement	R3	[35, 36]

3 Results and Discussion

This section summarizes the findings and outcomes derived from the investigation of bio-inspired cases of ecosystem service provision. By analyzing the case studies, we discover many services are available at all three urban scales. At the micro scale, the building envelope has an essential impact on climate regulation, purification, and aesthetic values, whether through green roofs and facades or other concepts such as dynamic facades and passive cooling technologies that work to reduce the urban heat island effect (UHI), which contributes to climate regulation. Modern materials, such as self-cleaning facades, also help with pollutant and air purification. When taken to the next level, the meso scale reveals many elements in outdoor spaces that are inspired by trees, many of which serve functions such as providing shade or purifying the air, as well as the process of photosynthesis to provide energy, which has a clear impact on adding aesthetic values and recreation. Furthermore, when we explore at the larger, macro-scale of cities, we can see several bio-inspired attempts and advances to provide water, energy, and climate regulation. Moreover, at this scale, the use of permeable pavements and green infrastructure to absorb water reduces extremes such as flooding more than previous scales. The variety of climates—tropical, temperate, and arid—showcases the possibility for developing inspired ideas in all regions.

When we examine the frequency of services in the case studies (see Fig. 3), we discover that climate regulation, purification, and aesthetic values are the most prominent, followed by energy and water, while food and raw material supply is quite low.

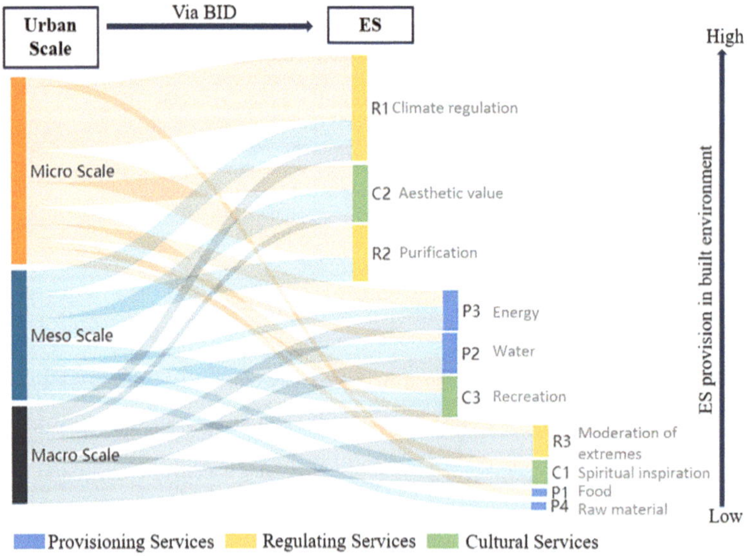

Fig. 3. Frequency of ES provided by BID in various urban scales. *Source* Authors.

Several studies are being conducted to develop more bio-inspired ideas, such as retro-reflective materials, inspired by Galanthus nivalis petals, effectively reduce urban heat islands and gain by reflecting solar radiation in the direction of incidence [37]. Additionally, a lot of research has been carried out on the usage of inspired facades, including hydro-active facades, which use evaporative cooling and rainwater harvesting to reduce runoff water, lowering the load on urban sewage systems and lowering the risk of flooding by absorbing wind-driven rain [38]. Further, kinetic facades use responsive elements to allow buildings to dynamically adjust to functional, aesthetic, and environmental requirements [39].

In this regard, numerous ideas are still being developed that could have a substantial impact on providing ecosystem services in the future. As a result, we conclude that BID plays a crucial role in providing ES in built environments over a wide range of climates. However, they necessitate more attention, a wider range of ideas, and a focus on delivering these services from the start. This should also focus on neglected services and broaden to include services other than those stated above to promote better balance and resilience in built environments.

4 Conclusion

This study examined case studies inspired by nature in diverse urban regions to determine their importance to ES provision. The findings revealed that the majority of applications focused on regulating ES, particularly climate and air purification, followed by energy and water, and food and raw materials received less attention. BID strategies were most commonly used at the building level, with fewer instances at the street or city level. By framing the investigation from ES lens, this study proved how BID contributes to urban sustainability.

Practically, this means that ES considerations should be integrated early in the design process to achieve multifunctional outputs. Also, the potential for street features or building envelopes to contribute to moderate extremes. Further empirical investigation into such projects is required to determine their true environmental benefits, particularly in terms of biodiversity, provisioning and cultural services, and other services not covered in this study. Collaboration across disciplines is also essential to fully realize the benefits of implementing BID principles in the built environment. Artificial intelligence tools such as machine learning and big data can be extremely useful at the first brainstorming stage.

Finally, combining BID with ES provides a pathway to more resilient and regenerative urban environments. Bridging current gaps can help nature reach its full potential as a paradigm for sustainable urbanization.

References

1. Deng, C., Liu, J., Nie, X., et al.: How trade-offs between ecological construction and urbanization expansion affect ecosystem services. Ecol. Indic. **122** (2021). https://doi.org/10.1016/j.ecolind.2020.107253
2. Yao, J., Xu, P., Huang, Z.: Impact of urbanization on ecological efficiency in China: an empirical analysis based on provincial panel data. Ecol. Indic. **129** (2021). https://doi.org/10.1016/j.ecolind.2021.107827

3. Peng, J., Tian, L., Liu, Y., et al.: Ecosystem services response to urbanization in metropolitan areas: thresholds identification. Sci. Total Environ. **607–608** (2017). https://doi.org/10.1016/j.scitotenv.2017.06.218
4. Breuste, J., Schnellinger, J., Qureshi, S., Faggi, A.: Urban ecosystem services on the local level: urban green spaces as providers. Ekol Bratislava. **32** (2013). https://doi.org/10.2478/eko-2013-0026
5. Grunewald, K., Bastian, O.: Maintaining ecosystem services to support urban needs. Sustainability. **9** (2017)
6. Yusuf, D.A., Zhu, J., Khaing, C.T., et al.: Regulating urban metabolism in semi-arid regions: classification and valuation of urban open spaces ecosystem services in metropolitan Kano. Environ. Dev. **54**, 101165 (2025). https://doi.org/10.1016/J.ENVDEV.2025.101165
7. Song, W., Deng, X.: Land-use/land-cover change and ecosystem service provision in China. Sci. Total Environ. **576** (2017). https://doi.org/10.1016/j.scitotenv.2016.07.078
8. Tolessa, T., Senbeta, F., Kidane, M.: The impact of land use/land cover change on ecosystem services in the central highlands of Ethiopia. Ecosyst. Serv. **23** (2017). https://doi.org/10.1016/j.ecoser.2016.11.010
9. Li, F., Liu, X., Zhang, X., et al.: Urban ecological infrastructure: an integrated network for ecosystem services and sustainable urban systems. J. Clean. Prod. **163** (2017). https://doi.org/10.1016/j.jclepro.2016.02.079
10. Pedersen Zari, M.: Ecosystem services analysis: mimicking ecosystem services for regenerative urban design. Int. J. Sustain. Built Environ. **4**, 145–157 (2015). https://doi.org/10.1016/J.IJSBE.2015.02.004
11. Landers, D.H., Nahlik, A.M.: Final ecosystem goods and services classification system (FEGS-CS) (2013)
12. Haines-Young, R., Potschin, M., Fish, R.: (2012) Classifying Ecosystem Services: An Ecosystems Knowledge Network Briefing Paper
13. CICES CICES, Towards a common classification of ecosystem services. http://cices.eu/. Accessed 8 Jan 2024
14. Ismail, R.M., Naglaa, A.M., Merhan, M.S., Eltarabily, S.: Bio-inspired design: leveraging nature for enhanced ecosystem services. Archit. Eng. Des. Manag. **21**(6), 1317–1348. https://doi.org/10.1080/17452007.2025.2480835
15. Stone, R.B., Goel, A.K., McAdams, D.A.: Charting a course for computer-aided bio-inspired design. In: Biologically Inspired Design (2014)
16. Ismail, A., Hosney, H.: The role of buildings envelope renovation to improve the visual image for existing buildings toward ecological balance. In: Advances in Science, Technology and Innovation (2022)
17. The impact of nanotechnology on exterior and interior finishing materials in hospitals. Int. J. Mech. Eng. (2021) https://doi.org/10.56452/2022-12-011
18. Hanafi, W.H.H.: Bio-algae: a study of an interactive facade for commercial buildings in populated cities. J. Eng. Appl. Sci. **68**, 37 (2021). https://doi.org/10.1186/s44147-021-00037-5
19. Elsakksa, A., Marouf, O., Madkour, M.: Biomimetic approach for thermal performance optimization in sustainable architecture. Case study: office buildings in hot climate countries. In: IOP Conference Series: Earth and Environmental Science (2022)
20. López, M., Rubio, R., Martín, S., Croxford, B.: How plants inspire façades. From plants to architecture: biomimetic principles for the development of adaptive architectural envelopes. Renew. Sust. Energ. Rev. **67** (2017)
21. Jamilu, G., Abdou, A., Asif, M.: Dynamic facades for sustainable buildings: a review of classification, applications, prospects and challenges. Energy Rep. **11**, 5999–6014 (2024). https://doi.org/10.1016/J.EGYR.2024.05.047

22. California Academy of Sciences Living roof. https://www.calacademy.org/exhibits/living-roof. Accessed 14 Aug 2025
23. Pritchard & Pritchard: The biggest and best biosolar roofs from around the globe (2024). https://green-roofs.co.uk/worlds-biggest-biosolar-roofs/. Accessed 19 Aug 2025
24. BOERI Vertical Forest Milan. https://www.stefanoboeriarchitetti.net/en/project/vertical-forest/. Accessed 15 May 2025
25. Solano, T., Bernal, A., Mora, D., Chen Austin, M.: How bio-inspired solutions have influenced the built environment design in hot and humid climates. Front. Built. Environ. **9** (2023). https://doi.org/10.3389/fbuil.2023.1267757
26. Hamdy, E.A.: Achieving thermal comfort by simulating the ecosystems of the desert environment BT—advances in architecture, engineering and technology. In: Rosso, F., Fabiani, C., Altan, H., Amer, M. (eds.) , pp. 227–242. Springer International Publishing, Cham (2022)
27. Beatley, T.: Handbook of Biophilic City Planning and Design (2017)
28. Sänger, P., Splittgerber, V.: The CityTree: a vertical plant filter for enhanced temperature management BT—innovation in climate change adaptation. In: Leal Filho, W. (ed.), pp. 75–85 Springer International Publishing, Cham (2016)
29. Januszkiewicz, K., Kowalski, K.G.: Air purification in highly-urbanized areas with the use of TiO_2. New approach in designing urban public space with beneficial human condition. IOP Conf. Ser. Mater. Sci. Eng. **603**, 032100 (2019). https://doi.org/10.1088/1757-899X/603/3/032100
30. del Guayo, P.M., Yannas, S.: Improving outdoor urban environments: three case studies in Spain. In: 30th International PLEA Conference: Sustainable Habitat for Developing Societies: Choosing the Way Forward—Proceedings (2014)
31. UrbiAtlas: Seeking shadow a cool fix for hot cities (2024). https://atlas.urbi.ae/seeking-shadow/. Accessed 21 Apr 2025
32. Chicago, C.: The Chicago Green Alley Handbook: An Action Guide to Create a Greener, Environmentally Sustainable Chicago. City of Chicago Department of Transportation (2010)
33. Fraker, H.: Vauban, Freiburg, Germany. In: The Hidden Potential of Sustainable Neighborhoods (2013)
34. East, M.: Rising cities: continuity, innovation and deliberation of Vauban District of Freiburg, Germany. Ecocycles. **4** (2018). https://doi.org/10.19040/ecocycles.v4i2.106
35. Slesson, C.: Floating houses, the Netherlands by Marlies Rohmer architects & planners. Archit. Rev. (2013). https://www.architectural-review.com/awards/ar-house/floating-houses-the-netherlands-by-marlies-rohmer-architects-planners. Accessed 20 Aug 2025
36. Ameh, H., Badarnah, L., Lamond, J.: Amphibious architecture: a biomimetic design approach to flood resilience. Sustainability. **16** (2024)
37. Wang, J., Liu, S., Meng, X., et al.: Application of retro-reflective materials in urban buildings: a comprehensive review. Energ. Buildings. **247** (2021)
38. Eisenbarth, C., Haase, W., Blandini, L., Sobek, W.: Potentials of hydroactive lightweight façades for urban climate resilience. Civ. Eng. Des. **4** (2022). https://doi.org/10.1002/cend.202200003
39. Razoki, F., Al-kazzaz, D.: Biomimetic strategies in kinetic architecture: a comparative analysis of nature-inspired roof and Façade designs. Int. J. Des. Nat. Ecodynamics. **20**, 1269–1282 (2025). https://doi.org/10.18280/ijdne.200608

Biomimetic Self-shading Buildings: An Analytical Approach to Improving Environmental Performance in Hot and Dry Climates

Mariam Salama, Merhan M. Shahda, Sara Eltarabily, and Naglaa A. Megahed(✉)

Architectural Engineering and Urban Planning Department, Faculty of Engineering, Port-Said University, Port Said, Egypt
Naglaaali257@eng.psu.edu.eg

Abstract. Biomimetic Self-Shading (BSS) is an effective passive design strategy to enhance energy efficiency and environmental performance in hot and dry climates. Inspired by natural systems, BSS techniques offer a solution by incorporating and designing building geometry (BG) or façade components (FC) that cast shades on their envelope. This study aims to present a comprehensive analytical framework that addresses the design methods of BSS buildings, their main variables, and the metrics used to evaluate their effectiveness. The study analyzes the potential of BSS techniques to lower heat gain and cooling consumption with a focus on solar radiation (SR) and self-shading calculations. The results highlight the design of FC and BG self-shading with their improvement methods, as well as their environmental performance improvement indicators. Furthermore, SR simulation tools and self-shading calculation methods are systematically classified and analyzed. Finally, the study concludes with a set of design guidelines for scalable and climate-responsive BSS buildings. The study also suggests future research directions that focus on integrating biomimetic principles into self-shading buildings. This approach aims to reduce energy consumption and serves as a practical pathway toward achieving sustainable and adaptive architecture.

Keywords: Biomimetic · Self-Shading · Environmental Performance Indicators · Solar Radiation (SR)

1 Introduction

Climate change represents an increasingly global challenge amid the significant rise in energy consumption and carbon dioxide emissions. The construction sector is one of the main contributors to the exacerbation of this environmental crisis [1]. Building envelopes act as "climate regulators" and "energy generators" through their ability to interact with SR via multiple mechanisms, including reflecting, collecting, storing, converting, and distributing light, in addition to providing shading, comfort, protection, and visibility [2]. In this context, biomimetic levels—drawing inspiration from biological forms and functions—emerge as a promising approach for improving EPII of building

Y. Zhu (Ed.): ISESIE 2025, LNCE 000, pp. 100–116, 2026.
https://doi.org/10.1007/978-981-92-2389-3_10

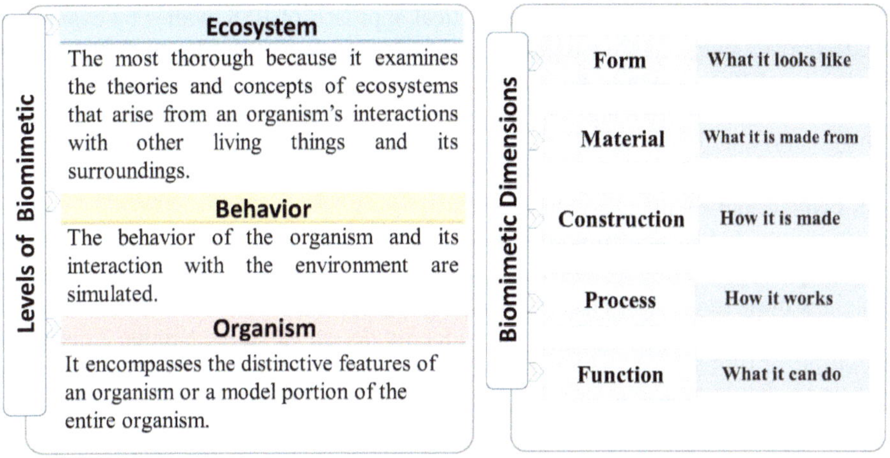

Fig. 1. The levels and dimensions of biomimetic. *Source* The authors based on [7, 8]

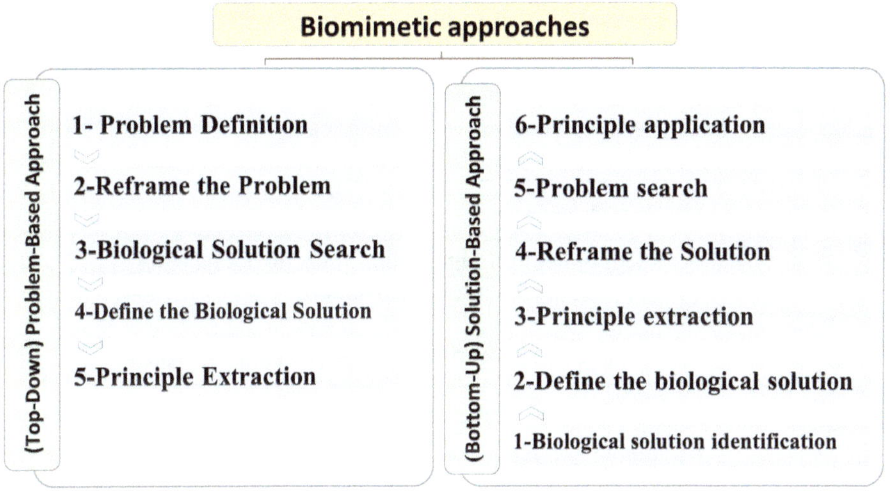

Fig. 2. Two main approaches of the biomimetic science. *Source* The authors based on [7, 8].

envelopes, as illustrated in Fig. 1. By studying organisms' millions of years of evolution to maximize their performance and survival, scientists and engineers can create novel solutions that resemble the elegance and efficiency of the natural world. Biomimetic science utilizes two primary approaches [3], as illustrated in Fig. 2. Among the nature-inspired strategies, self-shading mechanisms offer an effective model for passive solar design [4], especially in hot and dry climates. BSS is a design solution that utilizes BG or FC to create shading on specific areas of the envelope itself. Unlike traditional shading systems or reliance on surrounding buildings, this strategy aims to reduce direct SR on the envelope, contributing to lower cooling loads and enhancing thermal and visual comfort [5].

This research paper explores an analytical approach of BSS studies by examining the design variables of FC or BG. In addition, identifying EPII. The study focuses on the ability of BSS design to reduce SR, improve thermal comfort, and lower energy consumption [6]. By studying natural adaptations, thermal regulation mechanisms, and surface morphology, we propose design guidelines to apply these principles to self-shading design. The study also provides an in-depth understanding of various design BSS systems and simulation methods through comparative analysis of EPII.

The methods for calculating self-shading and the simulation tools for SR metrics are also analyzed, with the aim of identifying the most effective tools and methodologies for designing these studies [7]. The paper has been organized as follows: Sect. 2 presents the collection and analysis of BSS studies to identify design variables. Section 3 discusses the results of BSS analysis and the tools for evaluating various aspects of EPII. Finally, Sect. 4 presents conclusions with a future research direction on integrating biomimetic principles into self-shading design. Figure 3 presents the proposed research framework for this study.

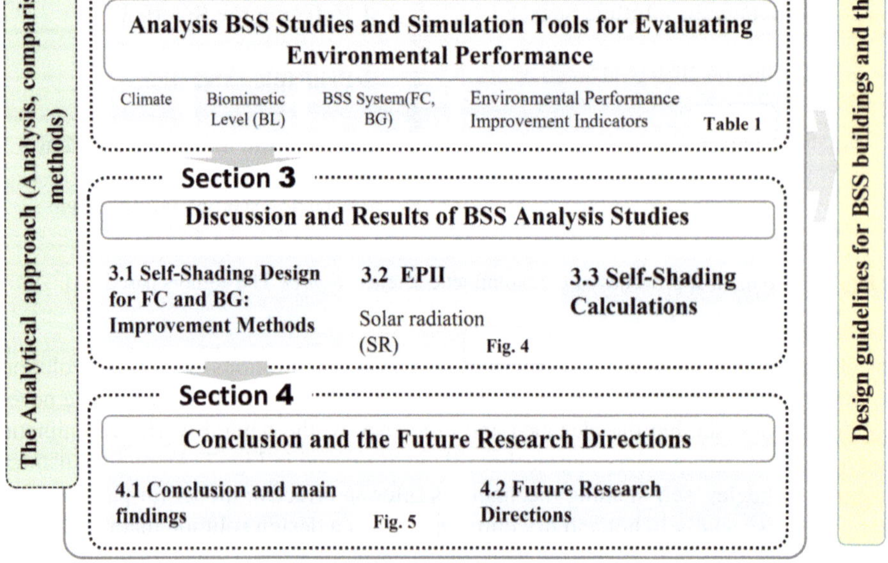

Fig. 3. A graphical representation of the methodological framework. *Source* Author's own creation.

2 Analysis BSS Studies and Simulation Tools for Evaluating Environmental Performance

This study relied on a systematic collection and analysis of studies related to BSS buildings through the Scopus, Web of Science, and Google Scholar databases. A limited search was conducted in each database using keywords (adjusted as needed), and the reference lists of all selected studies were scanned. In addition to selected studies addressing self-shading facades, this analysis was limited to studies that:

a) focused on building-related BSS and its impact on improving environmental performance.
b) discussed about self-shading techniques that can be used to the building shape as a whole or to facade components.
c) published between 2014 and 2025 to ensure relevance.
d) excluded studies that investigated self-shading provided by kinetic facades or photovoltaic panels. Table 1 provides a classification of BSS studies and their indicators for improving environmental performance. Finally, the total number at the end of the table indicates the prevalence and scientific use of this metric.

Table 1. Analysis of BSS studies categorized by climate, biomimetic level (BL), façade component (FC), building geometry (BG), and their environmental performance improvement indicators (EPII) with the program used.

References	Climate	(BL)	BSS system		Environmental performance improvement indicators (EPII)										
			FC	BG	Daylight	OTTV	Air flow	Surface temp	Air (Op) temperature	SR/gain	Energy consumption	Thermal comfort	Heating-cooling load	Sunlit hours/area	Program used& simulation tools
[8]	Hot climate	Organism level	X		X						X		X	X	Grasshopper plugin, Rhino 3D modeler and DIVA plugin
[9]	Hot-summer Mediterranean climate	Organism-behavior level	X		X					X			X	X	Rhino, Grasshopper, and Diva
[10]	(Hot Deserts, climates	Organism behavior level	X		X										Rhino 6, Grasshopper, and Diva simulation
[11]	Hot climate	Organism behavior level		X	X					X	X			X	"Grasshopper" in Rhino, ladybug & Honeybee plugins
[7]	Hot and dry climate	Organism behaviour level	X					X		X					Design Builder Program
[12]	(Hot-semi arid	NA		X						X	X				computer model SustArc ENERGY simulation
[13]	Tropical rainforest	—		X	X				X						HD35-data logger Lux meter sensor
[14]	Tropical climate	—		X	X					X	X		X		Design Builder simulation

(continued)

Table 1. *(continued)*

References	Climate	(BL)	BSS system		Environmental performance improvement indicators (EPII)										
			FC	BG	Daylight	OTTV	Air flow	Surface temp	Air (Op) temperature	SR/gain	Energy consumption	Thermal comfort	Heating-cooling load	Sunlit hours/area	Program used& simulation tools
[15]	Oceanic climate	—		X		X		X					X		Rhino, Grasshopper simulation, and Energy Plus
[16]	Hot deserts	—	X								X		X		Design Builder simulation
[17]	Hot summer	—		X						X			X		Energy Plus
[18]	Tropical climate	—		X	X					X					Rhino, Grasshopper and Octopus simulation
[19]	Hot summer	Behavior level	X			X	X	X				X			Energy Plus, Ladybug for grasshopper
[20]	Moderate temperatures &high humidity	—		X					X		X				Design Builder simulation
[21]	Hot summer	—	X							X			X		Design Builder
[22]	Hot summer Mediterranean climate	—	X		X						X		X		Design builder BM
[23]	Humid continental climate	—	X		X						X				Rhino, Grasshopper and Energy plus
[24]	Hot humid climate	—		X		X		X	X						Ladybug for Grasshopper
[25]	Hot summer and cold	—	X		X			X				X		X	Energy Plus

(continued)

Table 1. *(continued)*

References	Climate	(BL)	BSS system		Environmental performance improvement indicators (EPII)										
			FC	BG	Daylight	OTTV	Air flow	Surface temp	Air (Op) temperature	SR/gain	Energy consumption	Thermal comfort	Heating-cooling load	Sunlit hours/area	Program used& simulation tools
[26]	Cold semi-arid hot semi-arid mild-latitude	—		X	X						X				Grasshopper and Rhino &Radiance, Daysim, and Energy Plus
[27]	Hot Deserts climate	—		X						X					Utilizing design chart
[28]	Varies climate	—		X	X					X	X				Rhino + Grasshopper simulation
[29]	Tropical rainforest climate	—		X		X									OTTV manual calculation
[30]	Hot deserts climate	—		X	X										Rhino & Grasshopper simulation
[31]	Hot deserts climate	Behaviour level		X			X	X					X		Rhino, Octopus plugin & Energy Plus, and DIVA
[32]	Hot-summer Mediterranean	—		X	X			X		X					Rhino + Grasshopper simulation
[33]	Hot summer climate			X			X			X			X		DIVA for daylight simulations
[34]	Hot humid climate			X	X						X				Rhino & Grasshopper Simulation, DIVA
[35]	Hot humid climate		X		X						X				Energy Plus
[36]	Tropical climate			X					X	X				X	Sketchup, Open studio, and SPSS

(continued)

Table 1. *(continued)*

References	Climate	(BL)	BSS system		Environmental performance improvement indicators (EPII)										
			FC	BG	Daylight	OTTV	Air flow	Surface temp	Air (Op) temperature	SR/gain	Energy consumption	Thermal comfort	Heating-cooling load	Sunlit hours/area	Program used& simulation tools
[37]	Hot humid climate		X							X					Grasshopper, 3D FEA software
[38]	Hot deserts climate		X				X	X							Energy Plus
[39]	Tropical climate		X					X							Rhino & Grasshopper simulation
[40]	Hot summer and cold		X		X						X				BPS tools
[41]	Hot humid climate		X							X					-
[42]	Hot deserts climate		X					X						X	Energy Plus
[43]	Hot summer climate			X						X					"Grasshopper" in Rhino, ladybug & Honeybee plugins
[44]	Hot humid climate		X		X										Simulations in Energy Plus, and DIVA
[45]	Hot humid climate			X	X					X					Rhino, & Simulations in Energy Plus, and DIVA
[46]	Hot summer climate			X					X	X					Rhino + Grasshopper simulation
[47]	Hot deserts climate			X							X		X		Energy Plus
[48]	Mid-latitude climate			X						X					Design Builder Program
[49]	Hot summer climate			X							X		X	X	Rhino, Grasshopper and Energy plus

(continued)

Table 1. *(continued)*

References	Climate	(BL)	BSS system		Environmental performance improvement indicators (EPII)										
			FC	BG	Daylight	OTTV	Air flow	Surface temp	Air (Op) temperature	SR/gain	Energy consumption	Thermal comfort	Heating-cooling load	Sunlit hours/area	Program used& simulation tools
[50]	Hot summer climate			X						X			X		Design Builder BM
Percentage of each metric	Most climate: hot summer climate	15%	41%	59%	43%	9%	9%	23%	11%	45%	34%	5%	29%	15%	Most used: energy plus

3 Discussion and Results of BSS Analysis Studies

According to the comparative analysis shown in Table 1, there are many different methods for creating and assessing BSS buildings, as this comparative study has demonstrated. Even though there are standard methods for evaluating the strategies' effectiveness (such as calculating SR using building performance simulation), there appears to be a general lack of thorough investigation in some areas, which will be covered in more detail in this section.

3.1 Self-shading Design for FC and BG: Improvement Methods

In BSS studies, BG is the most common design study. It was also noted that few studies have combined BL with self-shading. Most of the limited studies in this field have applied a problem-based biomimetic approach primarily at the organism level. As for the improvement methods, many studies only used SR area values to examine self-shading buildings because SR could be calculated per unit area over a specified period of time. For example, various studies investigated how different folded pattern depths affected SR [14, 37]. However, they ignored the possible increase in surface area and, as a result, absolute (SR) values, instead using SR area as the only evaluation parameter. The area of surface patterns had not been taken into consideration because other FC research, such as the Alhuwayil WK study [21] and the Saifelnasr SS study [27], also used the SR area to comprehend changes in rotations and flips of patterns. Another study [18] evaluated how different window-to-wall ratios (WWR) and façade orientations impacted solar gains and natural daylight by utilizing SR area values. Regarding the simulation software used, EnergyPlus is the most widely used in the FC studies, while Grasshopper Simulation is the highest in the BG studies.

Twisted shapes [28] and tilted walls [20] greatly decreased SR area and increased energy efficiency for self-shading (BG) designs. Since self-shading might be quantified by simply computing the total SR values during a certain period for a particular surface without considering the surface area, few studies have just used absolute SR to assess self-shading. For instance, one or two research studies maximized natural daylighting [34] and minimized energy consumption intensity [27] by lowering the value of absolute SR to optimize self-shading. This metric has been utilized in several research studies to show the relationship between surface temperature [15], shaded surface area, and solar absorption in self-shading facades. By decreasing SR [34] and increasing natural daylight, other research attempted to strike a compromise between daylighting and solar heat gain reduction.

Numerous studies have examined the effect of shaded regions on facades together with other SR factors. A significant impact on cost, environmental impact, and thermal performance was observed when analyzing the SR area and shaded area of eight facade modules with varying forms and materials. Additionally, more shaded regions led to better thermal comfort for either the self-shading BG [49] or FC design [7], as well as decreased SR area and energy usage. These studies demonstrated that examining the facade using multiple metrics yielded a thorough evaluation. Additionally, it might aid researchers in precisely evaluating the effects of design modifications and improving the layout of an efficient facade.

3.2 Environmental Performance Improvement Indicators (EPII)

The studies used a variety of simulation tools, including EnergyPlus, DIVA, Ladybug, Honeybee, DesignBuilder, Vasari, Revit, Grasshopper, and others, to evaluate the EPII for BSS buildings, with each tool having different characteristics and advantages. Figure 4 illustrates the most commonly used software for each environmental metric.

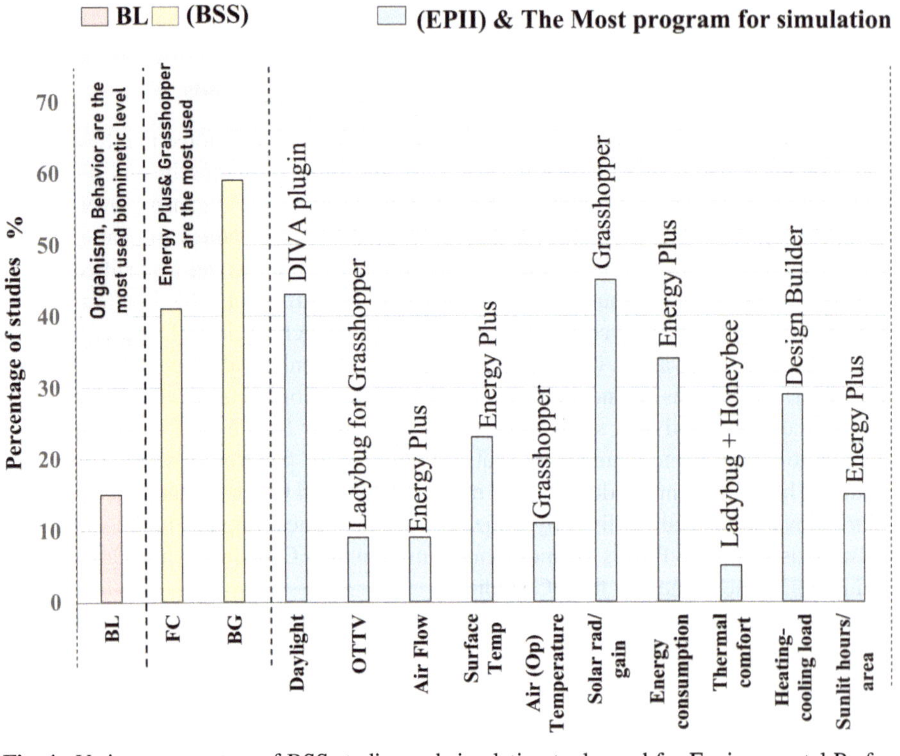

Fig. 4. Various parameters of BSS studies and simulation tools used for Environmental Performance Improvement Indicators (EPII). *Source* The authors based on references in Table 1

Solar radiation (SR)

SR is the most widely used performance indicator in studies and can be examined through solar heat gain and surface temperature [15], as shown in Fig. 4. A building's heat gain from external sources consists of contributions from solar radiation, thermal conduction through the envelope, and ventilation. When solar shading is applied to a building, it reduces heat gain from radiation. However, it does not directly affect heat gain through conduction, which is primarily affected by the temperature difference between the indoor and outdoor environments [7].

Regarding the simulation software, Grasshopper is the most widely used in evaluating SR studies. Studies indicate that the distribution of SR on shaded surfaces is significantly affected by the building's shape. Research has shown that adding a 10° tilt to the western wall reduces SR by approximately 50 W/m^2 during peak afternoon hours [17, 34, 43,

45]. Moreover, tilting building components resulted in a reduction of the shading rate SR by more than 10%. The values of incident SR are examined for BSS via four metrics: (a) absolute solar radiation, (b) SR per area, (c) shading area, and (d) sunlit hours [32, 37].

From a thermal performance perspective, self-shading strategies help reduce direct exposure to solar radiation, which in turn lowers thermal stress on building facades [15]. This, in turn, positively affects the reduction of cooling load in buildings, especially during the summer. The process of reducing heat gain and activating mechanisms for dissipating heat through thermal convection has become a central focus in recent studies [33]. For example, an applied study on a double-shelled facade inspired by the surface of a termite mound showed effectiveness in reducing the cooling load by up to 15% [31]. Studies by Al-Hweel, Abdul-Majeed, and Al-Jarni [16] and Farwa [49] also showed a similar 21% reduction in annual energy consumption, while Zerefos, Tsioumas, Koutsiopoulos, Fonda, and Kokkinis [33] observed an average decrease of 7.8%.

However, according to the study conducted by Adenugroho and Gadi [50], the decrease in average annual SR is not necessarily associated with an increase in heating load. The results of the model they studied showed that during winter days, higher SR is received despite the decrease in annual and monthly solar radiation. This indicates the necessity of analyzing SR data on a monthly or even daily basis to obtain an accurate assessment of the impact of radiation on heating load [33, 50].

3.3 Self-shading Calculations

Many studies have produced techniques for more precise shading calculations. For example, some studies have used shadow calculation methods to speed up the assessment of ambient shading [4, 51, 52]and improve shading geometry calculations within BIM tools [53]. Frameworks based on shadow projection principles have also been developed to enable more accurate analysis of energy consumption and reduce the gap between simulation and reality [54]. Furthermore, another study developed an approach to simulate complex surfaces using Grasshopper [18]. Overall, these methods help designers accurately assess shading and SR aspects.

4 Conclusion and Future Research Directions

This section further discusses the research's conclusion, main findings, and potential future directions.

4.1 Conclusion and Main Findings

This study makes a significant contribution by presenting an analytical approach to BSS building design and their associated EPII. Common design techniques for BSS systems were collected and studied, and they were classified into two main groups based on the biomimetic problem-based approach: (BG) design and (FC) design. The study also analyzed the indicators of environmental performance improvement, such as solar radiation, daylighting, energy consumption, and others. The study contributed the potential of BSS techniques to reduce heat gain and surface temperature, highlighting the critical role of SR and self-shading calculations. It further identified the most effective calculation methods and simulation tools for improving environmental performance. Figure 5 summarizes the design guidelines for BSS buildings and their EPII.

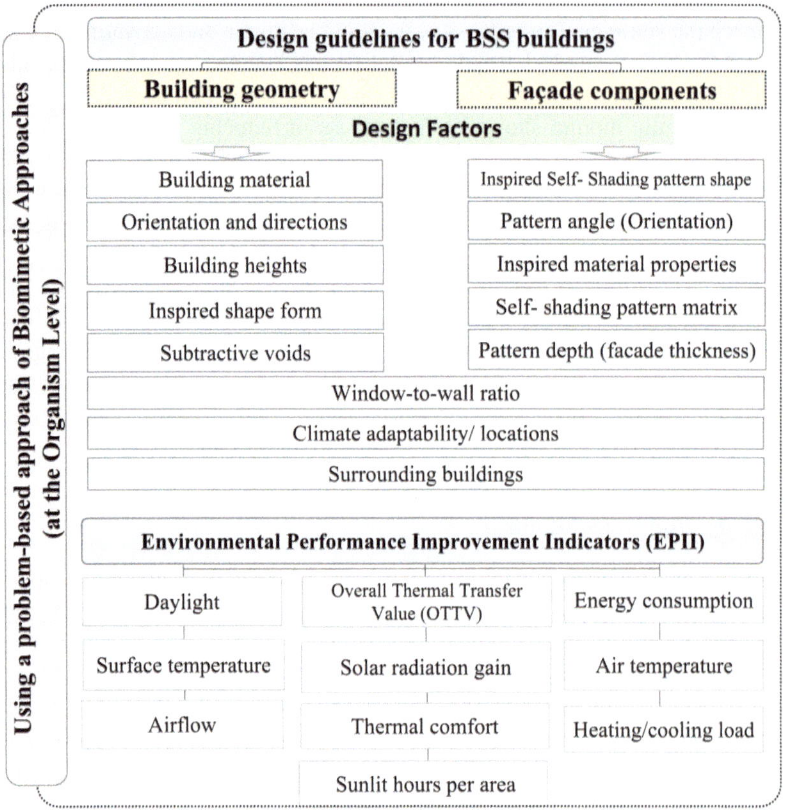

Fig. 5. Design guidelines for BSS buildings and their EPII. *Source* The authors based on [3, 23, 30, 55]

This systematic analysis yielded that BSS design has a significant impact on building energy consumption, especially in hot and dry climates. Architects can reduce glare while preserving natural light by using BSS design, which imitates organisms' natural defenses against direct sunlight. BSS improves lighting conditions in buildings by effectively balancing heat reduction with access to natural daylight. These designs can achieve daylighting efficiencies of approximately 2%, proving effective in both self-shading and natural lighting. Finally, BSS reduces solar heat gain, reducing cooling needs and supporting the building sector's transition to sustainability.

4.2 Future Research Directions

- The scientific literature shows a significant gap concerning the relationship between biomimetic and self-shading, with most BG studies over FC studies.
- Current environmental performance improvements predominantly focus on heat gain, energy consumption, and daylighting, neglecting other important crucial indicators.
- A lack of validation in many studies undermines their credibility and practical applicability.
- Designs with different surface areas need multiple self-shading measures for accuracy.
- Shading area and sunshine hours help assess thermal performance but have limits due to seasonal changes.in addition, combining these with a self-shading coefficient provides a more consistent evaluation, especially for complex building shapes.
- Many studies overlooked the potential increase in absolute surface area and self-shading values by solely relying on shading area as an evaluation metric.
- Many studies did not address building material efficiency, optimal design methods, or the impact of surrounding structures on self-shading.
- Some FC studies examined pattern rotations and flips using shading area, the actual surface pattern area was not considered.
- Ultimately, most studies concentrate on specific cases, highlighting the need for generalizable guidelines applicable across diverse climates, orientations, shapes, and materials.

References

1. Fu, S., Zhong, X., Zhang, Y., et al.: Bio-inspired cooling technologies and the applications in buildings. Energy Build **225**, 110313 (2020). https://doi.org/10.1016/j.enbuild.2020.110313
2. Kabir, M., Habiba, U.E., Khan, W., et al.: Climate change due to increasing concentration of carbon dioxide and its impacts on environment in 21st century; a mini review. J. King Saud Univ. Sci. **35**, 102693 (2023). https://doi.org/10.1016/J.JKSUS.2023.102693
3. Ismail, R., Megahed, N., Shahda, M., Eltarabily, S.: Bio-inspired design: leveraging nature for enhanced ecosystem services. Archit. Eng. Des. Manag. 1–32 (2025). https://doi.org/10.1080/17452007.2025.2480835
4. Liu, Z., Zhou, X., Shen, X., et al.: A novel acceleration approach to shadow calculation based on sunlight channel for urban building energy modeling. Energy Build. **315**, 114244 (2024). https://doi.org/10.1016/J.ENBUILD.2024.114244
5. Verbrugghe, N., Rubinacci, E., Khan, A.Z.: Biomimicry in architecture: a review of definitions, case studies, and design methods. Biomimetics **8**, 107 (2023). https://doi.org/10.3390/biomimetics8010107
6. Xing, Y., Jones, P., Bosch, M., et al.: Exploring design principles of biological and living building envelopes: what can we learn from plant cell walls? Intell. Build. Int. **10**, 78–102 (2018). https://doi.org/10.1080/17508975.2017.1394808
7. Shahda, M.: Self-Shading Walls to Improve Environmental Performance in Desert Buildings, pp. 1–14 (2020). https://doi.org/10.5923/j.arch.20201001.01
8. El Ahmar, S., Fioravanti, A.: Botanics and parametric design fusions for performative building skins an application in hot climates (2014)
9. Oukati Sadegh, S., Haile, S., Jamshidzehi, Z.: Development of Two-step biomimetic design and evaluation framework for performance-oriented design of multi-functional adaptable building envelopes. J. Daylighting **9**, 13–27 (2022). https://doi.org/10.15627/jd.2022.2

10. Hosseini, M., Fadli, F., Mohammadi, M.: Biomimetic kinetic shading facade inspired by tree morphology for improving occupant's daylight performance. J. Daylighting **8**, 65–82 (2021). https://doi.org/10.15627/jd.2021.5
11. El-Rahman, S., Esmail, S., Khalil, H., El-Razaz, Z.: Biomimicry inspired adaptive building envelope in hot climate. Eng. Res. J. **166**, 30–47 (2020). https://doi.org/10.21608/erj.2020.135274
12. Capeluto, I.G.: Energy performance of the self-shading building envelope. Energy Build **35**, 327–336 (2003). https://doi.org/10.1016/S0378-7788(02)00105-6
13. Kandar, M.Z., Nimlyat, P.S., Abdullahi, M.G., Dodo, Y.A.: Field study of thermal and visual performance of self-shading energy commission diamond building, Putrajaya, Malaysia. Indian J. Sci. Technol. **9** (2016). https://doi.org/10.17485/ijst/2016/v9i46/107120
14. Mohsenzadeh, M., Hedayati Marzbali, M., Maghsoodi Tilaki, M.J., Aldrin, A.: Building form and energy efficiency in tropical climates: a case study of Penang, Malaysia. urbe Rev Bras Gestão Urbana **13** (2021). https://doi.org/10.1590/2175-3369.013.e20200280
15. Fleckenstein, J., Molter, P., Chokhachian, A., Dörfler, K.: Climate-resilient robotic facades: architectural strategies to improve thermal comfort in outdoor urban environments using robotic assembly. Front. Built. Environ. **8**, 856871 (2022). https://doi.org/10.3389/fbuil.2022.856871
16. Alhuwayil, W.K., Abdul Mujeebu, M., Algarny, A.M.M.: Impact of external shading strategy on energy performance of multi-story hotel building in hot-humid climate. Energy **169**, 1166–1174 (2019). https://doi.org/10.1016/j.energy.2018.12.069
17. Chan, A.L.S., Chow, T.T.: Thermal performance of air-conditioned office buildings constructed with inclined walls in different climates in China. Appl. Energy **114**, 45–57 (2014). https://doi.org/10.1016/J.APENERGY.2013.09.048
18. Atthaillah, Mangkuto, R.A., Koerniawan, M.D., et al.: Optimization of daylighting design using self-shading mechanism in tropical school classrooms with bilateral openings. J. Daylighting **9**, 117–136 (2022). https://doi.org/10.15627/jd.2022.10
19. Su, M., Jie, P., Li, P., et al.: A review on the mechanisms behind thermal effect of building vertical greenery systems (VGS): methodology, performance and impact factors. Energy Build **303**, 113785 (2024). https://doi.org/10.1016/J.ENBUILD.2023.113785
20. Lavafpour, Y., Sharples, S.: Summer thermal comfort and self-shading geometries in Passivhaus dwellings: a pilot study using future UK climates. Buildings **5**, 964–984 (2015). https://doi.org/10.3390/buildings5030964
21. Alhuwayil, W.K., Almaziad, F.A., Abdul Mujeebu, M.: Energy performance of passive shading and thermal insulation in multistory hotel building under different outdoor climates and geographic locations. Case Stud. Therm. Eng. **45**, 102940 (2023). https://doi.org/10.1016/J.CSITE.2023.102940
22. Koç, S.G., Maçka Kalfa, S.: The effects of shading devices on office building energy performance in Mediterranean climate regions. J. Build. Eng. **44** (2021). https://doi.org/10.1016/j.jobe.2021.102653
23. De Luca, F., Sepúlveda, A., Varjas, T.: Multi-performance optimization of static shading devices for glare, daylight, view and energy consideration. Build. Environ. **217** (2022). https://doi.org/10.1016/j.buildenv.2022.109110
24. Kandar, M.Z., Nimlyat, P.S., Abdullahi, M.G., Dodo, Y.A.: Influence of inclined wall self-shading strategy on office building heat gain and energy performance in hot humid climate of Malaysia. Heliyon **5** (2019). https://doi.org/10.1016/j.heliyon.2019.e02077
25. Gou, S., Nik, V.M., Scartezzini, J.-L., et al.: Passive design optimization of newly-built residential buildings in Shanghai for improving indoor thermal comfort while reducing building energy demand. Energy Build **169**, 484–506 (2018). https://doi.org/10.1016/j.enbuild.2017.09.095

26. Mahdavinejad, M., Bazazzadeh, H., Mehrvarz, F., et al.: The impact of facade geometry on visual comfort and energy consumption in an office building in different climates. Energy Rep. **11**, 1–17 (2024). https://doi.org/10.1016/j.egyr.2023.11.021
27. Saifelnasr, S.S.: Design of a self-shading mass as a function of the latitude for automatic seasonal adjustment. IOP Conf. Ser. Earth Environ. Sci. **329**, 12050 (2019). https://doi.org/10.1088/1755-1315/329/1/012050
28. Jakica, N., Kragh, M.: Assessing self-shading benefits of twisting towers. J. Facade Des. Eng. **8**, 115 (2020). https://doi.org/10.7480/jfde.2020.1.5043
29. Nikpour, M., Kandar, M., Ghomeishi, M., et al.: Investigating the effectiveness of self-shading strategy on overall thermal transfer value and window size in high rise buildings. World Acad. Sci. Eng. Technol. **50**, 161–166 (2011)
30. Bhai, M.A.Y., Abdelkader, M., Neseem, A., Mustafa, A.: Impact of the geometric form of the building envelopes on the efficiency of natural lighting in the office space. IOP Conf. Ser. Earth Environ. Sci. **992**, 012001 (2022). https://doi.org/10.1088/1755-1315/992/1/012001
31. El Ahmar, S., Fioravanti, A.: Biomimetic-computational design for double facades in hot climates: a porous folded façade for office buildings (2015)
32. Maderspacher, J., Moosberger, S.: Design And Validation Of A Facade Zigzag-Shading Model Lucerne University of Applied Sciences and Arts, Department of Architecture & Technology, CH-6048 Horw ; Email : johannes.maderspacher@hslu.ch EQUA Solutions AG , CH-8934 Knonau ; Email : sven.moosb. Solutions 14–16 (2011)
33. Zerefos, S.C., Tessas, C.A., Kotsiopoulos, A.M., et al.: The role of building form in energy consumption: the case of a prismatic building in Athens. Energy Build. **48**, 97–102 (2012). https://doi.org/10.1016/J.ENBUILD.2012.01.014
34. Wang, L., Zhang, H., Liu, X., Ji, G.: Exploring the synergy of building massing and façade design through evolutionary optimization. Front. Archit. Res. **11**, 761–780 (2022). https://doi.org/10.1016/J.FOAR.2022.02.002
35. Song, J.Y., Lin, H., Shim, J.: LEAF (low energy adaptive façade) self-adapting micro shading façade design using responsive polymer sheets. In: World Congress: Facade Tectonics Institute, Los Angeles. Proceedings […], pp 145–155. Los Angeles, California (2016)
36. Srithongchai, T.: The role of layout in improving building thermal performance in bangkok. J. Sustain. Archit. Civ. Eng. **29**, 189–204 (2021)
37. Schultz, J., Katz, N.: Origami-inspired façade design: parametric studies for architectural and structural efficiency. In: Facade Tectonics 2018 World Congress, pp. 349–358 (2018)
38. Rupp, A., Gruber, P.: Bio-inspired evaporation from shaped interfaces: an experimental study. Bioinspir Biomim 16 (2021). https://doi.org/10.1088/1748-3190/abdd9e
39. Prawata, A.: A study of brick facade design as preventive measure for passive cooling. In: IOP Conference Series: Earth and Environmental Science. IOP Publishing, p. 12080 (2018)
40. Negendahl, K., Nielsen, T.R.: Building energy optimization in the early design stages: a simplified method. Energy Build. **105**, 88–99 (2015)
41. Mokhtar, S., Chronis, A., Leung, C.: Neighbourhood shading impacts on passive adaptive façade collective behaviour. In: 17th International conference, CAAD Futures 2017. Istanbul Technical University, pp. 199–210 (2017)
42. Molter, P.L., Chokhachian, A.: Experimental study of geometric configuration and evaporative cooling potential of brick elements. In: Eco-Efficient Materials for Reducing Cooling Needs in Buildings and Construction, pp. 101–115. Elsevier (2021)
43. Mamdoh Ali Youssef, A.: Optimizing building layouts for proper self-shading: a computational approach. MEJ-Mansoura Eng. J. **47**, 13–27 (2022)
44. Mainini, A.G., Speroni, A., Poli, T., Zinzi, M.: On the optical characterization of architectural three-dimensional skins and their solar control potential. Buildings **12**, 1103 (2022)

45. Liu, X., Wang, L., Ji, G.: Optimization-based design exploration of the mutual influence between building massing and façade design. In: 2022 Annual Modeling and Simulation Conference (ANNSIM). IEEE, pp. 487–499 (2022)
46. Lavafpour, Y., Sharples, S., Gething, B.: The impact of building form on overheating control: a case study of Larch House. Archit. Sci. Rev. **63**, 467–480 (2020)
47. Hinkle, L.E., Wang, J., Brown, N.C.: Quantifying potential dynamic façade energy savings in early design using constrained optimization. Build. Environ. **221**, 109265 (2022). https://doi.org/10.1016/J.BUILDENV.2022.109265
48. Hachem, C., Athienitis, A., Fazio, P.: Parametric investigation of geometric form effects on solar potential of housing units. Sol. Energy **85**, 1864–1877 (2011). https://doi.org/10.1016/J.SOLENER.2011.04.027
49. Freewan, A.A.Y.: Energy-efficient solutions depending on building forms design with tilted south and north facades. Buildings **12** (2022). https://doi.org/10.3390/buildings12060753
50. Adinugroho, T., Gadi, M.: Investigation on thermal performance of diverse innovative prismatic building models and establishment of the form indicator. Energy Procedia **152**, 407–412 (2018). https://doi.org/10.1016/j.egypro.2018.09.165
51. Elmalky, A.M., Araji, M.T.: Computational procedure of solar irradiation: a new approach for high performance façades with experimental validation. Energy Build. **298**, 113491 (2023). https://doi.org/10.1016/J.ENBUILD.2023.113491
52. Wen, J., Yang, S., Xie, Y., et al.: A fast calculation tool for assessing the shading effect of surrounding buildings on window transmitted solar radiation energy. Sustain. Cities Soc. **81**, 103834 (2022). https://doi.org/10.1016/J.SCS.2022.103834
53. Voivret, C., Bigot, D., Riviere, G.: A method to compute shadow geometry in open building information modeling authoring tools: automation of solar regulation checking. Buildings **13**, 3120 (2023). https://doi.org/10.3390/buildings13123120
54. Oliveira, M., Mendonça, P., Tenpierik, M., et al.: Shading calculation methods and regulation simplifications—the Portuguese case. Buildings **13** (2023). https://doi.org/10.3390/buildings13061521
55. Li, D.H.W., Lam, J.C.: Measurements of solar radiation and illuminance on vertical surfaces and daylighting implications. Renew. Energy **20**, 389–404 (2000). https://doi.org/10.1016/S0960-1481(99)00126-3

The Impact of Recreational Diving Participants' Marine Environmental Volunteer Literacy on the Development of Coastal and Rural Leisure Sports Environments

Chao-Yen Lin[1], Chin-Hsien Hsu[2], Hsiao-Hsien Lin[1,2], Po-Hsuan Wu[3], and Yong-Zhan Zheng[4](✉)

[1] School of Physical Education, Jiaying University, Meizhou, China
[2] Department of Healthcare Industry Technology Development and Management, National Chin-Yi University of Technology, Taichung, Taiwan
[3] Department of Electrical and Mechanical Technology, National Changhua University of Education, Changhua 50074, Taiwan
[4] Department of Recreation and Sports Management, Tajen University, No. 20, Weixin Rd., Yanpu Township, Pingtung County 90741, Taiwan
m04370001@tajen.edu.tw

Abstract. Background: Although humans have extensive experience in promoting ocean diving activities, tourism development continues to have negative impacts. Therefore, we aim to preserve marine ecological diversity and water quality by enhancing the knowledge and skills of marine environmental volunteers. Objective: This study investigates the influence of public marine environmental volunteer literacy on the development of coastal and rural leisure sports environments. Methods: A total of 300 participant responses were analyzed using Pearson's correlation analysis. Interview data were subsequently collected and triangulated through multivariate analysis. Conclusion: Public understanding of marine environmental issues—including cognition, attitudes, and behaviors—is inconsistent. Moreover, marine environmental literacy significantly influences the development of marine ecology, leisure sports environments, rural society, and community environments ($p < 0.05$). Behavioral engagement appears to exert a stronger influence than cognitive awareness. Recommendations: Sustained implementation of marine environmental literacy education, advocacy policies, and public conservation initiatives is essential for ensuring the long-term sustainability of marine ecology and leisure sports environments.

Keywords: Ocean · Environmental literacy · Leisure and sports environment · Cognition · Behavior

1 Introduction

Hainan Island covers an area of approximately 33,200 km^2 and is situated on the Leizhou Peninsula, nestled between the Qiongzhou Strait and the Eurasian continent. As the largest island within China's territorial waters, Hainan is classified as a continental

Y. Zhu (Ed.): ISESIE 2025, LNCE 000, pp. 117–129, 2026.
https://doi.org/10.1007/978-981-92-2389-3_11

island, located south of the Tropic of Cancer at approximately 110° east longitude. The island has a marine tropical monsoon climate, characterized by abundant sunshine, with over 2,000 h of sunshine annually [1]. This climate contributes to warm winters and cool summers, creating a pleasant living environment and fostering diverse ecological resources. Statistics indicate that Hainan Island is home to around 3,600 species of wild plants and more than 540 species of terrestrial vertebrates, including 145 species of key protected wild animals. The island's coastal zones contain 36 species of mangrove plants across 20 families, covering about 1,800 ha—representing 97% of China's total mangrove area [2]. Additionally, the surrounding ocean is rich in biodiversity, featuring over 300 species of coral reefs and fish, as well as six species of seagrass. Since 2016, the Chinese government has introduced a series of policies to promote the development of Hainan Island's tourism and economic sectors. These initiatives capitalize on the island's unique geographical, geological, ecological, and climatic characteristics to develop tourism resources, attract visitors, and establish a diverse coastal leisure and sports environment, thereby fostering local development [5–7]. Among these activities, recreational diving contributes to local tourism development and enhances the leisure sports environment due to its recreational, fitness, and tourism advantages. According to statistics, despite the impact of the pandemic, at least 1.2 million people participate annually following the local promotion of recreational diving, generating substantial tourism benefits [8]. This demonstrates that the promotion of local recreational diving activities has yielded significant results.

Scholars assert that to effectively cater to the needs of tourists, the development of tourism policies will inevitably attract external enterprises to invest, collaborate on development, promote industrial reform, and undertake infrastructure renovation projects [5–7]. Local governments aim to stimulate the economy, enhance social visibility, and improve the community environment through the growth of the tourism industry [8, 9]. Statistics indicate that since the implementation of Hainan Island's tourism policy, the number of tourists has peaked at approximately 8.34 million, with an average daily expenditure of around US$220 per person, resulting in an annual tourism foreign exchange of about US$185 million [11]. This data demonstrates that the growth of the tourism sector has a positive and significant impact on the local development of Hainan Island. However, research has highlighted that tourism development not only yields positive outcomes but also presents negative consequences. Economic challenges include rising prices, low-quality products, and inadequate service provision [12, 13]. From a social perspective, issues such as overconsumption, resource depletion, and interpersonal conflict have been observed [14, 15]. Environmental concerns include fuel emissions, noise pollution, and tourism-generated waste [16, 17]. Such factors may undermine policy effectiveness and reinforce negative public perceptions of local tourist destinations.

Scholars argue that improving citizens' marine literacy can help mitigate the adverse effects of tourism development while promoting positive engagement and environmental stewardship. Marine literacy encourages individuals to respect natural resources, such as the ocean and land, and raises public awareness about how daily behaviors affect marine ecosystems. It further promotes exploration and practical actions aimed at ocean conservation [18, 19]. Greater public awareness of marine literacy increases the likelihood of voluntary conservation efforts, reduced environmental harm, and enhanced concern for ecological well-being [20]. Consequently, this has a significant impact on reducing harmful human behaviors and practices, curbing excessive consumption, and enhancing collaboration in policy promotion [21, 22]. Thus, fostering marine literacy among citizens is crucial for preserving marine ecosystems, enhancing public leisure and sports environments, and laying the groundwork for a sustainable marine economy.

However, other scholars contend that various factors, including time, location, cultural context, and ecological conditions, can influence policy goals, established theories, and research findings. As a result, actual outcomes may diverge from initial expectations [23, 24]. Moreover, cultural diversity can lead to divergent perspectives regarding policy implementation and perceived effectiveness [23, 24]. Therefore, incorporating the experiences of directly affected individuals can help bridge the gap between policy design, historical precedent, and real-world outcomes. This study aims to utilize a case study approach to assess the impact of public marine environmental protection volunteer literacy on the coastal leisure sports environment. Additionally, it seeks to provide recommendations for policymakers in Hainan Island and similar island regions to support the formulation of sustainable development strategies for marine-based economies.

2 Methods

2.1 Process

This study employs a quantitative approach. The questionnaire was developed and refined based on a review of literature concerning marine leisure sports environments and social environment development theory. Further modifications were made using insights from studies on marine environmental literacy [18–22], marine leisure and environmental development [23–25], and social environment advancement [24, 26, 27]. The finalized questionnaire comprised three sections. The demographic section categorized respondents by gender (male and female) and by age group: under 20, 21–30, 31–40, 41–50, 51–60, and over 61 years. It includes four questions regarding marine leisure, sports, or marine environment development, and nine questions focused on social environment development. According to established criteria, a Kaiser-Meyer-Olkin (KMO) value exceeding 0.06 and a Bartlett's test p-value below 0.01 indicate the suitability of the scale for factor analysis [28–30]. Questions with a Cronbach's alpha greater than 0.60 demonstrate good reliability [29–31] and can be analyzed further.

2.2 Tools and Analysis

Four questionnaire items were designated to assess marine environmental literacy. Exploratory factor analysis yielded a KMO value of 0.932 and a Bartlett's test of sphericity ($\chi^2 = 7983.76$, df = 120, $p < 0.001$), indicating the dataset's suitability for factor analysis. The extracted components explained 77.375%, 5.048%, and 2.236% of the variance, respectively, with a cumulative variance of 84.66%. Based on factor analysis results, reliable items were retained and categorized into three dimensions: cognition (7 items), attitude (4 items), and behavior (5 items), with Cronbach's alpha values ranging from 0.979 to 0.980. The overall α coefficient for the total scale was 0.981.

The assessment of social environment development comprised eight items. Exploratory factor analysis yielded a KMO value of 0.930 and Bartlett's test of sphericity ($\chi^2 = 5382.21$, df = 36, $p < 0.001$), confirming the adequacy of the data for factor analysis. The two extracted components explained 90.013% and 1.811% of the variance, respectively, yielding a cumulative variance of 91.824%. Post-analysis, reliable dimensions were identified as social culture (7 items) and community environment (4 items), with Cronbach's alpha values ranging from 0.934 to 0.936. The overall α coefficient for the total scale is 0.986.

Four items were constructed to evaluate public cognition regarding the development of marine leisure sports environments. Factor analysis indicated a KMO value of 0.871 and Bartlett's test of sphericity ($\chi^2 = 2432.998$, df = 6, $p < 0.001$), confirming data suitability for further analysis. The total explained variance for this scale was 92.259%. Following analysis, items with strong reliability ($\alpha = 0.980$ to 0.984) were retained, with an overall Cronbach's alpha of 0.988. These results confirm that the KMO values, Bartlett's test statistics, and degrees of freedom for all scales—marine environmental literacy, social environment development, and marine leisure sports environment development—met established statistical thresholds. Therefore, the author concludes that this questionnaire demonstrates strong reliability.

Upon finalizing the questionnaire, a total of 450 copies were distributed between July and August 2024, resulting in 300 valid responses. Data were analyzed using SPSS version 27.0, employing descriptive statistics, Pearson's correlation analysis, and additional relevant statistical methods. Additionally, based on the quantitative findings, semi-structured interviews were conducted with experts in marine environmental management, policy analysis, leisure industry operations, ecology, and education, as well as with local residents aged 20–50. These interviews aimed to provide contextual insights, background information, and perspectives for triangulation, as detailed in Appendix 1. Finally, a multivariate verification method was employed to facilitate a discussion that considered multiple data sources and perspectives [32, 33]. The research framework is illustrated in Fig. 1.

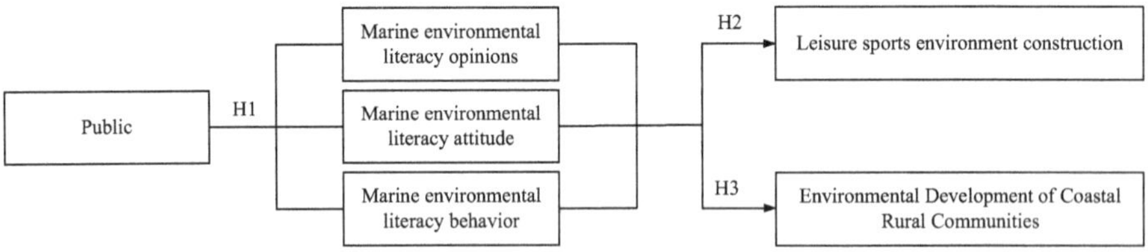

Fig. 1. Framework

2.3 Framework and Hypotheses

Environmental cleanliness, ecological diversity, and distinctive local cultural and landscape features are critical for the sustainable development of tourism. Enhancing public marine literacy can foster greater respect and appreciation for natural resources, including marine and terrestrial ecosystems. It also increases awareness of the impact of daily behaviors on marine ecosystems and encourages active engagement in conservation efforts [9, 10].

Research conducted by scholars Jiang, Fawehinmi, and Shin et al. highlights that positive and profound cognition can be embedded in individuals' subconscious and memory, forming the foundation for how they evaluate certain people, things, or objects [34]. Well-defined attitudes often influence initial impressions, which in turn affect individual behavior [35]. Such behavior represents an individual's response to practical interactions with specific people, things, or objects [36]. The interplay of these factors constitutes the steps and keys to human behavior and consumption patterns [34–38]. This dynamic effectively supports the development of local social culture and community environments [39, 40] while also enhancing the positive evolution of marine ecology, marine leisure sports, and the overall marine environment [41, 42]. Accordingly, this study posits that governmental initiatives promoting marine environmental literacy and public awareness are likely to positively impact the preservation of marine ecology and leisure environments, while also fostering social and community development in coastal rural areas. Based on this premise, the author proposes two hypotheses.

Hypothesis 1: The public exhibits a consistent level of understanding regarding marine environmental literacy.

Hypothesis 2: Public understanding of marine environmental issues significantly influences the development of marine ecology and leisure sports environments.

Hypothesis 3: Public understanding of marine environmental issues significantly influences the development of coastal rural societies and community environments.

3 Analysis

3.1 Analysis of Public Awareness of Marine Environmental Literacy

Descriptive statistical analysis was conducted using SPSS version 27.0 on data collected from 300 participants. The results are presented in Table 1. Among cognitive items, the highest mean score (M = 4.50) reflected participants' recognition of the importance of considering ecological carrying capacity during marine leisure activities. Conversely, the lowest cognitive score (M = 4.38) was associated with understanding that animal release may disrupt existing local ecosystems. In the attitudinal dimension, the highest score (M = 4.54) reflected a preference for low-pollution leisure activities and consumption patterns. The lowest score (M = 4.48) related to the belief that marine life should not be directly touched or interacted with during recreation. In the behavioral dimension, the highest score (M = 4.41) reflected the intention to refrain from touching or hunting marine organisms during leisure activities. The lowest behavioral score (M = 4.23) pertained to avoiding the purchase of souvenirs made from marine organisms.

Overall, participants demonstrated strong environmental literacy across cognitive, attitudinal, and behavioral dimensions—specifically acknowledging ecological carrying capacity, preferring low-pollution recreational practices, and avoiding interference with marine ecosystems. However, lower levels of agreement were observed for items concerning the ecological risks of animal release, avoidance of direct contact with marine life, and purchasing behavior related to marine life–based souvenirs.

☆Highest score; ○Lowest score

Table 1. Analysis of public awareness of marine environmental literacy

Facets (M)	Secondary facets (M)	Issue	Public
			M
Marine environmental literacy (4.45)	Marine environmental literacy opinions (4.45)	When participating in marine leisure sports, it is important to take into account the ecological carrying capacity	4.50☆
		To preserve the diversity of marine ecosystems, it is essential to protect marine life	4.43
		Improper fishing methods can adversely affect the marine environment for leisure sports	4.52
		Sea level rise is caused by excessive CO2 emissions and the premature greenhouse effect	4.39
		Mangroves play a crucial role in protecting coastlines and fostering ecological diversity	4.48
		Releasing animals into the wild can impact the ecosystem and the local organisms that already inhabit it	4.38○
		Improper fishing methods can adversely affect marine ecology.	4.47
	Marine environmental literacy attitude (4.51)	When participating in marine leisure sports, it is important to consider the impact of these activities on the marine environment	4.53
		When participating in marine leisure sports, it is important to choose low-pollution activities and consumption patterns	4.54☆

(*continued*)

Table 1. (*continued*)

Facets (M)	Secondary facets (M)	Issue	Public
			M
		When participating in marine leisure activities, it is important to avoid creating waste while diving	4.52
		When participating in marine leisure activities, it is important to avoid direct contact and interaction with marine life	4.48○
	Marine environmental literacy behavior (4.38)	Please avoid purchasing souvenirs made from marine life	4.23○
		Please refrain from touching or harming marine life while participating in water sports	4.41☆
		I will take the initiative to collect and remove trash from the ocean	4.30
		It is advisable to refrain from touching or interacting with marine life while participating in marine leisure activities	4.37

3.2 Analysis of the Impact of Public Marine Environmental Literacy on the Development of Marine Ecology and the Leisure Sports Environment

Pearson correlation analysis was conducted on data from 300 participants using SPSS version 27.0. The results of the data analysis are displayed in Table 2. Findings indicate that marine environmental literacy is significantly associated with the development of marine ecology and leisure sports environments ($p < 0.05$). Among the three dimensions, behavioral literacy demonstrated the strongest correlation with both marine ecology and leisure sports environment development ($r = 0.889$), while cognitive literacy showed the weakest correlation ($r = 0.828$).

These findings suggest that behavioral engagement has a greater influence than cognitive awareness on the development of marine ecology and leisure sports environments.

3.3 Analysis of the Impact of Public Marine Environmental Literacy on Marine Rural Society and Community Environmental Development

Pearson correlation analysis was conducted using SPSS version 27.0 on data collected from 300 participants. The results of the data analysis are presented in Table 2. The findings indicate a significant association between marine environmental literacy and the development of social culture and community environments in rural coastal areas ($p < 0.05$). Among the three dimensions, behavioral literacy exhibited the strongest correlation with social and cultural development ($r = 0.882$), whereas cognitive literacy showed the weakest correlation ($r = 0.816$). Additionally, behavioral literacy was most strongly correlated with the environmental development of rural coastal communities ($r = 0.889$), while cognitive literacy again showed the lowest correlation ($r = 0.819$).

These results suggest that behavioral engagement plays a more critical role than cognitive awareness in shaping social and environmental development in rural coastal communities.

Table 2. The impact of marine environmental literacy on the development of marine ecology, leisure and sports environment, rural social culture, and community environment

Facets	Marine conservation literacy	Marine environmental literacy opinions	Marine environmental literacy attitude	Marine environmental literacy behavior
Rural social and cultural development	0.823**	0.816**○	0.836**	0.882**☆
Rural community environmental development	0.825**	0.819**○	0.844**	0.889**☆
Marine ecology and the development of leisure sports environments	0.835**	0.828**○	0.848**	0.889**☆

* $P < 0.05$; ** $P < 0.01$; ☆ Highest score; ○ Lowest score

4 Discuss

4.1 Public Awareness of Marine Environmental Literacy

Government initiatives to promote marine environmental literacy aim to enhance public appreciation of natural ecological resources—such as oceans and land—and to raise awareness of how daily behaviors affect marine ecosystems [10, 35]. Such initiative encourages individuals to explore and actively engage in marine conservation efforts, fostering a positive experience that promotes the recognition of sustainable marine resources. Additionally, the public can access valuable information through documentaries on marine pollution produced by the government, marine conservation organizations, and the media. These resources contribute to reinforcing public belief in and commitment to marine environmental protection. Consequently, individuals are more likely to consider low-impact recreational activities, environmentally conscious consumption patterns, and ecological carrying capacity when participating in marine-based leisure. They also express willingness to refrain from disturbing or exploiting marine ecosystems. This tendency may explain public acknowledgment of ecological carrying capacity and the preference for low-pollution leisure activities during marine recreation. This perspective aligns with the findings of researchers such as Jayasekara and others.

The traditional Chinese practice of purchasing fish and releasing birds reflects a cultural belief in the equality of all living beings and the value of ecological coexistences. Tourists participating in marine tourism or water-based recreation often perceive these activities as consumption-oriented experiences, asserting their right to enjoyment. Additionally, the marine ecosystem is incredibly diverse, with a vast number of species. Consequently, the negative impact of a few individual actions on the overall health of the marine environment is relatively minor. This perception may account for why certain participants do not view activities such as interacting with marine life, purchasing marine-based souvenirs, or releasing animals during recreation as ecologically harmful. Such perspectives may also explain inconsistencies observed in previous studies, including those conducted by Jayasekara et al.

4.2 The Impact of Public Marine Environmental Literacy on the Development of Marine Ecology and the Leisure Sports Environment

Government initiatives to promote marine environmental literacy aim to enhance public moral awareness and understanding of marine ecosystems, along with associated conservation practices [10, 35]. These efforts also seek to reinforce individuals' commitment to marine resource conservation, minimize environmental degradation, and support ecological protection. The dissemination of these policies and knowledge plays a crucial role in curbing environmentally harmful behaviors that may arise during marine recreational activities. Consequently, these initiatives may encourage individuals to engage in marine recreation while subconsciously adopting environmentally responsible consumption and leisure behaviors. This tendency may explain why the public perceives marine environmental literacy as a contributing factor in the development of marine ecology and leisure environments. This perspective aligns with the findings of Jayasekara and other researchers [9, 10].

Additionally, individuals' perceptions of tourism and education policy implementation and effectiveness may vary depending on cultural background and social context [9, 10]. Cognitive and subconscious differences in environmental knowledge may shape initial perceptions of people, objects, or situations, thereby influencing individual behaviors [11, 24]. As such, psychological expectations may affect real-world decision-making processes. The public often perceives tangible conservation actions as more effective than abstract plans in promoting sustainable marine ecology and leisure development. This perception reinforces the view that behavioral actions have a greater impact than cognitive awareness. Such discrepancy may also explain the inconsistencies found in the research results of Jayasekara and other scholars [9, 10].

4.3 The Impact of Public Marine Environmental Literacy on Coastal Rural Society and Community Environmental Development

Beyond promoting marine ecological awareness and sustainability, marine environmental literacy plays a pivotal role in fostering individuals' sense of responsibility toward community environmental stewardship [9, 10]. This awareness encourages sustained commitment to environmentally responsible behaviors during tourism and recreational activities, including low-impact consumption practices aimed at ecosystem preservation. Consequently, individuals are motivated to safeguard coastal rural communities and their environments through informed policies and knowledge. As a result, marine environmental literacy is increasingly recognized as a significant factor influencing the development of rural coastal social culture and community environments. This may explain the discrepancies in research findings from scholars such as Wu [11, 24].

Cultural background and individual developmental differences may also shape cognition and attitudes toward environmental and societal issues [9, 10]. These cognitive factors influence individual perceptions and serve as the basis for decision-making in environmental behavior. Consequently, delays or cognitive dissonance during decision-making may lead to unpredictable behavioral outcomes, which can be either beneficial or detrimental [9, 10]. Therefore, behavioral action is generally regarded as having a greater impact than cognitive awareness in achieving sustainable development goals related to marine ecology and environmental protection.

5 Conclusions

Although governmental efforts have prioritized marine environmental literacy to support sustainable development in marine ecology and rural leisure sports, they primarily succeed in fostering public consensus on the importance of ecological carrying capacity during marine recreational activities. This includes choosing low-pollution leisure options and refraining from disturbing or hunting marine life. However, policy efforts are less effective at discouraging direct interaction with marine life, the purchase of marine-derived souvenirs, and animal release during recreational activities. Moreover, despite ongoing policy advocacy, individual perceptions, attitudes, and behaviors vary considerably across cultural and experiential contexts. Such variability introduces uncertainty regarding the actual behaviors and benefits associated with environmentally friendly consumption and leisure activities. Consequently, the public tends to believe that actual behavior is more significant than mere awareness when it comes to achieving sustainable goals for marine ecology and the leisure sports environment.

Acknowledgements. We thank all author members and survey respondents.

Funding. This study was supported by the Science Research Department of the College level Development Program (324E1531) at Jiaying University, which aims to explore "related research on competition and sports psychology". Funding agencies do not participate in the design or implementation of research, data collection, management, analysis or interpretation, manuscript preparation, review or approval, nor do they participate in deciding to submit manuscripts for publication.

References

1. Liu, M.: Handbook of World Physical Geography. Knowledge Press, Beijing, China (1984)
2. Wang Y. International day for biological diversity themed interview—a snapshot of the seagrass special reserve. https://res.cenews.com.cn/h5/app/html/new-article.html?id=1132812&openInApp=1. Accessed 28 Aug 2024
3. Zou, C., Song, X., Jiang, H., Chen, K.: Wetland types and characteristics of Hainan Island. Trop. Geogr. **19**(3), 204–207 (1999)
4. Zhu, H.: On the biogeographic origin of Hainan Island, China. J. Plant Sci. **38**(6), 839–843 (2020)
5. National Development and Reform Commission of the People's Republic of China. Hainan Province's 14th Five-Year Plan for National Economic and Social Development and Outline of Long-term Goals for 2035. https://www.ndrc.gov.cn/fggz/fzzlgh/dffzgh/202104/P020210428646025031658.pdf. Accessed 28 Aug 2024
6. National Development and Reform Commission of the People's Republic of China. Outline of the Thirteenth Five-Year Plan for National Economic and Social Development of Hainan Province. https://dtfz.ccchina.org.cn/archiver/LowCD/UpFile/Files/Default/20160711161307319386.pdf. Accessed 28 Aug 2024
7. Qianzhan Enterprise Research Institute (Tourism Research Institute of the People's Republic of China). Analysis report on Hainan Province's pillar industries, advantageous industries, emerging industries, and key enterprises in 2022
8. Chen, W.: National Diving Members Congress held in Wanning: Suggestions for the development of diving in Hainan (2023). https://baijiahao.baidu.com/s?id=1775096080128189308&wfr=spider&for=pc. Accessed 31 Aug 2024
9. Ferreira, J.C., Vasconcelos, L., Monteiro, R., Flávia, F.Z., Duarte, C.M., Ferreira, F.: Ocean literacy to promote sustainable development goals and agenda 2030 in coastal communities. Educ. Sci. **11**(2), 62 (2023)
10. National Marine Educators Association. Ocean Literacy Scope and Sequence for Grades K–12. https://www.marine-ed.org/ocean-literacy/scope-and-sequence Accessed 31 Aug 2024
11. Wu, C.H.: Research on the coastal marine environment and rural sustainable development strategy of island countries—taking the Penghu Islands as an example. Water. **13**(10), 1434 (2021)
12. Streimikiene, D., Svagzdiene, J.E., Simanavicius, A.: Sustainable tourism development and competitiveness: the systematic literature review. Sustain. Dev. **29**(1), 259–271 (2021)
13. Zheng, M.Y., Chen, C.C., Lin, H.H., Tseng, C.H., Hsu, C.H.: Research on the impact of popular tourism program involvement on rural tourism image, familiarity, motivation and willingness. Sustainability. **13**(9), 4906 (2021)
14. Sadiqe Ali Mohsen, L.: The Relationship between Marketing Deception and Customer Repurchase Intention (Master Thesis). Gelisim University, Türkiye (2023) https://acikerisim.gelisim.edu.tr/xmlui/bitstream/handle/11363/4779/791469.pdf?sequence=1&isAllowed=y
15. Pressey, A.D., Harris, L.C.: When we practice to deceive: service worker manipulation and (Mis) behavior on cruise ships. J. Travel Res. **63**(2), 447–465 (2024)
16. Wondirad, A., Ewnetu, B.: Community participation in tourism development as a tool to foster sustainable land and resource use practices in a national park milieu. Land Use Policy. **88**, 104155 (2019)

17. Chandel, S.P.K.: Impacts of tourism on environment. Cent. Asian J. Innov. Tour. Manag. Financ. **3**(10), 90–98 (2022)
18. Kelly, R., et al.: Connecting to the oceans: supporting ocean literacy and public engagement. Rev. Fish Biol. Fish. **32**(1), 123–143 (2022)
19. Schoedinger, S., Tran, L.U., Whitley, L.: From the principles to the scope and sequence: a brief history of the ocean literacy campaign. NMEA Spec. Rep. **3**, 3–7 (2010)
20. McAteer, B., Flannery, W., Murtagh, B.: Linking the motivations and outcomes of volunteers to understand participation in marine community science. Mar. Policy. **124**, 104375 (2021)
21. Jayasekara, K.D.D.S., Rajapaksa, D., Gunawardena, U.A.D.P.: Impacts of environmental knowledge, motives, and behavior on eco-tourism. Sustainability. **16**(11), 4724 (2024)
22. Ozbey, D.O. et al.: How does environmental education moderate local People's commitment to the environment and intention to support tourism development? Sustainability. **16**(13), 5646 (2024)
23. Wu, C.H.: A study on the current impact on island tourism development under COVID-19 epidemic environment and infection risk: a case study of Penghu. Sustainability. **13**(19), 10711 (2021)
24. Lin, H.H. et al.: Adolescents' consumption intentions toward leisure tourism in high-risk leisure environments in riverine areas. Open Geosci. **16**(1), 1–10 (2024)
25. Martínez Vázquez, R.M., Milán García, J., De Pablo Valenciano, J.: Analysis and trends of global research on nautical, maritime and marine tourism. J. Mar. Sci. Eng. **9**(1), 93 (2021)
26. Baloch, Q.B. et al.: Impact of tourism development upon environmental sustainability: a suggested framework for sustainable ecotourism. Environ. Sci. Pollut. Res. **30**(3), 5917–5930 (2023)
27. Alamineh, G.A., Hussein, J.W., Endaweke, Y., Taddesse, B.: The local communities' perceptions on the social impact of tourism and its implication for sustainable development in Amhara regional state. Heliyon. **9**(6) (2023)
28. Fornell, C., Larcker, D.F.: Evaluating structural equation models with unobservable variables and measurement error. J. Mark. Res. **18**, 39–50 (1981). https://doi.org/10.2307/3151312
29. Wang, Z., Yin, L., Qin, X., Wang, S.: Integrated assessment of sediment quality in a coastal lagoon (Maluan Bay, China) based on AVS-SEM and multivariate statistical analysis. Mar. Pollut. Bulleti. **146**, 476–487 (2019)
30. Devellis, R.F.: Scale Development: Theory and Applications. Sage, Newbury Park, CA, USA (1991)
31. Chen, S.: Structural Equation Model: Amos Operation. Psychological General Distribution, Taiwan, Taipei (2007)
32. Janesick, V.J.: The choreography of qualitative research design: minuets, improvisations, and crystallization. In: Denzin, N.K., Lincoln, Y.S. (eds.) Handbook of Qualitative Research, pp. 379–399. Sage, Thousand Oak, CA (2000)
33. Gursoy, D., Jurowski, C., Uysal, M.: Resident attitudes: a structural modeling approach. Ann. Tour. Res. **20**, 79–105 (2002). https://doi.org/10.1016/S0160-7383(01)00028-7Get
34. Jiang, Y.: The internal logic and quantitative characteristics of Learners' subconscious learning from the perspective of psychology. Front. Educ. Res. **6**(31) (2023)
35. Fawehinmi, O., Yusliza, M.Y., Tanveer, M.I., Abdullahi, M.S.: Influence of green human resource management on employee green behavior: the sequential mediating effect of perceived behavioral control and attitude toward corporate environmental policy. Corp. Soc. Responsib. Environ. Manag. **31**(3), 2514–2536 (2024)
36. Shin, D.: The actualization of meta affordances: conceptualizing affordance actualization in the metaverse games. Comput. Hum. Behav. **133**, 107292 (2022)
37. Verplanken, B., Orbell, S.: Attitudes, habits, and behavior change. Annu. Rev. Psychol. **73**(1), 327–352 (2022)

38. Halkiopoulos, C., Antonopoulou, H., Gkintoni, E., Aroutzidis, A.: Neuromarketing as an indicator of cognitive consumer behavior in decision-making process of tourism destination—an overview. In: Transcending Borders in Tourism through Innovation and Cultural Heritage: 8th International Conference, IACuDiT, pp. 679–697, Hydra, Greece (2021)
39. Huang, H., Te Hsin, C.: Environmental literacy education and sustainable development in schools based on teaching effectiveness. Int. J. Sustain. Dev. Plan. **18**(5) (2023)
40. Georgiou, G.K., Inoue, T., Parrila, R.: Developmental relations between home literacy environment, reading interest, and reading skills: evidence from a 3-year longitudinal study. Child Dev. **92**(5), 2053–2068 (2021)
41. Lin, H.H. et al.: From the perspective of high-risk leisure environment: a study of the impact of green education literacy on river water, ecology, surrounding rural economic development, and youth well-being. In: International Symposium on Water Resource and Environmental Management, pp. 229–243. Springer Nature Switzerland, Cham (2023)
42. Costa, D.D.A., de Lucena, R.F.P., Christoffersen, M.L., Piñeiro-Corbeira, C., Dolbeth, M.: Improving environmental awareness and ocean literacy through hands-on activities in the tropics. Appl. Environ. Educ. Commun. **21**(2), 120–139 (2022)

Challenges and Solutions in Using Green Hydrogen for Vehicles to Improve the Environment and Public Health

Maryam S. Hamed[1], Mohamed H. Mohamed[1], and Salah H. R. Ali[2(✉)]

[1] Faculty of Oral and Dental Medicine, Ahram Canadian University (ACU), 6th of October, Egypt

[2] Precision Engineering Division, National Institute for Standards (NIS), Al-haram, Giza, Egypt
salahali20@yahoo.com

Abstract. *Background* The use of green hydrogen (GH) has gained significant global attention as a promising and clean energy source. GH plays a key role in maintaining a sustainable environment and, consequently, improving public health. *Introduction* Several types of hydrogen fuel cells are available for vehicle power, but they are still in the research and development stages. This calls for further research to find scientific solutions to maximize hydrogen utilization at affordable prices and high efficiency. *Methodology* This paper focuses on monitoring and understanding the current status and challenges facing automotive hydrogen fuel cells. The cell consists of a cathode, an anode, and an electrolyte as a medium. *Objective* This paper aims to present examples of the current progress in hydrogen fuel cells operating using water electrolysis. In addition, offers solutions based on advanced engineering materials and appropriate measurement techniques to improve the performance of the cathode material and its surface in a scientific and useful manner. Furthermore, the use of scanning electron microscopy and atomic force microscopy is proposed to improve the characterization of the cathode material, its surface, and the electrolyte in hydrogen fuel cells. *Conclusion* Adopting this approach may be a necessary step toward helping provide clean and sustainable energy, reducing pollution, enhancing doctors' ability to perform their medical duties, and maximizing patient treatment and recovery in a healthy environment. In addition, it will enhance the understanding of researchers, students, and those interested in alternative energy sources, highlighting the importance of the environment and public health.

Keywords: Hydrogen · Fuel cell · Vehicle transportation · Environment · Healthcare and public health

1 Introduction

The environment and public health have received significant attention from all international organizations and bodies. The environment is the surroundings in which humans live and interact, including water, the air on Earth, and space [1]. Public health, as defined by CEA Winslow in 1887, is "*the science and art of preventing disease, prolonging life,*

Y. Zhu (Ed.): ISESIE 2025, LNCE 000, pp. 130–144, 2026.
https://doi.org/10.1007/978-981-92-2389-3_12

and promoting health and well-being through the organized social effort and informed choices of society, organizations, public and private communities, and individuals" [2]. CEA Winslow, a university professor and one of the pioneers of modern public health in the US and the world, founded the Yale School of Public Health in 1915. Public health is also defined as a combination of sciences, skills, and beliefs aimed at maintaining and improving the health of populations through collective action [3]. Public health is a collective social effort to promote health and prevent disease [4]. Healthcare is a set of services and procedures that aim to improve the health of individuals and communities, and to prevent, treat, manage, and maintain diseases [5, 6]. Therefore, it can be said that a clean environment is one of the fundamental pillars of improving healthcare, by enhancing the role of doctors in successfully performing their medical duties and providing a suitable environment for work systems in various areas of life, which improves public health, enhances human well-being, and promotes economic prosperity.

On the other hand, back in 1839, Sir William Robert Grove (1811–1896) invented the first primitive model of a fuel cell, which he called a "*Gas Battery*" [7]. He combined hydrogen and oxygen to produce an electric current and water from a chemical reaction (the same idea as modern fuel cells for electric cars). They were then developed in the 1960s for use as simple models in space programs, and have been used in small cars since the 1990s [8]. Now, hydrogen fuel cells (HFCs) are being used on a modest scale in transportation, especially for small cars, such as trucks, buses, trains, and even ships and airplanes. However, the vast majority of cars still run on fossil fuels. About a century ago, the number of conventional cars on the planet was estimated at thousands. Today, there are approximately one and a half billion cars on our planet—the equivalent of one car for approximately eight people. As the natural population increases, the rate of car use has increased, along with the rate of environmental pollution. It is expected that the number of cars will exceed 2 billion by the next 15 years. So, let's imagine for a moment that our Earth will become like a giant gas station with a limited supply of fuel, which will quickly make us realize that our children and grandchildren will inevitably face a real issue called "energy shortages." Some geologists predict that we will soon reach a point called "peak oil," at which everything will be made from oil, and consequently, the supply will dwindle. If that happens, where will everyone get fuel for their cars? This is on the one hand. The other, more important aspect lies in the environmental pollution produced by fossil fuels. Studies have indicated that air pollution in urban areas reaches about 27% of land transportation alone [9–12]. Air pollution from railways also increases as a result of the transportation of iron ore, coal, or mineral soil, which may lead to the release of nitrogen oxide, carbon monoxide, sulfur dioxide, and hydrocarbons into the atmosphere [13].

Therefore, the scarcity of energy resources and the increase in environmental pollution have become two strategic issues facing scientists and politicians worldwide for more than twenty years. Therefore, we are currently going through a difficult phase, especially as everyone is now exploring the need for a delicate balance at the centre of the triangle whose sides are economy, nature, and operational efficiency. The short-term solution is to improve the operational efficiency and fuel consumption of traditional cars with internal combustion engines (ICEs) that run on gasoline or diesel. This solution is constantly evolving, but it is not sufficient in light of the expected energy crisis and environmental pollution. The optimal long-term solution may only lie in diversifying energy sources. That is, providing cars that run on non-fossil fuels, whether using electric or hydrogen vehicles, or a combination of both. Although there are research opinions that are not optimistic about the continued use of electric cars, due to their significant risks, both during manufacturing and use, whether announced or unannounced so far, except in special circumstances [11, 14]. Although anyone following global politics will notice the US government's role in financially supporting and promoting the spread of electric cars in recent years, some, and we agree, believe that hydrogen or hybrid cars may be the optimal solution and alternative, given their advantage in maintaining a clean environment and protecting public health. This is because hydrogen cars run on hydrogen, while hybrid cars run on two energy sources together, and there are many types, such as cars that run on HFCs and electricity, or hydrogen fuel cells and gasoline together, etc. Green hydrogen cells are similar to batteries but run on hydrogen gas, which never runs out. Thanks to the silence and theoretical lack of pollution, hydrogen cells may be considered the cleanest cars ever, as they use cleaner energy sources and are more environmentally friendly, public health, and life-saving.

All relevant entities, both governmental and non-governmental, have placed great importance on supporting everything that provides a clean and suitable environment for all living beings, especially humans. There are even specialized periodicals and scientific conferences devoted to environmental science and environmental industries. Our current conference is a prime example of this. All of this will undoubtedly support the exploration of the challenges we face and provide new opportunities for innovation in the environmental industry, to continue promoting global environmental protection. This is evident in light of the significant and growing population growth, the diversification of transportation methods, and industries that rely on energy sources. In light of the increasing environmental challenges resulting from heavy reliance on fossil fuels, it has become imperative to explore alternatives, even if only relative, to provide a clean and sustainable energy source for the present and the future. Clean energy can be either green hydrogen to produce green electricity or green electricity to produce green hydrogen. Green hydrogen is therefore one of the most prominent of these alternatives, as it is produced through the electrochemical analysis of water using renewable energy sources, making it carbon-free. Green hydrogen is also used via electrochemical decomposition as a clean fuel to generate electricity through fuel cells using oxygen in the air. However, when HFCs are used to power vehicles, only water vapor is produced as a byproduct, making it an ideal environmental choice. It has shown increasing global interest, particularly in the transportation sector, as it is considered a major contributor to reducing environmental pollution and thus reducing the spread of infections for many diseases, which supports a reduction in the spread of diseases and a decrease in the number of deaths [15, 16]. Studies have shown that environmental pollution contributes to the spread of many diseases, such as respiratory diseases, heart disease, some types of cancer, and oral diseases [14–21]. Air pollution resulting from fossil-fuel vehicle exhaust contains dangerous pollutants, such as nitrogen oxides (NO_x), carbon monoxide (CO), and fine particulate matter (PM_{10} and $PM_{2.5}$). $PM_{2.5}$ refers to particulate matter with a diameter of less than 2.5 μm, which is very fine and penetrates deep into the lungs and even the bloodstream. PM_{10} refers to particulate matter with a diameter of less than 10 μm, and even smaller, penetrates the respiratory system. These pollutants directly affect oral and dental health, and their most prominent effects [21–25] include:

- Increased gingival inflammation: Inhaling pollutants leads to inflammatory responses in the body, increasing the likelihood of gingivitis and periodontitis.
- Weakened local immune system: Pollutants weaken the immune system response within the mouth, increasing the chances of bacteria that cause tooth decay multiplying.
- Dry mouth (Xerostomia): Some components of polluted air lead to dry mouth, which reduces saliva production and increases the growth of harmful bacteria. Increased risk of oral cancer: Prolonged exposure to carcinogens in the air (such as benzene and formaldehyde) may increase the risk of mutations in oral cells.

In light of these serious issues of energy scarcity, increased pollution, and public health risks, researchers must seek new, clean energy sources. It is also imperative to comply with United Nations regulations requiring the gradual transition of vehicle ICEs to alternative energy sources, such as fuel cells and electric vehicle technology [26].

This research paper aims to monitor and understand the current status of using green hydrogen fuel cells to produce clean energy for vehicle operation. It also provides a simplified and documented systematic overview of the latest related technologies, with the aim of enhancing the concepts and knowledge of researchers, students, and those interested in alternative energy sources and their role in environmental protection. The paper also seeks to explore the technical and material challenges facing the production and use of green hydrogen, and propose scientific and practical solutions based on advanced engineering materials and precise characterization techniques. This work contributes to improving the efficiency of alternative energy sources and raising environmental awareness to achieve a clean environment and sustainable public health.

2 Hydrogen Fuel Cells (HFCs)

Hydrogen is used to provide clean energy for vehicles in one of two main ways: (1) a proton exchange membrane fuel cell (PEMFC) or (2) hydrogen engines. Regarding hydrogen engines, some researchers and major companies have attempted to develop conventional ICE systems to run on hydrogen fuel, but the results so far have been modest in terms of performance, efficiency, and emissions [27–30]. There is also significant interest in this field from some researchers in producing and using hydrogen mixed with other fuels, including methane, gas, diesel, methanol, and gasoline, to improve the operating performance of vehicles with ICEs [30–32]. Recent developments in hydrogen storage (up to 700 bar) and its production using solar and wind energy have raised interest in hydrogen as a future fuel for ICEs [31, 32] due to its emission-free properties. However, development is still underway to improve storage systems and reduce costs (so-called green ammonia), and successful trials are expected to be completed and achieve a suitable economic return by the beginning of 2050 [32]. On the other hand, HFCs have begun to make their way into the automotive world. HFCs are a promising technology for powering cars, and their use is expected to become widespread in the near future. Given the importance of this global trend at the present time, we will review this topic in detail, in terms of types and reaction mechanisms during use.

2.1 Types of HFCs

There are six basic types of hydrogen fuel cells. Depending on the need and application, the most appropriate type can be selected and implemented. They are as follows:

1. Proton Exchange Membrane Fuel Cell (PEMFC): The most commonly used in vehicles. It's operates at low temperatures (≈80 °C). Operates quickly and has a high energy density.
2. Alkaline Fuel Cell (AFC): Historically used in space applications. Operates efficiently but is sensitive to carbon dioxide.
3. Molten Carbonate Fuel Cell (MCFC): Operates in large industrial applications. It's Operates at high temperatures (600–700 °C).
4. Solid Oxide Fuel Cell (SOFC): Relies on a solid ceramic electrolyte. It's Operates at high temperatures of 1000 °C. It's suitable for industrial use and stationary electricity generation.
5. Phosphoric Acid Fuel Cell (PAFC): Used to generate electricity for facilities and hospitals. Operates at high temperatures of 200 °C
6. Direct Methanol Fuel Cell (DMFC): Used to generate electricity for portable and small-sized devices.

The PEMFC is the most important and most common type for use in vehicles and transportation in general, due to its unique properties such as high efficiency, rapid response at suitable temperatures, and environmental cleanliness, which makes it superior to other types. The PEMFC cell operates as an electrochemical device in one of two modes: either converting hydrogen to electricity or vice versa. The PEMFC cell converts the chemical energy stored in hydrogen (H_2) and oxygen (O_2) directly into electrical energy. This is done through a chemical reaction that does not involve combustion. The reaction produces only electricity, water, and heat, thus is making this cell a clean energy source, free of harmful emissions. This type of cell is used in hybrid vehicles (hydrogen to electricity). The diagram shown in Fig. 1 illustrates the general framework of one of the most popular modern vehicle types, called hybrid vehicles or hydrogen fuel cell electric vehicles (HFCEVs), which generate electricity from a hydrogen fuel cell. Figure 1 illustrates the importance of a hydrogen fuel cell in the process of converting hydrogen to electrical energy, as is the case in operating some hybrid vehicles. Hydrogen gas travels from the tank to the fuel cell with air, and electrons flow continuously from the fuel cell, providing the electrical voltage used to power the electric motor that drives the car's wheels. Inside the fuel cell, hydrogen atoms are split into protons and electrons. As the PEMFC fuel cell produces electricity, the hydrogen protons combine with oxygen from the air to produce water, as detailed in Fig. 2.

HFCEVs have a lower risk of fire or explosion than conventional non-hybrid electric vehicles. These vehicles are equipped with a wide range of sensors to ensure safety. In the event of a fuel leak or collision, HFCEVs are automatically programmed to shut down in a safety mode. This means that the fuel system will not operate normally, and therefore, the vehicle cannot be restarted [33]. Although hydrogen is a stable fuel source, its properties

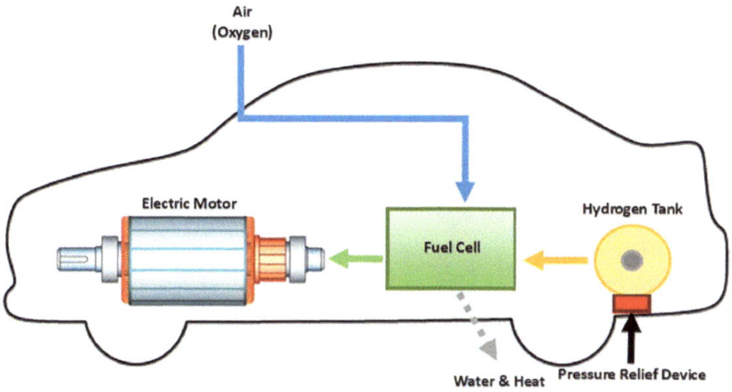

Fig. 1. Simplified diagram of the operation of hydrogen-powered hybrid electric vehicles [33]

are well known thanks to its extensive use in industrial processes. However, for added safety, these vehicles are equipped with a thermal pressure relief device (TPRD) on the hydrogen fuel tank, as shown in Fig. 1. This device rapidly discharges the hydrogen fuel into the open air in the event of a vehicle fire [33]. In the event of a catastrophic accident, assessing the system may become more difficult due to the lack of time for the safety shutdown mechanism. In such a situation, extreme caution is essential. In this case, too, hydrogen is likely to escape into the open air, potentially igniting the area. Depending on the power lines being down, there is a risk of electric shock and collision with metals and charged equipment. This makes planning for these situations difficult, and they must be handled carefully and on a case-by-case basis [33]. However, this is where the matter requires a pause, as the authors sense an unexpected danger not to hydrogen but to the electrical power of cars, which requires further studies and research to predict and provide appropriate solutions to this issue.

In detail, Fig. 2 illustrates the working principle of a PEMFC, which consists of an anode, a cathode, a proton exchange membrane, catalyst layers, a gas diffusion layer, and bipolar plates. This type of HFC uses a proton-conducting polymer membrane as an electrolyte, consisting of:

1. Anode: Hydrogen gas (H_2) is introduced into the anode, where hydrogen is decomposed into protons and electrons using a catalyst (usually made of platinum).
2. Cathode: Oxygen from the air is introduced into the anode, where electrons and protons are accepted to form water.
3. Electrolyte membrane: This is a polymer membrane that only allows protons to pass through. The most common material used is Nafion.
4. Catalysts: These are typically made from noble metals such as platinum (Pt), and accelerate redox reactions.
5. Bipolar plates: These distribute gases and collect the resulting electrical current.

2.2 Reaction Mechanism

As shown in Fig. 2, the mechanism of the electrochemical reaction inside the cell is as follows:

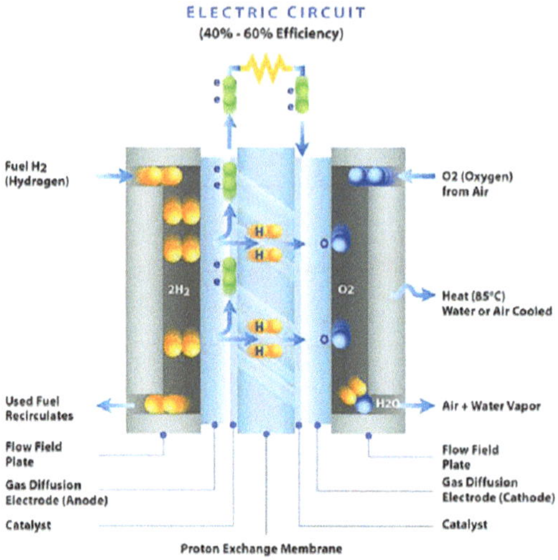

Fig. 2. Simplified diagram of the working principle of a PEMFC [34]

1. At the anode: Hydrogen dissociates into protons and electrons, as in Eq. (1), where:

$$H2 \rightarrow 2H + +2e \tag{1}$$

2. Electrons and the external circuit: Electrons pass through an external electrical circuit to power electrical devices, generating an electrical current used to power loads.
3. At the cathode: Oxygen combines with the electrons and protons (as a result of the hydrogen dissociation) to form water (H_2O) as in Eq. (2), where:

$$O2 + 4H + +4e- \rightarrow 2H2O \tag{2}$$

Figure 3 shows a three-dimensional image of a PEMFC hydrogen fuel cell used in hybrid electric vehicles. Figure 4 shows a real fuel cell stack of PEMFC-type HFCs, used in many transportation systems such as cars, airplanes, ships, and buses to reduce environmental pollution. This cell is capable of producing 5 kW of power.

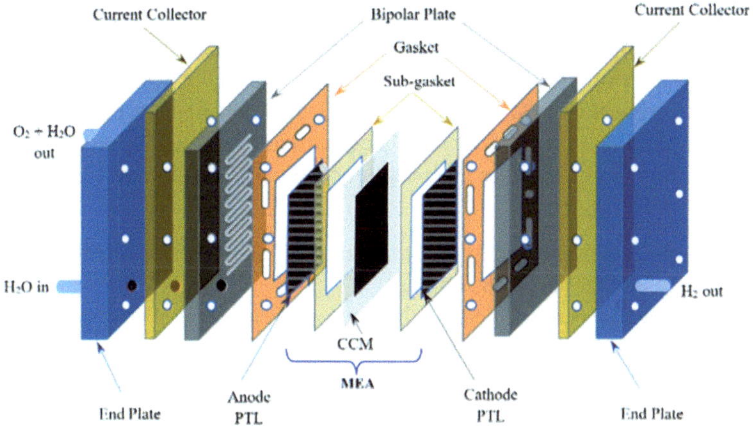

Fig. 3. 3D schematic of a PEMFC hydrogen fuel cell in a hybrid electric vehicle [35, 36]

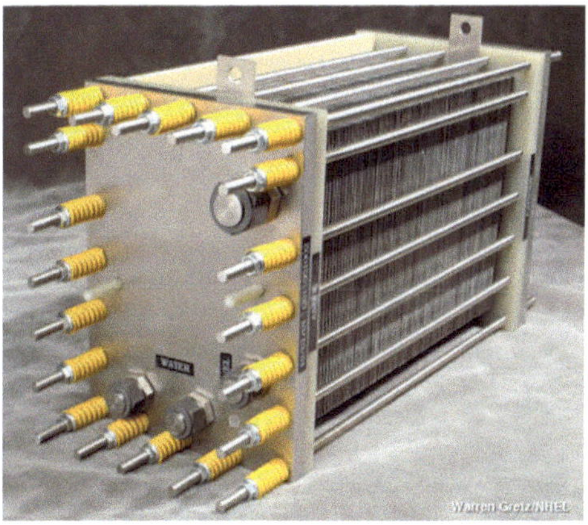

Fig. 4. A cell module assembled from several PEMFC-type HFCs for vehicle use [37]

3 Progress in the Use of Green HFCs in Vehicles

Currently, the automotive industry is witnessing a radical shift in progress toward the use of new energy sources. Hydrogen fuel cell technology has emerged as a promising alternative for cars, particularly the PEMFC type, due to its unique features, as we mentioned earlier. Hydrogen fuel cells are used in one of two ways in vehicles. In other words, one cell produces green hydrogen from electricity and vice versa. Here, we will present two distinct experiences of green hydrogen fuel cell technologies in powering vehicles, one at the international level and the other at the local level.

3.1 Globally

Figure 5 shows a fuel cell stack (in a single cell unit called a Fuel Cell Stack) used in modern hydrogen automobile produced by several global companies, such as Toyota

(Mirai), Honda (Clarity Fuel Cell), and Hyundai (Nexo). This cell unit design is characterized by its compactness, allowing it to be integrated into the vehicle's chassis. The unit includes a large number of PEMFC fuel cells connected electrically, mechanically, and hydraulically, including cooling, water, and gas management elements. Figure 5 illustrates a comprehensive picture of the latest technologies for a hydrogen cell assembly assembled together to produce electricity for use in hybrid electric vehicles. The unit consists of the following:

- Cell assembly (stack): A group of hydrogen fuel cells stacked together to generate sufficient voltage to operate the engine.
- MEA components: These include the proton membrane, electrodes (anode and cathode), and gas diffusion layers.
- Advanced catalysts and improved materials: These are used to improve efficiency and reduce cost, and are also used in development of Engineering in Shape-Controlled Catalysts and Others.
- Automotive-Friendly Installation: The design reflects the evolution of PEMFC modules to become more integrated and compatible with the automotive environment.

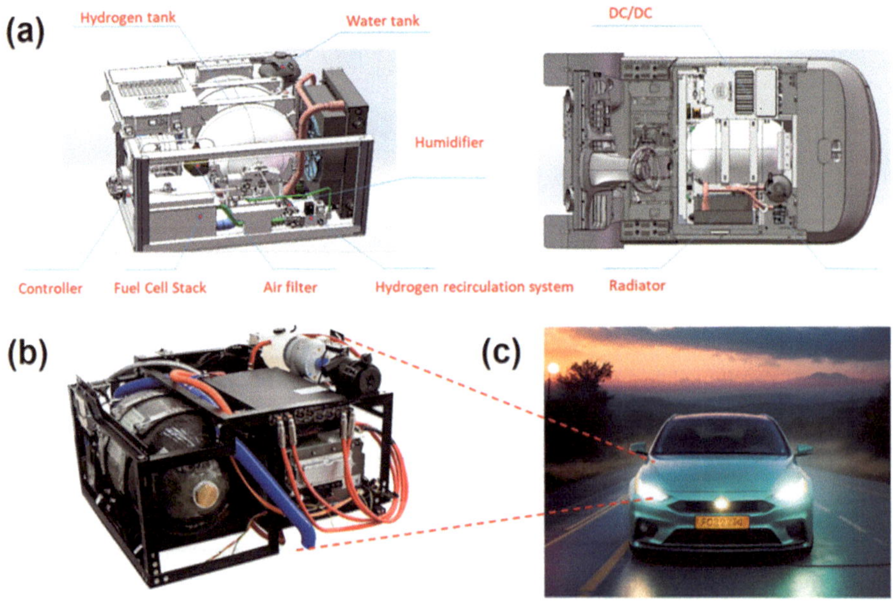

Fig. 5. A PEMFC hydrogen fuel cell module used in modern vehicles [38].

3.2 At the National Level

Egypt has achieved significant progress in exploring and producing green hydrogen for a variety of applications, including operating power plants. There are currently approximately 36 green hydrogen and green ammonia production projects in Egypt, with the goal of producing approximately 1.5 million tons annually by 2030 [39]. After extensive cooperation, effort, and perseverance, Egypt has also succeeded in exporting green hydrogen in the form of green ammonia to several countries, including India [40]. In April 2025, the latest partnership agreement was signed between Egypt and France, with Emirati cooperation worth seven billion euros ($7.7 billion), to finance and operate a new green hydrogen production project in the Ras-Shukeir area at Egypt. This project will begin producing 300,000 tons of green ammonia annually in 2029 and aims to produce one million tons annually in three phases, providing clean fuel for supplying ships, in addition to exporting to European countries and global markets [41, 42]. These projects represent a step toward achieving sustainable development goals, diversifying energy sources, and reducing carbon emissions. They also contribute to strengthening Egypt's position as a regional hub for renewable energy and green hydrogen. We also aspire to use it in the automotive sector. Efforts have been made to build a handmade cell to produce very small quantities of green hydrogen. The cell is then mixed and injected simultaneously with gasoline in a special circuit to improve the operation of ICEs, enhancing performance and reducing pollution. There are two circuits: one electrical to connect the cell to the vehicle's circuit, and the other hydraulic to the vehicle's engine. However, its feasibility remains incomplete and requires further development and improvement. It is still under development, see Fig. 6. The dry hydrogen generation cell was designed and implemented for use with engines with a displacement of approximately 2000 cc, reducing fuel consumption by up to 35% and reducing emissions, while prolonging engine life by cleaning internal components. If the automobile has a 2-l engine, it requires a 10-plate electrolysis cell, or bipolar plates, selected from stainless steel such as 304 L or 316 L, 1 mm thick. The surfaces of the plates must be chemically treated (by boiling them in water with citric acid) to remove any fingerprints from the manufacturing process, including cutting and finishing, to maintain electrical conductivity. Two 10 mm thick from acrylic or Bakelite plates, and a 1/4 in. diameter polyamide (PA6) hose are also required. The electrolyte is prepared as a mixture of 800 mg of cold distilled water and 200 mg of sodium hydroxide (a caustic substance). The required voltage is 14.5 to 15 V. The total cell weight is approximately 1.5 kg, and it produces approximately 0.5 l/min of hydrogen fuel.

Fig. 6. A real image of a stacked cell unit of HFCs to operate with gasoline automotive engines

4 Most Important Current Challenges

Despite the significant progress made in the recent years in the production and use of green hydrogen to power automobiles, electrochemical hydrogen fuel cells, especially the PEMFC type, still face several challenges in improving their efficiency [40, 43]. The most important of these challenges are the following:

1. The high cost of using platinum metal due to its scarcity, which affects the cost of building the cell and its economic feasibility.
2. The rapid deterioration and corrosion of the peritoneal membrane over time: Membranes of the type Nafion (trade name: co-fluoropolymer) deteriorate rapidly with time, heat, and humidity. Nafion is primarily composed of polytetrafluoroethylene (trade name: Teflon) and a series of other chemical components.
3. High sensitivity to polluting gases: This type is negatively affected by the presence of carbon dioxide or carbon monoxide gases, whether in the fuel or the air.
4. Short lifespan in some industrial applications, particularly in carbonated environmental conditions: These require special care and periodic maintenance, along with the replacement of some essential components.

This is in addition to the difficulty of providing infrastructure due to the difficulty of storing and transporting ICE vehicles. This has made hydrogen distribution networks limited to date, hindering the spread of hydrogen-powered vehicles. Because of the low density and explosive nature of hydrogen, its storage and transportation require special and safe technologies. It requires very special tanks due to the low temperatures and high pressures required to store it, making it difficult to deploy hydrogen fueling stations to date.

5 Progress Research and Development and the Most Important Proposed Solutions

A large number of previous studies and research have focused on improving the issue of hydrogen fuel cell production. However, they are still insufficient to improve production, especially for PEMFC hydrogen cells. These studies require further research and interpretation to provide innovative solutions to this issue in the pursuit of optimal results. The operation of hydrogen cells is considered a challenging electrochemical process and also requires further efforts and appropriate solutions. Therefore, we will present two categories here: (1) the findings of previous research on development, and (2) the proposals we have reached in this paper to maximize the development of green hydrogen production, highlighting how to facilitate the implementation of these appropriate solutions.

5.1 Research Trends for the Development of PEMFC Hydrogen Cells

Research has diversified and focused on hydrogen cell development over the past years. However, although these trends have been promising and some have been implemented over the past six years, the most important of these are:

- Replacing platinum with cheaper alternative catalysts, such as cobalt, nickel, or carbon nanocomposites supported by metal oxides, by Wang et al. in 2021 [44].
- Designing new polymer membranes with improved permeability, using highly conductive and degradation-resistant membranes, by Zhao et al. in 2020 [45].
- Using nanomaterials, to increase the surface area of catalysts and improve electrochemical performance, by Zhang et al. in 2019 [46].
- Three-dimensional electrode structures, to improve gas distribution and reduce efficiency loss, by Liu et al. in 2022 [47].

However, previous research trends have not achieved the desired development and require further effort and ideas for development. Therefore, we may need to participate in providing proposed solutions that complement and are more effective in maximizing the development of hydrogen fuel cell production.

5.2 The Most Important Proposed Solutions for HFCs Development

As we noted in Sect. 3 above, there are various practical applications around the world for the use of green HFCs for vehicles. In other words, there are cells that produce electricity from green hydrogen and vice versa, and both require development and improvement to maximize operating efficiency. Therefore, the focus here is on the proposals in this paper that enhance hydrogen fuel cell production, as follows:

- Developing and using advanced engineering materials for electrodes: Using new alloys and nanocomposite materials in the manufacture of cathodes (electrodes) to achieve reasonable prices while improving performance and reducing electrical resistance, such as platinum or cobalt compounds supported on carbon substrates or nanoscale metal oxides.
- Improving cathode surface design using GD&T systems and nanotechnology: This helps improve the geometric features of the cathode surface, thus increasing the effective interaction area, thus enhancing production efficiency.
- Using advanced metrological characterization techniques: Scanning electron microscopy (SEM) and atomic force microscopy (AFM) are proposed to study surface structure and interactions at the atomic level, providing accurate measurements that enhance design and application concepts for reducing corrosion rates during reaction processes.
- Attempting to provide a more suitable working environment within the cell: This will reduce the presence of harmful gases in the production process, such as carbon monoxide and carbon dioxide. This will enhance the sensitivity of handling within the cell.
- Reducing costs through modern manufacturing techniques: Innovative technologies such as 3D printing of components and nanofabrication techniques can contribute to reducing the overall cost of the system.

6 Conclusion

- The use of green hydrogen represents a fundamental pillar in the global transition to a low-carbon economy as a clean energy source. Adopting this type of energy not only contributes to raising environmental awareness and improving environmental quality, but also reduces pressure on healthcare systems, supporting public health and enhancing quality of life. Despite significant technical and economic challenges, the proposed scientific solutions provide innovative practical paths to improving the efficiency and effectiveness of hydrogen fuel cells.
- Hydrogen fuel cells, particularly PEMFCs, are among the most promising technologies for meeting future clean energy needs, given their ability to produce electricity without harmful emissions and green hydrogen. They are also expected to play a pivotal role in sustainable transportation technologies, particularly in the automotive sector.
- Continued advances in the fields of engineering materials and engineering metrology will increase the chances of these technologies achieving wider deployment in the coming years, particularly for the diagnosis and improvement of electrochemical systems.
- In addition, blending green hydrogen with conventional fuels such as gasoline is a promising option for reducing emissions and complying with international and local environmental standards.

Eventually, the move toward developing hydrogen fuel cells, whether for hydrogen production or generating electricity from hydrogen opens new horizons for overcoming the challenges of the environmental industry, creating a more sustainable environment. It also supports medical efforts by reducing pollution to improve healthcare and the overall health climate. This is in addition to raising awareness among researchers and students about the importance of clean energy and its role in protecting the environment and public health.

References

1. Buongiorno, F., Chiaramonte, X.: Environment. In: Wallenhorst, N., Wulf, C. (eds.) Handbook of the Anthropocene. Springer, Cham (2023). https://doi.org/10.1007/978-3-031-25910-4_8
2. Visel, A.J., Tear, C.-E.A.: Winslow and the early years of public health at Yale. Yale J. Biol. Med., 1915–1925 (1982) https://pmc.ncbi.nlm.nih.gov/articles/PMC2596005/
3. Acheson, R.M., Payne, A.M.: Preventive medicine at the Yale School of Medicine, 1950-1965. Milbank Mem. Fund Q. **45**(3), 287–301 (1967)
4. Culyer A.J.: Encyclopedia of Health Economics, Book Reference. Science Direct, New York, (2014)
5. Rohani, M., Marquart-Pyatt, S.T.: The costs, health and economic impact of air pollution control strategies: a systematic review. Glob. Health Res. Policy. **9**(1), 15 (2024). https://doi.org/10.1186/s41256-024-00373-y
6. Balakrishnan, K., Ghosh, S., Kumar, P.: Effectiveness of community and household level interventions for reducing impact of air pollution on health outcomes: A systematic review. Environ. Health Perspect. **130**(10), 106001 (2020). https://doi.org/10.1289/isee.2022.P-0949
7. Larminie, J., Dicks, A.: Fuel Cell Systems Explained, 2nd edn. Wiley (2023)

8. Appleby, A.J., Foulkes, F.R.: Fuel Cell Handbook. Van Nostrand Reinhold (1989)
9. Heydari, S., Tainio, M., Woodcock, J., de Nazelle, A.: Estimating traffic contribution to particulate matter concentration in urban areas using a multilevel Bayesian meta-regression approach. Environ. Int. **141**, 105800 (2020)
10. Bajwa, A.U., Sheikh, H.A.: Contribution of road transport to Pakistan's air pollution in the urban environment. Air. **1**(4), 237–257 (2023). https://doi.org/10.3390/air1040018
11. Ali, S.H.R., A-Razek, M.M., Ali, A.S.H.: Evaluation of environmental impact for electric automobiles. In: Zhu, Y. (ed.) Proceedings of the 4th International Symposium in Environmental Science and Industrial Ecology. ISESIE 2024. Springer, Singapore (2025). https://doi.org/10.1007/978-981-96-1578-0_14
12. Ali, A.S.H.R., Ali, S.H.R.: Entrepreneurship in the Automotive Industry using Computational Modeling Technology. In: Advances in Transdisciplinary Engineering, vol. 61, pp. 392–400. IOS Press, Netherlands (2024). https://doi.org/10.3233/ATDE240784
13. Carpenter, T.G.: The Environmental Impact of Railways. Wiley, New York (1994)
14. Ali, S.H.R., Elsayed, G.A.I., Ali, A.S.H.: Strategic solutions to provide a sustainable environment and automotive use in modern underground cities. In: Zhu, Y. (ed.) Proceedings of the 4th International Symposium in Environmental Science and Industrial Ecology. ISESIE 2024. Springer, Singapore (2025). https://doi.org/10.1007/978-981-96-1578-0_13
15. Ali, M.S.: Proposed solutions to reduce the impact of air pollution on oral and dental public health. Int. J. Community. Dent. **9**, 107–112 (2021). https://doi.org/10.4103/ijcd.ijcd_14_22. 2021. www.ijcommdent.com
16. Wired, air pollution is deadlier than you think (2023)
17. World Health Organization: Ambient (outdoor) air pollution and health (2018). Retrieved from https://www.who.int/news-room/fact-sheets/detail/ambient-(outdoor)-air-quality-and-health
18. Stępień, Z.: Analysis of the prospects for hydrogen-fuelled internal combustion engines. Combustion Engines. **197**(2), 32–41 (2024). https://doi.org/10.19206/CE-174794
19. Du, W. et al.: Environmental particulate matter—one of the culprits in the development of caries. Front. Public Health **13**, Article 1559384 (2025). https://www.ncbi.nlm.nih.gov/pmc/articles/PMC
20. Sinjari, B., et al.: The correlation between oral health and air pollution: a systematic review. MDPI. (2024)
21. Verywell Health: 8 Ways Environmental Pollutants Can Harm the Body. Verywell Health (2021)
22. Lin et al.: Prolonged exposure to air pollution increases periodontal disease risk: a nationwide, population-based, Cohort study. MDPI (2021)
23. Crystal Marruganti et al.: Air pollution as a risk indicator for periodontitis. PMC (2023). Website at: https://www.ncbi.nlm.nih.gov/pmc/articles/PMC
24. Zhu, L., Tang, M., Cai, Y., Wang, P.: Association between exposure to environmental pollutants and increased oral health risks: a comprehensive review. Frontiers in Public Health. **12**, 1482991 (2025). https://doi.org/10.3389/fpubh.2024.1482991
25. Mendelian Randomization Study: Air Pollution and Oral Health: An Overall Insight from Genetic Causality. Pubmed (2025)
26. Wang, Z., et al.: Effect of pre-injection on combustion and emission characteristics of a diesel engine fueled with diesel/methanol/n-butanol blended fuel. PRO. (2021)
27. Verhelst, S., Wallner, T.: Hydrogen-fueled internal combustion engines. Prog. Energy Combust. Sci. **35**(6), 490–527 (2009). https://doi.org/10.1016/j.pecs.2009.08.001
28. Fukai, J., et al.: Development of a hydrogen internal combustion engine and its vehicle. Toyota Technical Review. **50**(1), 59–64 (2001)
29. Verhelst, S., Wallner, T.: Hydrogen-fueled internal combustion engines: A review of efficiency and emissions performance. Int. J. Hydrog. Energy. **34**(7), 3213–3226 (2009)

30. Cao, Y., et al.: Oxygen-enriched air gasification of biomass materials for high-quality syngas production. Energy Convers. Manag. **199**, 111628 (2019)
31. Muhssen, H.S., Zöldy, M., Bereczky, Á.: A Comprehensive Review on the Hydrogen–Natural Gas–Diesel Tri-Fuel Engine Exhaust Emissions. Energies. **17**(15), 3862 (2024). https://doi.org/10.3390/en17153862
32. Ampah, J.D., Jin, C., Afrane, S., Yusuf, A.A., Liu, H., Yao, M.: Race towards net zero emissions (NZE) by 2050: reviewing a decade of research on hydrogen-fuelled internal combustion engines (ICE). Green Chem. **26**(16), 9025–9047 (2024)
33. Schmidt K., Schmidt J., Rivkin C.: Hydrogen Fuel Cell Vehicles—What First Responders Need to Know. Firehouse. Website at: https://www.firehouse.com/rescue/article/12385113/hydrogen-fuel-cell-vehicles-what-first-responders-need-to-know-firehouse
34. Richard.: Thicker Electrodes for Smarter Batteries (2023). Website at: https://blog.upsbatterycenter.com/thicker-electrodes-denser-batteries/
35. Araújo, H.F., Gómez, J.A., DMF, S.: Proton-exchange membrane electrolysis for green hydrogen production: fundamentals, cost breakdown, and strategies to minimize platinum-group metal content in hydrogen evolution reaction Electrocatalysts. Catalysts. **14**(12), 845 (2024). https://doi.org/10.3390/catal14120845
36. Yuan, X.-Z. et al.: The porous transport layer in proton exchange membrane water electrolysis: perspectives on a complex component. Sustain. Energy Fuels. **6**, 1824–1853 (2022)
37. Woodford C.: Fuel cells (2024). Website at: https://www.explainthatstuff.com/fuelcells.html
38. Mo, T., Li, Y., Luo, Y.: Advantages and technological Progress of hydrogen fuel cell vehicles. World Electr. Veh. J. **14**(6), 162 (2023). https://doi.org/10.3390/wevj14060162
39. Amar A.: The largest Arab countries in clean hydrogen production targets (report). Energy Research Unit (2024). Website at: https://attaqa.net/
40. Hamed M.S., Ali S.H.R.: Green hydrogen production: progress, challenges and proposed solutions. SAE paper, WCX SAE World Congress Experience (2025). https://doi.org/10.4271/2025-01-8601
41. Magdy Y.: With a €7 billion investment, an Egyptian-French-Emirati alliance signed a contract to launch an integrated green hydrogen production plant. Amwal Al-ghad News (2025). Website at: https://amwalalghad.com/
42. Huaxia: Egypt, France sign 7-bln-euro deal for green hydrogen complex on Red Sea. Source: Xinhua News, China (2024). Website: https://english.news.cn/20250409/8b8d7de0d36747099dc6b5aead0fb09e/c.html
43. Barbir, F.: PEM Fuel Cells: Theory and Practice. Academic (2013)
44. Wang, Y., et al.: Non-precious metal catalysts for oxygen reduction reaction in PEM fuel cells. Energy Environ. Sci. **14**(2), 992–1023 (2021)
45. Zhao, Y., et al.: Recent advances in PEMFCs with new polymer electrolyte membranes. J. Power Sources. **448**, 227456 (2200)
46. Zhang, J., et al.: Nanostructured catalyst layers for PEMFCs. Nat. Catal. **2**(5), 377–386 (2019)
47. Liu, D., et al.: 3D printed electrodes for high-performance PEM fuel cells. Adv. Funct. Mater. **32**(5), 2106789 (2022)

Applying ES and IE Concepts to Reduce the Risks of Medical Waste in the Dental Sector

Maryam S. Hamed[1], Alaa W. Al-Hor[2], Mohamed H. Mohamed[1], and Salah H. R. Ali[3](✉)

[1] Faculty of Oral and Dental Medicine, Ahram Canadian University (ACU), 6th of October, Egypt

[2] Medical Devices Department, Faculty of Eng. and Info Tech, Al-Azhar University (AUG), Gaza, Palestine

[3] Precision Engineering Division, National Institute for Standards (NIS), Giza, Egypt

salahali20@yahoo.com

Abstract. Dealing with the waste issue poses a major challenge to the healthcare and public health sectors, especially with increasing population density and increased patient attendance. It has become necessary to constantly search for innovative and effective solutions to reduce environmental and health risks. ES and IE play a pivotal role in designing sustainable strategies for managing the issue, including medical waste and recycling, applying circular economy concepts, and improving the design of medical devices to reduce waste. The paper aims to present an appropriate environmental strategy for managing the risk of medical waste in a scientifically safe and sound manner, ensuring the safety and health of workers, patients, and the community. It focuses on monitoring and understanding the current situation and identifying the challenges facing the medical sector in implementing environmental measures to ensure smooth and safe medical operations. It indicates that integrating sustainable environmental strategies significantly contributes to reducing harmful emissions and improving the performance of the medical sector and hospitals, particularly in the dental sector, ensuring the safety of patients and workers. This research highlights the importance of utilizing administrative, engineering, and environmental approaches together to develop sustainable strategies for medical waste management in hospitals in general, and in the dental sector in particular. This enhances understanding of how IE can be applied to maximize sustainability in the healthcare sector, enhances the healthcare and promotes environmental awareness among workers and members of the community.

Keywords: Medical waste · Dental Sector · ES and IE concepts · Environment · Healthcare and public health

1 Introduction

Medical waste is considered one of the most dangerous types of waste due to its content of biological, mineral, chemical, and radioactive materials, which makes it a direct threat to human health and the environment. According to the World Health Organization

Y. Zhu (Ed.): ISESIE 2025, LNCE 000, pp. 145–161, 2026.
https://doi.org/10.1007/978-981-92-2389-3_13

(WHO), improper disposal of medical waste leads to toxic pollutants and dangerous pathogens that leak into the environment through air, water, and soil pollution, posing a significant threat to human health and increasing the likelihood of infectious diseases and chemical poisoning [1]. The medical sector includes many subsectors, the most important of which is dentistry, due to the high frequency of patient visits. The dental sector is considered a major source of medical waste and is also one of the medical sectors that produce the most waste, due to its frequent use of various materials and tools, such as amalgam used in dental fillings, sharp instruments such as needles and scalpels, and chemical solutions used in sterilization and dental imaging. If tools and materials are not disposed of properly after use in a safe and sustainable manner, they pose a direct health risk to patients and workers, in addition to the risks they pose to the environment [2]. For example, amalgam contains mercury, which can leach into water and soil, causing long-term contamination of all living organisms, including humans [3]. Sharps can also be a major vector for transmitting infectious diseases, particularly HIV (Human Immunodeficiency Virus), Hepatitis B Virus (HBV), and Hepatitis C Virus (HCV), if not disposed of safely [4]. These and other risks make it imperative to adopt waste management strategies that combine careful environmental assessment through Environmental Science (ES) with practical and managerial solutions through industrial Ecology (IE) to ensure public health protection and environmental conservation. ES and IE are both fields focused on understanding and mitigating human impact on the environment, but they approach the problem from different perspectives. ES broadly studies the natural environment, its processes, and how humans affect it. IE, on the other hand, focuses specifically on industrial systems and their relationship with the environment, aiming to make them more sustainable.

Therefore, this research represents an important step towards monitoring, understanding, and managing medical waste in the dental sector. By integrating ES with IE, it can provide sustainable strategies that reduce environment and health risks. The research also seeks to provide a scientific framework applicable in both developing and developed countries, contributing to the enhancement of healthcare and the protection of public health. It also contributes to the conservation of environmental resources and supports evidence-based decision-making in enhancing dental waste management.

2 Previous Literature and Global Concerns

International organizations such as the WHO and the United Nations Environment Programme (UNEP) have paid extensive attention to the safe management and disposal of medical waste, given its increasing threat to the environment and public health. A large body of academic studies has also emerged that address the global volume of medical waste, its characteristics, and its extended impacts on air, water, and soil. Given the importance of this, this literature review will address three main themes: First, the global problem of medical waste and the specific challenges facing the dental sector. Third, the applications of the concepts of ES and IE as presented in previous studies.

2.1 The Global Problem of Medical Waste

Medical waste is one of the most pressing environmental and health issues globally, given the direct and indirect risks it poses to humans and the environment. According to WHO estimates, healthcare activities generate millions of tons of waste annually, approximately 85% of which is general waste that poses no significant risk, while 15% is classified as hazardous waste, including metallic, chemical, pharmaceutical, and radioactive waste [1]. Studies have also shown that the disparity between developed and developing countries in medical waste management remains large. While developed countries have advanced sorting and treatment systems (such as steam treatment and thermal sterilization), many developing countries still rely on open burning or unsanitary landfilling, which exacerbates harmful emissions and increases environmental risks [5], and other disparities are shown in Table 1.

Table 1. Comparison between developed and developing countries in medical waste management

Comparison	Developed countries	Developing countries
Collection and sorting mechanism	Accurate sorting systems at the source (color coded) [1]	Mixing medical waste with waste public [1]
Processing methods	Steam sterilization, heat treatment, recycling [5]	Open burning or burial health [5]
Legislation	Strict laws and control mechanisms [5]	Weak legislation or lack of oversight effective [5]
Financial resources	Government funding and advanced technologies [7]	Limited funding and heavy dependence on donor countries [7]
Environmental impact	Lower emissions and continuous monitoring [6]	High emissions and water pollution/soil [6]

The COVID-19 pandemic has also contributed to an unprecedented increase in medical waste volumes. Some major producing countries (such as China) have recorded daily production rates of hundreds of tons of protective equipment and disposable medical supplies [6]. In addition, mercury from certain medical practices—most notably the use of dental amalgam—has posed a global environmental challenge. The United Nations Environment Programme (UNEP) has estimated that mercury emissions from the healthcare sector amount to tons annually, prompting the inclusion of this issue in the Minamata Convention on Mercury, which aims to reduce emissions and protect human health and the environment [7]. This data confirms that the medical waste problem is not local or sectoral, but rather a transboundary issue that requires the integration of international policies, environmental assessments, and industrial technologies to achieve sustainable management that reduces health and environmental risks on a global scale.

2.2 Specific Challenges Facing the Dental Sector from Medical Waste

The dental sector is considered one of the medical specialties that produce the most hazardous waste. While this waste does not constitute a significant proportion of the total medical waste in hospitals, its hazardousness exceeds its numerical proportion. This is due to the heavy reliance on sharp instruments, metal materials, chemical solutions, and radiological imaging devices [8]. Given that the waste generated from these materials and instruments contains environmental and health pollutants, careful management is required. Table 2 shows the types of this waste, their examples, and their environmental and health risks.

Table 2. Types of waste in the dental Sector, their examples, and their environmental and health risks

Type of waste	Environmental risks	Health risks	Common examples in dental
Metallic waste	Long-term contamination of water and soil with mercury [9]	Neurological disorders and mercury poisoning upon direct exposure [9]	Amalgam fillings (mercury, silver, tin)
Sharps waste	Biological contamination of soil and water	Transmission of infectious diseases (HBV, HIV, HCV)	Needles, scalpels, surgical instruments
Chemical waste	Wastewater contamination with toxic chemical compounds	Skin and respiratory irritation among workers; chronic toxicity risks [10]	Disinfectant solutions, sterilization agents, dental adhesives
Radiological waste	Soil and groundwater contamination with silver [10]	Secondary contamination from improper reuse	X-ray imaging solutions containing silver and lead
Plastic waste	Increase in solid waste and high carbon footprint	Transmission of harmful bacteria and viruses	Plastic cups, protective equipment, gloves
Biological waste	Decomposition produces toxic gases (e.g., methane) [11]	Risk of infection and disease spread	Tissue residues, blood, body fluids

The above demonstrates a clear gap between developed and developing countries in this field, a topic that will be addressed more extensively later in this research. Developing countries have clear legislation and plans for managing dental waste according to sustainability standards, while developing countries suffer from a lack of infrastructure and weak legislation, which often results in the disposal of this waste through primitive methods such as random burning or uncontrolled sanitary landfills.

While numerous studies have addressed medical waste in the healthcare sector in general, specialized studies in the dental sector remain relatively limited and often focus on a single aspect, such as waste containing mercury or sharp instruments, without providing a comprehensive vision for managing all types of waste. Furthermore, most research is limited to quantitative descriptions (the amount of waste produced daily or annually) without analyzing its life cycle or assessing its long-term environmental impact using tools such as life cycle analysis (LCA).

Hence, there is an urgent need for scientific research addressing the problem of medical waste in the dental sector. It is desirable that the proposed research combines quantitative and qualitative waste characterization with the application of analytical tools derived from ES and IE, contributing to the development of strategies applicable in various environments, particularly in developing countries. To bridge this gap, this topic is focused on in our current study.

2.3 ES and IE Strategies

With the increasing environmental and health risks associated with dental waste, it has become necessary to adopt scientific strategies to assess these risks and provide practical solutions to mitigate them. In this context, the concepts of Environmental Sustainability (ES) and Industrial Ecology (IE) have played a pivotal role in analyzing the environmental impact of waste and developing effective and sustainable management methods.

This section aims to review previous studies that have used ES to assess the environmental impact of dental waste, as well as studies that have adopted IE to provide practical and technical solutions, such as steam treatment, amalgam filters, and recycling.

2.3.1 Application of ES in Dental Waste Management

ES is defined as the ability to meet the resource needs of current generations without compromising the ability of future generations to meet their future needs, by reducing the environmental impact and enhancing the efficient use of natural resources [11]. In the healthcare context, particularly the dental sector, ES represents an analytical framework for assessing waste generation, its distribution pathways, and associated environmental and health impacts, while developing practical strategies for reducing them.

- Life Cycle Analysis (LCA) for Dental Practices: LCA has become a crucial tool for understanding the environmental impacts of dental practices, allowing analysis of every stage of care, from the procurement and use of supplies to the management of the resulting waste. A study using OpenLCA version 1.10.3 with the Ecoinvent database version 3.8 demonstrated the tool's ability to assess the environmental impact of key elements in a dental clinic, including water systems, consumables, and waste disposal methods. The results showed that using rainwater instead of mains water could reduce the carbon footprint by approximately 30 g of CO_2 equivalent per patient, equivalent to a reduction of up to 90% compared to standard practices [12]. In a related context, another study based on a life cycle analysis of routine dental examinations revealed that simple materials, such as soap, paper aprons, and disinfectants, may represent critical points with disproportionate environmental impact. The results indicated that these materials affect specific environmental categories, including human poisoning, water scarcity, and freshwater eutrophication. Despite its simplicity, the study, using Open LCA software and the Ecoinvent database, demonstrated that these materials have an environmental impact that exceeds initial estimates, highlighting the need for conscious and effective management of these materials within dental clinics [13].

- *Environmental Impact Assessment of Dental Materials*: The study focused on comparing the environmental impact of three common restorative materials in dentistry: dental amalgam, resin composite (RBC), and glass-polyacrylate cement (GIC). This study used a life cycle analysis (LCA) and a "cradle-to-gate" methodology with the Ecoinvent database and the ReCiPe methodology. This approach enabled the assessment of each material's impact across nine major environmental categories, including global warming, ocean acidification, and toxic effects on humans and living organisms. As shown in Table 3.

Table 3. The environmental impact of dental restorative materials based on LCA analysis

Restorative material	Main sources of environmental impact	Most affected environmental categories	Main environmental advantage/disadvantage
Dental Amalgam	Mercury and heavy metals (Hg, Ag, Sn, Cu)	Human and environmental toxicity, water and soil contamination, long-term ecosystem impact	Highly toxic due to mercury, persistent environmental burden
RBC (Resin-Based Composite)	Energy consumed during manufacturing, plastic monomers	Global warming (CO_2 emissions), aquatic toxicity	Less toxic than amalgam but high carbon footprint
GIC (Glass Ionomer Cement)	Production and packaging phase, fluoride use	Resource consumption, moderate global warming	Lower environmental impact during use, but depends on production efficiency

The results shown in Table 3 above show that:

- Dental amalgam is the most environmentally problematic, as it is the most prominent source of mercury in dental medical waste. Due to mercury's stability in the environment, its long-term effects have prompted the adoption of international policies and legislation, such as the Minamata Convention, which recommends the phasing out of amalgam use.
- Resin composite (RBC) is considered a less toxic alternative to amalgam, but its manufacturing process is highly energy-intensive, increasing its carbon footprint and global warming impact [14].

- Glass ion cement (GIC) exhibits moderate environmental impact, particularly during the in-use phase. However, its environmental efficiency depends on the manufacturing and packaging processes, which opens the door to improvements through the adoption of more sustainable production practices [15].

These results highlight that adopting ES through LCA not only measures environmental impacts, but also helps formulate sustainable procurement strategies for dental materials and guide international policies to reduce the use of highly toxic materials. This demonstrates the direct link between scientific-level environmental impact assessment and global legislation and policies [14].

- Analysis of Pollution Pathways in Dental Waste: Dental waste, particularly that containing dental amalgam, is a significant environmental source of toxic mercury. This waste poses a significant environmental risk, as it can enter sewage systems or spread through the air and water, requiring strict management measures. First, the discharge of amalgam waste via wastewater is one of the most significant non-industrial sources of mercury release to the environment. An estimated 5.1 tons of mercury are sent annually by dental practices to public wastewater treatment plants (POTWs). This contaminates the resulting sludge, which can be used as fertilizer or incinerated, contributing to the re-release of mercury to the air and soil [16]. Studies have also shown that dental wastewater contains elevated concentrations of mercury, along with other metals such as silver, tin, copper, and zinc, often exceeding local permissible limits. These metals have also been detected in treatment plant pre-water, classifying dental waste as an environmentally hazardous waste requiring careful treatment before discharge [17]. Furthermore, evidence suggests that other contamination pathways include evaporation of mercury from fillings during placement or incineration, as well as emissions from human or corpse cremations containing amalgam fillings, which release mercury into the air. Inhalation of aerosols or their dispersal into the environment is one of the factors contributing to long-term environmental impacts [18].

Finally, the environmental impact of dental amalgam is partly linked to poor dental waste management, as large quantities of mercury are wasted due to improper disposal or the absence of amalgam separators. The installation of such units could significantly reduce mercury release to the environment and enhance environmental safety [3].

2.3.2 The Role of IE in Dental Waste Management

IE provides measurement and assessment tools; industrial ecology focuses on implementing practical technical and management solutions to transform environmental assessments into enforceable policies and procedures within dental practices, as follows:

- Amalgam Separators: Amalgam filters are an essential tool for reducing mercury contamination in dental wastewater. Data from the U.S. Environmental Protection Agency (EPA) indicate that installing these filters can remove between 95% and 99.99% of mercury, reducing its migration to treatment plants or the surrounding environment [19]. Studies on ISO 11143-compliant filters have shown a removal capacity of up to 97.05%, reflecting their high effectiveness in preventing the release

of amalgam particles into the environment. The use of these filters is an essential part of strategies to reduce secondary mercury emissions from wastewater treatment [20].
- Autoclaving: Steam sterilization is a sustainable alternative to traditional incineration for dental waste, particularly infectious and non-infectious waste. Studies indicate that this technology effectively reduces greenhouse gas emissions such as carbon dioxide and limits the production of toxic pollutants such as dioxins and furans associated with incineration [1]. In addition, this technology offers greater economic benefits and safety for workers, with the ability to process various types of medical waste without adverse environmental impact [21].
- Recycling and Industrial Symbiosis: Some clinics have begun implementing systematic recycling strategies, including the recycling of metal materials such as hand instrument handles and NiTi rotary files, in addition to plastic materials used in dental procedures. A recent study has demonstrated the potential to convert these metal materials into alloys or raw materials for industrial use, enhancing the economic value of dental waste [22, 23]. The concept of "Industrial Symbiosis" also represents an advanced model in the industrial environment, where healthcare facilities and industrial facilities collaborate to redirect waste to other production chains, reducing the volume of final waste and improving the environmental efficiency of the system as a whole [24].

The analysis shows that the application of the ES and IE concepts has provided the dental sector with two integrated paths: the first enables a precise scientific assessment of environmental impact, and the second transforms this assessment into practical technical and management strategies. This integration of assessment and solutions is not limited to reducing pollution and emissions alone, but also forms the basis for reshaping dental practices within a global sustainability framework driven by environmental legislation and standards.

3 Risks and Challenges Facing Dental Waste Management

There are numerous risks and challenges facing dental waste management. We will highlight the most important of these in detail here:

3.1 Environmental Risks

The dental sector is a significant non-industrial source of environmental pollution due to its reliance on mercury-containing amalgams, in addition to chemicals used in clinical and diagnostic procedures. Evidence indicates that the environmental risks associated with this waste are relatively greater in quantity than other types of medical waste [11].

- Mercury: Dental amalgam poses one of the greatest environmental threats in this sector due to its mercury content. According to estimates by the EPA, dental clinics release approximately 5.1 Tons of mercury annually into wastewater treatment plants [16]. These quantities settle into the sludge from treatment plants, which is often reused in agriculture or incinerated, resulting in the re-release of mercury into the soil and air. Dental clinics are responsible for approximately 14% of total mercury emissions into the environment, making them a major source of mercury pollution [25].

- Water and Air Pollution: Surface and groundwater are most vulnerable to contamination due to improper disposal of dental waste. A study in Pakistan revealed that mercury concentrations in dental clinic wastewater ranged from 86.7 to 343.3 mg/L, exceeding the permissible limit (0.01 mg/L) according to public health standards [26]. In the air, the burning of dental waste or even the evaporation of mercury from amalgam fillings within the clinic can release significant concentrations of mercury vapor. Studies have indicated that mercury levels in the air of some clinics have exceeded 25 micrograms/m^3, the permissible limit for occupational exposure according to the WHO [27].
- Other Chemicals: In addition to mercury, the dental sector produces significant amounts of chemical waste, including radioactive acid solutions, disinfectants, and phenols. If these materials are not disposed of according to health and safety standards, they will lead to increased risks of environmental pollution. These materials can poison aquatic organisms, disrupt aquatic ecosystems, and increase levels of toxic organic compounds in the surrounding environment [28].
- Weak Waste Management Infrastructure: One of the most prominent environmental challenges in developing countries, to varying degrees, is the lack of integrated systems for treating dental waste. Clinics are not fully equipped with amalgam separation units, steam treatment, or adequate health and environmental protection measures. The remaining waste is disposed of in unqualified landfills, contributing to long-term environmental pollution.

3.2 Health Risks Associated with Sharps

Handling sharps (such as needles, scalpels, rotary files, and microsurgical instruments) is a major challenge in dental clinics, due to the direct risk of viral and bacterial infection. Although sharps waste in dental clinics typically does not constitute a significant proportion of the total medical waste, its danger far outweighs its numerical weight due to its potential to cause penetrating wounds that could expose staff to infection.

It is estimated that more than 3 million needlestick injuries occur annually among healthcare workers worldwide [29], a significant proportion of which occur in dental clinics due to the nature of procedures that require direct contact with bodily fluids and blood. These injuries are associated with the potential transmission of dangerous viruses, such as: HIV, with a transmission risk of approximately 0.3% per needlestick injury; hepatitis B virus (HBV), with a transmission risk that may exceed 30% in the absence of vaccination; and hepatitis C virus (HCV), with a transmission rate estimated at 1.8%. These rates are according to the Centers for Disease Control and Prevention (CDC). One study of the dental sector shows that the rate of needlestick injuries among dentists and dental students ranges from approximately 44% annually in some countries, a higher rate than in other medical specialties due to the frequent use of local injections, the removal of metal fillings, and the use of fine sharp instruments in a limited work environment [30]. The effects of repeated sharps injuries are not limited to the risk of viral transmission. They also extend to psychological and social burdens on workers, such as constant anxiety about infection, reluctance to perform certain clinical procedures, and high rates of absenteeism. Thus, it can cat: The environmental challenge in dental hospitals and clinics not lies only in reducing injuries, but also in implementing strict protocols for handling sharps. These protocols include ongoing training, the use of puncture-resistant and secure containers, and appropriate vaccinations against viruses, as deemed appropriate by health organizations.

3.3 Administrative Challenges in Dental Waste Management

Administrative challenges represent a core component of the dental waste management crisis. The absence of strict policies or their weak implementation exacerbates environmental and health risks, as illustrated by the following points:

- Weak Infrastructure: Dental clinics in developing countries lack specialized systems for treating medical waste. It is often disposed of with municipal waste and is also disposed of by burying it in vehicles near population centers and burning it without proper treatment. This defect leads to the release of mercury and chemicals into the environment, increasing the likelihood that waste collectors will be exposed to the risk of infection and injury [31, 32].
- Lack of funding: The cost of installing and maintaining amalgam filters and the preventative measures required to dispose of dental waste for health and environmental safety are very high, an amount many small clinic owners are reluctant to pay. In developing countries, where funding for waste management is scarce, many clinics resort to unsafe methods such as open burning or uncontrolled burial.
- Poor compliance with legislation: Poor compliance with legislation on medical waste management within dental clinics is one of the most prominent regulatory and environmental challenges in many countries. In Nigeria, for example, a study in Lagos State showed that only 17.2% of dental workers had a good understanding of medical waste management legislation, while the actual practice rate was much lower, at only 4.1%. Regarding mercury management, it was noted that 92.2% did not follow proper mercury conservation practices, such as proper waste disposal or the use of appropriate separation devices [33]. This stark gap between theoretical knowledge and practical application clearly reflects a systematic flaw in the training of environmental personnel and the enforcement of judicial oversight.

In the West Bank and Jerusalem, the level of adherence to infection prevention measures among dentists was assessed. The results indicated that only 18.5% of the sample adhered to good compliance, despite most adhering to basic practices such as the use of gloves and sterilization of instruments. The study also showed that adherence varied according to age and years of experience, in addition to a positive association with working in the public sector or in the northern West Bank. Meanwhile, exposure to needlestick injuries was associated with a lower level of adherence [34].

Based on the above, it is clear that dental clinics face numerous challenges, not only in the quantity of waste, but also in its type and its pathways in the environment and throughout the workflow within the facility. Mercury emitted from amalgam waste can migrate with wastewater to treatment plants and then into sludge or waterways, creating chronic, difficult-to-reverse hazards. Solvents and solutions add additional toxic and climate-related loads when incinerated or landfilled. Sharps puncture wounds and poor waste sorting increase the burden of infection and occupational injuries, while administrative and regulatory gaps, lack of funding and training lead to system closure when performance falls below desired levels. This finding suggests that combining ES as a measurement tool and IE as an implementation tool is the most appropriate practical approach to reducing risks in the dental sector. Based on this, and in this context, we will present a case study comparing developed and developing countries, with the aim of highlighting real-world applications and identifying gaps and successful experiences that can be generalized.

4 Case Study

International experiences demonstrate significant variations in the management of dental waste between developed and developing countries. This is why we will focus on this important issue and discuss it to understand it and provide appropriate innovative solutions.

4.1 Case Study of a Developed Country

Healthcare waste management, including dental waste, in the United Kingdom and European countries is based on clear and specific standards. These standards are part of an integrated national policy adopted by ministries of health through Technical Guide HTM 07-01. This guide, updated in 2024, provides a clear framework for waste sorting by color, storage, transportation, and treatment in ways that align with sustainability principles and reduce the carbon footprint. For example, infectious waste is often treated with steam treatment (autoclaving) or other alternative methods instead of relying on incineration, which reduces the emissions of dioxins and furans associated with inefficient incineration [35]. Chemical waste, particularly amalgam residues, is subject to strict regulatory requirements under European Regulation 2017/852, which requires clinics to install amalgam separators with an efficiency of at least 95% according to ISO 11143. Evaluation studies have shown that these separators are capable of removing the largest proportion of amalgam particles, reducing the entry of mercury into wastewater treatment plants and enhancing long-term environmental protection [36, 37]. According to strict standards.

4.2 Case Study of a Developing Country

Select one of the worst cases of developing countries currently exists, the Gaza Strip would be the worst of them. The Gaza Strip represents a harsh reality characterized by the absence of integrated infrastructure and the scarcity of financial and technical resources. Studies estimate that the total medical waste in Gaza exceeds tons per day, of which more than one ton per day is infectious waste, including sharps by Al-Khatib et al., 2009 [32]. Despite these large quantities, healthcare facilities lack safe waste storage facilities. These facilities also rely on inefficient open burning, which releases massive amounts of fine particles and toxic pollutants into the surrounding air. Specifically, field studies in Khan Yunis city, for example, revealed that amalgam residues are often disposed of in the general trash or flushed down the drains after rudimentary filtration, with the near-total absence of standard separators. Furthermore, the use of puncture-resistant containers for sharp instruments is low, increasing the risk of injury and infection among staff and patients [38, 39]. This demonstrates a clear difference between waste management in the dental sector in developed countries such as the United Kingdom and European countries, and in developing countries such as the Gaza Strip, as detailed in Table 4.

Table 4. Comparison between a model from a developed country and one from a developing country, focusing on dental clinics

Comparison aspect	Developed country, UK	Developing country, Gaza
Governance framework	HTM07-01 (updated 2024), specifies segregation, transport, treatment, and environmental impact reduction	Lack of widely adopted detailed operational guidelines, significant variability among facilities
Amalgam	Amalgam separators required under Regulation 852/2017 and ISO 11143 standard	Field evidence of improper discharge/disposal of amalgam residues, widespread absence of separators
Treatment	Priority for alternative treatment of infectious waste, high-temperature incineration only when necessary	Extensive reliance on incineration and open burning due to service constraints
Storage and transport	Secure, color-coded storage and licensed, monitored chain of custody	Most facilities lack secure storage, frequent mixing with municipal waste during transport
Expected impact	Significant reduction of Hg discharge to wastewater and air pollutants, improved occupational safety	Higher community and occupational exposure to infections and airborne/Gy pollutants
Cost and technology	Higher costs but high efficiency, regular maintenance	Limited resources
Integration of ES, IE	Life cycle approach + advanced industrial solutions	Weak IE, ES partially implemented or limited

This disparity, as evident in Table 4, clearly highlights the vast difference when comparing the two systems. The UK is considered to offer a model based on strict governance, adherence to environmental standards, and the systematic use of separation and treatment. In Gaza, however, the absence of standardized frameworks and basic technologies is evident, leading to significant gaps in occupational safety and environmental protection. From an ES assessment perspective, the most critical issues in developed countries are primarily amalgam and high-emission treatment technologies, which have been addressed through strict policies and effective separators. In the developing context, fundamental challenges such as the lack of storage infrastructure, the mixing of medical waste with municipal waste, and the reliance on open burning are compounded by gaps in implementation capacity rather than scientific knowledge. In this context, implementing IE becomes more complex with the simple installation of amalgam separators, the provision of secure sharps containers, and the gradual transition from incineration to steam treatment where resources permit.

The purpose of this comparative presentation is not only to highlight the gap between the two contexts, but also to illustrate how ES and IE applications can constitute a feasible transitional tool in resource-limited settings. Conducting a periodic environmental assessment enables the monitoring of the most hazardous flows (such as mercury), while industrial environmental solutions provide practical pathways to reduce emissions and risks through gradual technological interventions. Thus, it becomes possible to build a safer and more sustainable waste management system in Gaza and other developing countries, even within available resources.

5 Analysis and Discussion

5.1 Linking Challenges to Solutions

The results of the literature review and case study revealed that the dental sector faces an intertwined set of environmental, health, and administrative challenges, which go beyond simple technical issues to structural issues related to governance and sustainability. For example, mercury contamination from dental amalgam cannot be isolated from surrounding water and air pollution, as UNEP reports have shown that dental hospitals and clinics represent one of the main sources of mercury leakage into water [40]. This problem cannot be solved with a single measure (such as amalgam banning), but rather requires a dual framework of the following:

- Through ES concepts, where LCA analysis allows for an understanding of the full impacts of amalgam use, from extraction to disposal.
- Through IE, where ES findings are translated into practical practices, such as the installation of amalgam spacers, the recycling of dental materials, or the use of safer alternatives.

This link between scientific diagnosis using ES and technical implementation using IE represents the cornerstone for addressing challenges and demonstrates that effective solutions must be complementary, not partial.

5.2 Comparison of Implementation Between Developed and Developing Countries

In developed countries, experiences such as those in the United Kingdom indicate that the integration of ES and IE has become an essential part of health and environmental policies. Since 2018, the government has mandated the installation of amalgam spacers in all clinics, resulting in a reduction of mercury emissions by more than 95%. Life cycle assessments have also been incorporated into policies for the gradual replacement of amalgam with alternative fillings. In developing countries, however, the situation is more challenging. Environments such as the Gaza Strip suffer from a lack of infrastructure and limited funding, with waste often disposed of with general waste or through indiscriminate burning. Despite growing awareness of the dangers of amalgam and chemicals, the lack of IE tools (such as industrial interlocking or recycling) makes theoretical solutions derived from ES limited in their impact. This comparison demonstrates that the gap is not in scientific knowledge as much as it is in legislation, first, and second, in the ability of health systems to translate knowledge into practical practice.

5.3 Strengths of Technical and Environmental Solutions

ES provides accurate tools for measuring environmental impact over an entire life cycle, allowing for estimation of the actual environmental cost. It provides a scientific language that can be used in formulating policies and legislation, such as the Minamata Convention. Where, IE provides proven technical solutions such as amalgam separation, autoclaving as an alternative to incineration, and industrial mesh for recycling. These solutions increase safety in clinics and reduce the likelihood of infection and exposure to toxic substances.

However, operational costs and maintenance challenges remain barriers to their implementation in low-resource countries, requiring international support or public-private partnerships.

6 Conclusions

1. Dental waste is considered one of the most dangerous types of medical waste, despite its smaller quantity compared to other hospital sectors. However, its danger lies in the frequency and increase in visits, as well as its working components: such as mercury-containing amalgams, sharp instruments, chemical solutions, metal, plastic, biological waste, and others. These elements create an environmental, health, and administrative challenge that requires systematic, strategic, scientific intervention.
2. Mercury as a Major Source of Pollution: Studies have shown that dental clinics are responsible for global emissions of mercury into the environment, making it the most serious threat to groundwater, air, and soil. The Minamata Convention established a framework to reduce these emissions, but implementation remains uneven across countries.
3. Sharps and Infection: In some countries, the incidence of sharps injuries among dental workers reaches 44% among dental students, posing a continued risk of transmission of serious diseases such as hepatitis B, HCV, and HIV.
4. The gap between developed and developing countries is large: Developed countries (such as the UK and Germany) have advanced separation and treatment systems, and implement amalgam separators and ISO 11143 standards, in addition to steam treatment as an alternative to incineration. Developing countries (such as the Gaza Strip) lack infrastructure and often resort to open burning or random landfilling, which increases health and environmental risks.

5. It can recommend the importance of applying the ES approach: Tools such as LCA (Life Cycle Analysis) and Pollution Pathway Analysis have proven effective in accurately characterizing environmental impact and have helped introduce legislative reforms and international policies.
6. It can also recommend the importance of applying the IE approach: The application of practical technologies such as amalgam separators (which remove a significant percentage of mercury) and waste autoclaving instead of incineration, in addition to recycling and industrial networking programs, have contributed to transforming theoretical assessments into implementable practical solutions.
7. The integration of ES assessment and IE implementation is the foundation for sustainable dental waste management. Without this integration, efforts remain either diagnostic without solutions or random solutions without scientific evaluation.

Adopting this approach and continuing it on a large scale could radically transform waste management in the dental sector and lay a solid foundation for healthcare, public health, and environmental protection for future generations in both developed and developing countries.

References

1. World Health Organization (WHO): Health-care waste: fact sheet. https://www.who.int/news-room/fact-sheets/detail/health-care-waste. Last accessed 21 March 2025
2. Khanna, R., et al.: An overview of dental solid waste management and associated environmental impacts: a materials perspective. Sustainability. **15**(22), 15953 (2023)
3. Treasure, E.T., Treasure, P.: An investigation of the disposal of hazardous wastes from New Zealand dental practices. Community Dent. Health. **18**(3), 150–155 (2001)
4. Author unknown: Mercury exposure in dental profession. J. Dent. (1998)
5. Yin, J., Ma, H.: Regional medical waste management in China: a case study of Shenyang. Waste Manag. **33**(3), 710–715 (2013)
6. Abu-Qdais, H., Al-Ghazo, M., Al-Ghazo, E.: COVID-19 pandemic and its impact on solid waste management in the healthcare sector. Waste Manag. **124**, 448–455 (2021)
7. United Nations Environment Programme (UNEP): Global Mercury Assessment 2018. UNEP, Geneva (2019)
8. Azhar, A., Yi, Y.: Management of dental waste in dental hospitals of Pakistan. Pak. J. Med. Sci. **34**(6), 1482–1487 (2018)
9. Encyclopédie de l'Environnement: The Minamata convention on mercury. https://www.encyclopedie-environnement.org/en/society/the-minamata-convention-on-mercury-2/. Last accessed 22 March 2025
10. Al-Rajhi, D., Jamal, H.: Management of dental radiographic waste. Environ. Res. J. **12**(4), 225–233 (2018)
11. World Health Organization (WHO): Safe Management of Wastes from Health-Care Activities, 2nd edn. WHO Press, Geneva (2014)
12. Kumar, S., Sharma, J.: Dental waste management: an insight. Int. J. Clin. Dent. **12**(2), 110–117 (2024)
13. Choudhary, R., Singh, A.: Dental solid waste management: current practices and future challenges. J. Environ. Health. **83**(5), 410–418 (2021)
14. Patel, M., Desai, H.: Mercury pollution from dental amalgam waste: a global perspective. J. Hazard. Mater. **431**, 128–134 (2022)

15. White, S., Jones, R.: Sustainable dental practice: waste reduction strategies. Int. Dent. J. **72**(6), 591–600 (2022)
16. United States Environmental Protection Agency (EPA): Dental effluent guidelines. https://www.epa.gov/eg/dental-effluent-guidelines. Last accessed 23 March 2025
17. Kümmerer, K., Al-Ahmad, A.: Wastewater from dental offices: a source of environmental contaminants. Water Res. **44**(2), 393–402 (2011)
18. European Commission Scientific Committees: Environmental risks and impacts of dental amalgam. https://ec.europa.eu/health/scientific_committees/opinions_layman/en/dental-amalgam
19. American Dental Association (ADA): Amalgam separators. https://www.ada.org/resources/ada-library/oral-health-topics/amalgam-separators. Last accessed 25 March 2025
20. Stone, M.E., Cohen, M.: Dental office wastewater characterization and treatment alternatives. J. Am. Dent. Assoc. (JADA). **145**(9), 1020–1029 (2014)
21. Ramalingam, K., Sarkar, S.: Nanomaterials in dentistry: environmental implications. Appl. Sci. **10**(15), 5056 (2020)
22. Maheshwari, P., Khan, S.: Dental waste generation and disposal practices: a systematic review. Environ. Monit. Assess. **197**(12), 101–109 (2024)
23. Ozturk, M., Altun, I.: Evaluation of dental waste management in Turkey. European Journal of Science and Technology. **19**, 259–265 (2020)
24. Sharma, A., Gupta, R.: A comprehensive review of healthcare waste management in dentistry. Sustainability. **16**(22), 10047 (2024)
25. Rother, H.A., Hall, D.: Mercury exposure pathways in dental professionals. Occup. Med. **48**(2), 99–105 (1998)
26. Singh, R.D., Yadav, S.: Biomedical waste management: current scenario and future prospects. J. Environ. Biol. **39**(5), 765–770 (2018)
27. Lakshmi, V., Rao, A.: Dental health-care waste management practices in India. Indian J. Occup. Environ. Med. **21**(2), 79–84 (2017)
28. Zhang, L., Chen, X.: COVID-19 and healthcare waste management challenges. Environ. Sci. Technol. **59**(8), 4221–4230 (2025)
29. Ali, S., Mahmood, T.: Dental amalgam waste and mercury release: a review. Environ. Monit. Assess. **193**(4), 218–225 (2021)
30. Ahmed, R., Qureshi, N.: Nanoplastics in dental materials: risks and regulations. J. Hazard. Mater. **430**, 127–132 (2022)
31. Rathore, M., Singh, A.: Dental waste management: need for strict protocols. J. Environ. Health. **72**(6), 60–65 (2009)
32. Al-Khatib, I.A., Sato, C.: Solid health care waste management status at health care centers in the West Bank-Palestinian territory. Waste Manag. **29**(8), 2398–2403 (2009)
33. Khan, H., Malik, Z.: Environmental impact of dental amalgam and waste management. J. Dent. **101**, 103–110 (2021)
34. WHO Regional Office for the Eastern Mediterranean: Assessment of adherence to infection prevention and control guidelines among dentists in the West Bank and Jerusalem. East Mediterr. Health J. **29**(3), 193–200 (2023)
35. European Union: Regulation (EU) 2017/852 on mercury. https://eur-lex.europa.eu/legal-content/EN/TXT/?uri=CELEX%3A32017R0852. Last accessed 30 April 2025
36. International Organization for Standardization (ISO): ISO 11143: Dentistry—Amalgam separators, 2nd edn., ISO Online Browsing Platform. https://www.iso.org/obp/ui/en/#iso:std:iso:11143:ed-2:v1:en. Last accessed 30 April 2025
37. Lee, H., Park, J.: Medical waste management in dental clinics: comparative study. J. Mater. Cycles Waste Manag. **23**, 567–576 (2021)
38. Al-Khatib, I.A., Monou, M.: Evaluation of dental waste management in two cities in Palestine. East Mediterr. Health J. **12**(Suppl 2), S217–S223 (2006)

39. Oliveira, T., Silva, R.: Dental solid waste: environmental and health risks. Environ. Sci. Pollut. Res. **26**(7), 6755–6764 (2019)
40. Care Quality Commission (CQC): Dental Mythbuster 1: use of disposal dental amalgam. https://www.cqc.org.uk/guidance-providers/dentists/dental-mythbuster-1-use-disposal-dental-amalgam. Last accessed 21 May 2025

Environmental Impacts and Costs of Electric Vehicle Battery Recycling

Thanwadee Chinda(✉) and Anatawat Chayutthanabun

School of Management Technology, Sirindhorn International Institute of Technology, Thammasat University, Bangkok 131,, Pathumthani 12000, Thailand
Thanwadee@siit.tu.ac.th

Abstract. End-of-life management of electric vehicle batteries (EVBs) is crucial to sustainable development. This study utilizes the lifecycle assessment (LCA) method to examine environmental impacts and environmental-related costs of the EVB recycling process. The calculated impacts are normalized and compared with global scales to determine the suitability of the recycling process. The results reveal that EVB recycling significantly reduces human toxicity (i.e., avoiding 45,934 kg 1,4-DB_{eq}), terrestrial acidification (i.e., avoiding 425 kg SO_{2eq}), and metal depletion (i.e., avoiding 20,129 kg Fe_{eq}), but highly incurs CO_{2eq} emissions from energy consumption (i.e., releasing 1689 CO_{2eq}). Using renewable energy in electricity generation may reduce the CO_{2eq} emissions in the EVB recycling process. The environmental-related costs of the EVB recycling process are also calculated, where avoiding impacts are converted to benefits and incurred impacts are costs. The results show the environmental-related benefits of $3388.3 per battery, proving the suitability of the recycling process for EVB's end-of-life management. The study results suggest that EVB recycling should be encouraged and supported for concrete implementation to achieve sustainable development.

Keywords: Electric Vehicle Battery · Environmental Cost · Environmental Impact · Recycling

1 Introduction

The rapid global adoption of electric vehicles (EVs) is driving enormous demand for lithium-ion batteries, which serve as the primary power source for EVs. Global EV sales surpassed 14 million units in 2023, representing approximately 18% of total vehicle sales worldwide [1]. Consequently, the number of end-of-life EVBs (EVBs) will increase significantly, creating both an opportunity and a challenge for sustainable resource management. Sustainable recycling has become a significant process to prevent environmental harm, recover raw materials, and reduce dependence on the extraction of raw materials [2]. EVBs can contribute to hazardous waste, releasing heavy metals and toxic electrolytes, contaminating soil and water. Uncontrolled disposal increases environmental risks and represents a loss of raw materials that could be reused in the production cycle, supporting a more circular economy [3]. Despite these benefits, EVB recycling is energy-intensive and involves complex mechanical, thermal, and chemical processes, which may

Y. Zhu (Ed.): ISESIE 2025, LNCE 000, pp. 162–172, 2026.
https://doi.org/10.1007/978-981-92-2389-3_14

generate environmental impacts. The economic feasibility of recycling is also influenced by volatile raw material prices, high capital investment requirements, and operational costs [4]. This paper investigates the environmental impacts and costs of EVB recycling. By integrating environmental life cycle assessment (LCA) with economic cost analysis, the study aims to provide a comprehensive understanding of trade-offs, identify expected environmental benefits, and offer recommendations to improve environmental performance and financial feasibility.

2 EVB Recycling Process

EVB recycling involves processes designed to recover valuable materials for reuse in manufacturing, particularly for producing new EVBs [5]. An EVB comprises four major units: EVB cell, battery management system (BMS), cooling system, and battery pack (with battery frame) [6]. In this study, the inventory data for the EVB cell production follows the Thai EVB specifications, such as a 75-kWh battery, a cell weight of 312 kg, and a battery case weight of 42 kg [7]. The materials are expected to recover by up to 95% using industrial recycling processes: pyrometallurgical and hydrometallurgical [5]. Pyrometallurgy is first employed to concentrate cobalt, nickel, and copper into a metal alloy, while other elements are retained in the slag. The slag or partially refined products are subjected to a hydrometallurgical process to extract lithium and any remaining metals [8, 9]. Once EVBs are received from collection points, they are fully discharged to eliminate residual electrical energy. The casings and external components, such as wiring harnesses, electronics, and cooling systems, are removed manually. Modules and cells are separated and shredded to produce a mixture of active materials, metals, and polymeric components [5]. During smelting, metallic components are recovered in an alloy phase [10]. Once smelting is complete, the molten slag is tapped from the furnace and cooled. This slag is then crushed and milled to increase surface area, making it more reactive and suitable for subsequent hydrometallurgical leaching [11]. Sulfuric acid combined with a reducing agent such as hydrogen peroxide is used to leach metals and slag into solution [4]. These dissolved metals are then selectively recovered using solvent extraction, precipitation, or ion exchange [12]. Hydrometallurgy can achieve recovery rates above 90% for cobalt, nickel, and lithium, and produce high-purity compounds [13]. Residual solids are filtered for safe disposal, and the purified products are sent for reuse in EVB manufacturing [3].

3 Environmental Impacts Related to EVB Recycling

Prior to conducting LCA, the scope of the study, system boundaries, material flows (including consumption and recovery), and functional units must be identified. This study focuses on the most common EVBs used in the Thai market (i.e., 75-kWh NMC batteries due to their high energy density and stability). EVBs comprise four main components: EVB cells, battery management system (BMS), cooling system, and battery pack [6, 7]. Each cell weighs approximately 70 g and consists of a cathode made from nickel, manganese, and cobalt (NMC) and an anode containing graphite [6]. The functional unit is the recycling of one end-of-life EVB, which weighs 466 kg. Each category is expressed in its respective standardized unit, such as kilograms of CO_{2eq} for climate change, kilograms of SO_{2eq} for terrestrial acidification, m^3 for water depletion, and kilograms of Fe_{eq} for metal depletion. The system boundary includes the complete life cycle of EVB recycling, from raw material extraction to using recycled materials for new EVB production [7]. This reflects a cradle-to-cradle approach, where the recycled materials are looped back into the production cycle, reducing the demand for virgin resources [9]. The system boundary is shown in Fig. 1. It begins with extracting virgin raw materials and their subsequent processing for EVB production. Once in use, EVBs eventually reach their lifecycles, where they go through the recycling process. The recycled materials are reintegrated into the production chain, partially replacing virgin materials in the new EVB production [12]. While recycling reduces the demand for raw material extraction and associated environmental degradation, the recycling operations consume energy, affecting the environment, and generating waste that requires careful management. The steps of EVB recycling produce environmental impacts. For example, the pyrometallurgical process has high environmental footprints due to its reliance on high-temperature furnaces, resulting in greenhouse gas (GHG) and CO_{2eq} emissions, particularly if fossil fuels or carbon-intensive electricity grids are used [4]. On the other hand, the hydrometallurgical process produces a lower carbon footprint, with GHG emissions ranging from 2–5 kg CO_{2eq} per kg of battery material processed [5]. The production of high-purity compounds reduces the environmental impact of downstream battery manufacturing. However, hydrometallurgy requires significant chemical inputs, causing resource depletion, toxicity potential, and energy consumption [7]. The environmental impacts of EVB recycling are extracted using SimaPro software version 8.0, applying the ReCiPe method to cover 18 midpoint impacts [7].

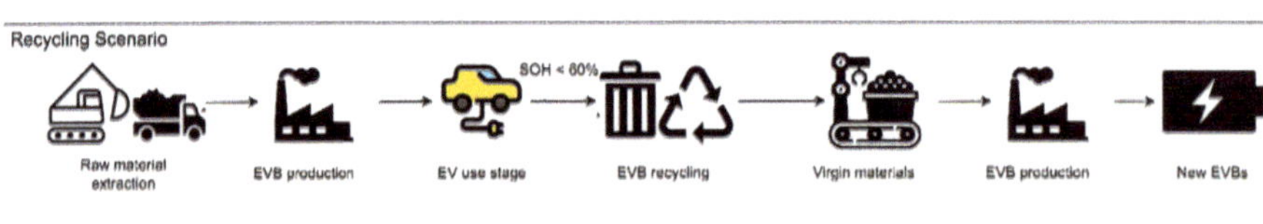

Fig. 1. System boundary of the EVB recycling process

- CO_{2eq} emission: Recycling avoids CO_{2eq} emissions by reducing the need for mining virgin materials, though energy-intensive processes like pyrometallurgy may generate CO_{2eq}.
- Human toxicity: Recycling reduces exposure to toxic material mining but may introduce risks via acid use in hydrometallurgy if not correctly managed.
- Terrestrial acidification: Recycling avoids virgin material extraction, reducing this impact.
- PM formation: Avoiding mining and smelting reduces particulate emissions; however, pyrometallurgy can still generate PMs.
- Photochemical oxidant formation: Volatile organic compound emissions from recycling can contribute to smog, whereas avoiding virgin material extraction can also reduce this impact.

- Water depletion: Hydrometallurgical process consumes water, but significantly less than that used in primary lithium (Li) and cobalt (Co) extraction.
- Metal depletion: Recycling directly avoids metal depletion by recovering virgin materials (e.g., Li, Co, nickel).
- Ozone depletion: Negligible in recycling due to the ban on ozone-depleting substances.
- Fossil depletion: Recycling reduces fossil fuel usage by avoiding primary material production.
- Freshwater eutrophication: Recycling avoids this impact by limiting runoff and wastewater discharge.
- Marine eutrophication: Wastewater from material mining and recycling may contribute nutrients to marine environments.
- Terrestrial ecotoxicity: Recycling reduces leachates through material mining.
- Freshwater ecotoxicity: Material mining (e.g., Li and Co) produces much acid and heavy metal leakage during mining.
- Marine ecotoxicity: Recycling avoids marine discharge from material mining.
- Ionizing radiation: Negligible in recycling, unless the electricity is generated from nuclear sources.
- Agricultural land occupation: Recycling facilities occupy less land than large-scale mining operations, thus reducing land-use impacts.
- Urban land occupation: Recycling facilities in industrial zones may occupy urban land but prevent further land expansion for mining.
- Natural land transformation: Recycling facilities avoid using land and reduce the need to clear natural habitats for new virgin material mining

4 Environmental Costs Related to EVB Recycling

While recycling offers clear benefits over landfilling by reducing the need for raw material extraction, the recycling operations create their own environmental footprint [15]. According to Holland [16], emissions from heavy metals, solvents, and chemical residues during recycling can be translated into human health-related costs. Water depletion is calculated using the Thai water pricing rate of \$0.58 per m^3. Since recycling reduces the need for mining and associated land disturbance, higher recovery rates translate to less land costs (see Table 1).

Table 1. Environmental-related costs of EVB recycling

Impact category	Cost per unit	References
CO_{2eq} emission	\$0.0052/kg CO_{2eq}	[15, 17]
Human toxicity	\$0.0169/kg 1,4-DB_{eq}	[15, 16]
Terrestrial acidification	\$1.22/kg SO_{2eq}	[15, 19]
PM formation	\$0.45/kg PM_{eq}	[15–19]
Photochemical oxidant formation	\$3.57/kg $NMVOC_{eq}$	[15–19]

(*continued*)

Table 1. (*continued*)

Impact category	Cost per unit	References
Water depletion	\$0.58/m^3 H_2O	[20]
Metal depletion	\$0.25/kg Fe_{eq}	[21]
Ozone depletion	\$24,387/kg CFC-11$_{eq}$	[22]
Fossil depletion	\$0.26 /kg oil$_{eq}$	[23]
Freshwater eutrophication	\$19.4/kg P_{eq}	[24]
Marine eutrophication	\$3.85/kg N_{eq}	[24]
Terrestrial ecotoxicity	\$0.0169/kg 1,4-DB$_{eq}$	[18, 19]
Freshwater ecotoxicity	\$0.0169/kg 1,4-DB$_{eq}$	[18, 19]
Marine ecotoxicity	\$0.0169/kg 1,4-DB$_{eq}$	[18, 19]
Ionizing radiation	\$0.035/kBq U235$_{eq}$	[18, 19]
Agricultural land occupation	\$15.7/m^{2}a	[25]
Urban land occupation	\$15.7/m^{2}a	[25]

5 Results and Discussion

5.1 Environmental Impact Results

The environmental impacts of EVB recycling are calculated (see Table 2) and compared with the global benchmark values to normalize the magnitude of each impact. Table 3 presents the normalized values for each environmental impact category, presenting the relative significance of recycling benefits in a normalized format. The results reveal that EVB recycling significantly reduces metal depletion (i.e., avoiding 16,941.2 kg Fe_{eq}), human toxicity (i.e., avoiding 7007.8 kg 1,4- DB_{eq}), water depletion (i.e., avoiding 535.8 m3 H^2O), and terrestrial acidification (i.e., avoiding 424 kg SO_{2eq}), but highly incurs CO_{2eq} emissions from energy consumption (i.e., releasing 1697.2 CO_{2eq}). Metal depletion represents a significant environmental reduction, equivalent to approximately 55 times the annual metal utilization per individual. This is primarily due to the high material recovery rates achievable in pyrometallurgical and hydrometallurgical recycling systems. These recovered materials directly displace the need for new raw materials. Raw material extraction generates significant environmental impacts and EVB recycling significantly reduces human toxicity by 140 times the annual individual toxicity potential generated due to high concentrations of heavy metals in EVBs. The final normalization impacts from EVB recycling is − 225.4, indicating that the recycling process yields net environmental benefits. The negative value reflects how the impacts avoided by material recovery outweigh the potential impacts generated during recycling. The study results prove that EVB recycling reduces environmental damage and resource scarcity. The high reduction in human toxicity and metal depletion impacts comes from the material composition of EVBs. They contain large amounts of valuable but hazardous metals, and recovering these metals through recycling prevents pollution and reduces the need for new extraction. This not only cuts health risks but also saves natural resources. Other impacts like acidification or water use also improved. EVB recycling still generates CO_{2eq} emissions, as the primary fuel used in electricity generation in Thailand is coal and natural gas (i.e., non-renewable energy). Utilizing renewable energy may reduce this impact further.

Table 2. Environmental impact results

Data		Electricity	Data	Raw materials					
Detail		D[a]	P[a]	Copper		Aluminum		Nickel	
				C[a]	R[a]	C[a]	R[a]	C[a]	R[a]
(Environmental impacts)	CO_2eq emission	– 7	3662	7	– 149	0.2	– 3	15	– 314
	Human toxicity	0	1	42	– 900	16	– 230	189	– 3000
	Terrestrial acidification	0	5	1	– 12	0.0	– 0.03	5	– 92
	PM formation	– 0.4	20	0	– 6	0	0	1	– 34
	PO formation	0	5	0	– 3	0	0	1	– 10
	Water depletion	0.1	66	11	– 83	48	– 347	12	– 85
	Metal depletion	0	0	212	– 4025	0	0	121	– 2299
	Ozone depletion	0	0	0	0	0	0	0	0
	Fossil depletion	1	684	3	– 62	0.1	– 2	6	– 125
	Freshwater eutrophication	0	0	0	0	0	0	0.4	– 2
	Marine eutrophication	0.04	0.8	0	– 0.1	0	– 0.1	0.1	– 0.3
	Terrestrial eutrophication	0	0	0	– 0.1	0	0	0.1	– 0.4
	Freshwater ecotoxicity	0	0	2	– 14	1	– 4	14	– 83
	Marine ecotoxicity	0	0	3	– 15	1	– 5	15	– 89
	Ionizing radiation	0	0	1	0	2	0	12	– 1
	Agricultural land occupation	0	0	0	– 1	0	– 3	1	– 6
	Urban land occupation	0	0	0	– 1	1	– 3	1	– 7
	Natural land transformation	0	0	0	0	0	0	0	0

Data		Raw materials								
Detail		Manganese		Cobalt		Lithium		Steel		Total
		C[a]	R[a]	C[a]	R[a]	C[a]	R[a]	C[a]	R[a]	
(Environmental impacts)	CO_2eq emission	1	– 14	4	– 84	71	– 1344	8	– 155	1697
	Human toxicity	1	– 18	90	– 1710	– 43	– 855	31	– 620	– 7008
	Terrestrial acidification	0	– 0.1	0.1	– 1	17	– 315	2	– 31	– 424
	PM formation	0	– 0.2	0	– 1	4	– 72	0	– 7	– 96
	PO formation	0	– 0.1	0.1	– 1	1	– 25	0	– 3	– 34
	Water depletion	1	– 5	4	– 28	12	– 96	2	– 48	– 536

(continued)

Table 2. *(continued)*

Data		Raw materials								
Detail		Manganese		Cobalt		Lithium		Steel		Total
		C[a]	R[a]	C[a]	R[a]	C[a]	R[a]	C[a]	R[a]	
	Metal depletion	77	– 1461	1	– 23	719	– 10486	11	212	– 16941
	Ozone depletion	0	0	0	0	0	0	0	0	0
	Fossil depletion	0.3	– 6	2	– 36	19	– 404	2	– 37	45
	Freshwater eutrophication	0.2	– 1	0.4	– 2	0	– 1	0	0	– 6
	Marine eutrophication	0	– 0.3	0.1	– 0.4	0	– 0.2	0	– 0.1	– 1
	Terrestrial eutrophication	0	– 0.1	0.1	– 0.5	0	– 0.3	0	0	– 1
	Freshwater ecotoxicity	4	– 24	18	– 106	10	– 59	0	– 2	– 242
	Marine ecotoxicity	5	– 30	19	– 113	11	– 65	0	– 2	– 266
	Ionizing radiation	6	0	14	– 1	9	– 0.4	0.3	0	41
	Agricultural land occupation	1	– 4	1	– 7	1	– 5	0	0	– 22
	Urban land occupation	1	– 4	1	– 8	1	– 6	0	– 1	– 24
	Natural land transformation	0	– 0.001	0	0	0	– 0.001	0	0	– 0.004

[a] *Note* D for discharged, P for purchase, C for consumed, R for recovery

Table 3. Normalized environmental impacts

Environmental impact	Impact value	Global value	Normalized impact
CO_{2eq} emission	1697.2 kg CO_{2eq}	4,700	0.36
Human toxicity	−7007.8 kg 1,4-DB_{eq}	50	−140.16
Terrestrial acidification	−424 kg SO_{2eq}	29.41	−14.42
Particulate matter formation	– 95.5 kg PM_{eq}	9.08	– 10.52
Photochemical oxidant formation	– 34.4 kg $NMVOC_{eq}$	21.13	– 1.63
Water depletion	– 535.8 m^3 H^2O	1385	– 0.39
Metal depletion	– 16,941.2 kg Fe_{eq}	305.53	– 55.45
Ozone depletion	– 0.0001 kg CFC-11_{eq}	0.04	– 0.0025
Fossil depletion	44.6 kg oil_{eq}	2400	0.02
Freshwater eutrophication	– 6.0 kg P_{eq}	2.50	– 2.41
Marine eutrophication	– 1.2 kg N_{eq}	28.00	– 0.04

(continued)

Table 3. (*continued*)

Environmental impact	Impact value	Global value	Normalized impact
Terrestrial ecotoxicity	– 1.2 kg 1,4-DB_{eq}	5.95	– 0.19
Freshwater ecotoxicity	– 242.4 kg 1,4-DB_{eq}	1000	– 0.24
Marine ecotoxicity	– 266.3 kg 1,4-DB_{eq}	1100	– 0.24
Ionizing radiation	41 kBq $U235_{eq}$	18750	0.0022
Agricultural land occupation	– 22 m^2a	1700	– 0.01
Urban land occupation	– 24.4 m^2a	200	– 0.12
Natural land transformation	– 0.0041 m^2	10	– 0.0004
Final environmental impact			– 225.4

5.2 Environmental Cost Results

The environmental impacts in Table 2 are calculated with the environmental-related costs (see Table 1) to achieve the total environmental costs in Table 4. The results show that most impact categories are effectively converted into environmental-economic benefits from the reuse of recovered materials of EVBs. The most significant benefit is attributed to avoided metal depletion since recycling directly offsets the need for extracting new raw materials. This reduction in primary resource demand translates into economic savings and environmental preservation. EVB recycling generates an estimated $3388.3 in environmental-economic savings, enhancing their contribution to environmental sustainability.

Table 4. Total environmental costs

Environmental impact	Impact value	Cost ($)	Total cost ($)
CO_{2eq} emission	1697.2 kg CO_{2eq}	0.0052	8.8
Human toxicity	– 7007.8 kg 1,4-DB_{eq}	0.0169	– 118.4
Terrestrial acidification	– 424 kg SO_{2eq}	0.61	– 257.8
Particulate matter formation	– 95.5 kg PM_{eq}	0.45	– 43.0
Photochemical oxidant formation	– 34.4 kg $NMVOC_{eq}$	3.57	– 122.8
Water depletion	– 535.8 m^3 H^2O	0.58	– 310.8
Metal depletion	– 16,941.2 kg Fe_{eq}	0.10	– 1694.1
Ozone depletion	– 0.0001 kg CFC-11_{eq}	24387	– 2.4
Fossil depletion	44.6 kg oil_{eq}	0.26	11.6
Freshwater eutrophication	– 6.0 kg P_{eq}	19.40	– 116.6

(*continued*)

Table 4. (*continued*)

Environmental impact	Impact value	Cost ($)	Total cost ($)
Marine eutrophication	– 1.2 kg N_{eq}	3.85	– 4.7
Terrestrial ecotoxicity	– 1.2 kg 1,4-DB_{eq}	0.02	0.0
Freshwater ecotoxicity	– 242.4 kg 1,4-DB_{eq}	0.02	– 4.1
Marine ecotoxicity	– 266.3 kg 1,4-DB_{eq}	0.02	– 4.5
Ionizing radiation	41 kBq $U235_{eq}$	0.0015	0.1
Agricultural land occupation	– 22 m^2a	15.70	– 345.8
Urban land occupation	– 24.4 m^2a	15.70	– 383.2
Natural land transformation	– 0.0041 m^2	137.00	– 0.6
Final environmental cost			– 3388.3

6 Conclusion

This study evaluated the environmental costs and benefits of EVB recycling by integrating environmental assessment with environmental valuation. The analysis covered key categories including CO_{2eq} emission, human toxicity, terrestrial acidification, particulate matter formation, photochemical oxidant formation, water depletion, metal depletion, ozone depletion, fossil depletion, freshwater eutrophication, marine eutrophication, terrestrial ecotoxicity, freshwater ecotoxicity, marine ecotoxicity, ionizing radiation, agricultural land occupation, urban land occupation, natural land transformation, with impacts converted to economic terms using Thai and global reference values. The need for this study arises from the rapid growth in EV adoption and the increase in EVBs, which contain high-value but hazardous materials. By quantifying the environmental impacts of recycling, this study provides clear evidence for decision-making in industry and policy. The results show that EVB recycling reduces several environmental impacts, specifically metal depletion, human toxicity, water depletion, and terrestrial acidification. The normalized impact reveals that human toxicity is reduced by 140 times of an individual's annual reduction impact. In contrast, metal depletion is reduced by 55 times, proving that EVB recycling prevents large-scale pollution and resource loss from landfilling. Nevertheless, the EVB recycling process generates CO_{2eq} emissions mainly from electricity generation. The government should promote using renewable energy to minimize this impact and enhance the environmental benefits of EVB recycling. From an environmental-economic perspective, EVB recycling can generate savings of approximately \$3388 per battery. The most significant benefit is from saving in metal depletion, which is \$1649 per battery, followed by water depletion, which is \$310 per battery, and terrestrial acidification, which is \$257 per battery. These savings arise because recovered metals directly replace raw materials, reducing the environmental costs of mining and refining.

These findings highlight the value of EVB recycling. It is both an environmentally protective and economically feasible process. The results show the industry's feasibility of integrating recycling into battery manufacturing supply chains to reduce environmental liabilities and improve resource security. To maximize benefits, recycling systems should focus on achieving high recovery rates, integrating renewable energy sources into operations, and adopting processing methods that combine the strengths of pyrometallurgical and hydrometallurgical techniques. The strategies will enhance environmental performance and strengthen the circular economy and long-term sustainability of the EVB industry. This study has some limitations. The data used in the analysis are from the literature, not specifically from the Thai context. To achieve accurate results, future studies should examine EVB recycling from a sustainable perspective that covers economic, environmental, and social contexts.

References

1. Baum, Z.J., Bird, R.E., Yu, X., Ma, J.: Lithium-ion battery recycling—overview of techniques and trends. ACS Energy Lett. **7**(2), 712–719 (2022). https://doi.org/10.1021/acsenergylett.1c02602
2. Fan, X. et al.: Separation and recovery of valuable metals from spent lithium-ion batteries via concentrated sulfuric acid leaching and regeneration of $LiNi_{1/3}Co_{1/3}Mn_{1/3}O_2$. J. Alloys Compd. **863**, 158775 (2021). https://doi.org/10.1016/j.jallcom.2021.158775
3. Chen, Z., Yildizbasi, A., Wang, Y., Sarkis, J.: Safety concerns for the management of end-of-life lithium-ion batteries. Global Challenges. **6**(12) (2022). https://doi.org/10.1002/gch2.202200049
4. Kopsidas, O.N., Giakoumatos, S.D.V.: Economics of recycling and recovery. Nat. Resour. **12**(04), 73–90 (2021). https://doi.org/10.4236/nr.2021.124007
5. Ali, A., Bahrani, M.A., Ahmed, S., Islam, M.T., Qadir, S.A., Shahid, M.: Sustainable recycling of end-of-life electric vehicle batteries: EV battery recycling frameworks in China and the USA. Recycling. **10**(2), 68 (2025). https://doi.org/10.3390/recycling10020068
6. Zhang, X., Zhang, L., Fung, K.Y., Bakshi, B.R., Ng, K.M.: Sustainable product design: a life-cycle approach. Chem. Eng. Sci. **217**, 115508 (2020). https://doi.org/10.1016/j.ces.2020.115508
7. Chayutthanabun, A., Chinda, T., Papong, S.: End-of-life management of electric vehicle batteries utilizing the life cycle assessment. J. Air Waste Manage. Assoc., 1–13 (2025). https://doi.org/10.1080/10962247.2024.2430325
8. Fu, Y. et al.: Effective leaching and extraction of valuable metals from electrode material of spent lithium-ion batteries using mixed organic acids leachant. J. Ind. Eng. Chem. **79**, 154–162 (2019). https://doi.org/10.1016/j.jiec.2019.06.023
9. Pagliaro, M., Meneguzzo, F.: Lithium battery reusing and recycling: a circular economy insight. Heliyon. **5**(6), e01866 (2019). https://doi.org/10.1016/j.heliyon.2019.e01866
10. Huang, F., Liao, Y., Zhou, J., Wang, Y., Li, H.: Selective recovery of valuable metals from nickel converter slag at elevated temperature with sulfuric acid solution. Sep. Purif. Technol. **156**, 572–581 (2015). https://doi.org/10.1016/j.seppur.2015.10.051
11. Qu, Y., Tan, K., Zhao, B., Xie, S.: Recovery of Cu-Fe alloy from copper smelting slag. Meta. **13**(2), 271 (2023). https://doi.org/10.3390/met13020271
12. Kauppinen, T., Vielma, T., Salminen, J., Lassi, U.: Selective recovery of manganese from anode sludge residue by reductive leaching. ChemEngineering. **4**(2), 40 (2020). https://doi.org/10.3390/chemengineering4020040
13. Seyrankaya, A.: Pressure leaching of copper slag flotation tailings in oxygenated sulfuric acid media. ACS Omega. **7**(40), 35562–35574 (2022). https://doi.org/10.1021/acsomega.2c02903

14. Brückner, L., Frank, J., Elwert, T.: Industrial recycling of lithium-ion batteries—a critical review of metallurgical process routes. Meta. **10**(8), 1107 (2020). https://doi.org/10.3390/met10081107
15. Chayutthanabun, A., Chinda, T.: Long-term benefits of remanufactured electric vehicle batteries: a case of Thailand. Suranaree J. Sci. Technol. **31**(5), 010336, 1–13 (2024)
16. Holland, M. (2020). Reducing the health risks of the copper, rare earth and cobalt industries. OECD Green Growth Papers. https://doi.org/10.1787/88ce1db4-en
17. TGO: Carbon market (2025). Available online: https://carbonmarket.tgo.or.th/index.php?lang=EN&mod=Y2N0X3ByaWNl
18. WHO: The Public Health Impact of Chemicals: Knowns and Unknowns. World Health Organization, Geneva, Switzerland (2025) https://apps.who.int/iris/bitstream/handle/10665/206553/WHO_FWC_PHE_EPE_16.01_eng.pdf
19. OECD: Saving Costs in Chemicals Management. The Organization for Economic Co-operation and Development, Paris, France (2025) https://www.oecd.org/content/dam/oecd/en/publications/reports/2019/01/saving-costs-in-chemicals-management_g1g9e0e5/9789264311718-en.pdf
20. PWA: PEA Water Tariff Rate (2025). Available online: https://en.pwa.co.th/contents/service/table-price
21. DMP: Daily Metal Prices (2025). Available online: https://www.dailymetalprice.com/
22. Simpkins, G.: Ozone damage costs billions. Nat. Rev. Earth Environ. **3**(2), 102 (2022). https://doi.org/10.1038/s43017-022-00266-z
23. Ministry of Energy: Fuel prices (2025). Available online: https://www.eppo.go.th/index.php/th/petroleum/price/structure-oil-price
24. Pretty, J.N., Mason, C.F., Nedwell, D.B., Hine, R.E., Leaf, S., Dils, R.: Environmental costs of freshwater eutrophication in England and Wales. Environ. Sci. Technol. **37**(2), 201–208 (2002). https://doi.org/10.1021/es020793k
25. Empire Estates: How expensive is land in Thailand? (2025). Available online: https://empire-estates-phuket.com/how-expensive-is-land-in-thailand/

Accelerating Transformation: Price Wars, Innovation, and Restructuring in the New Energy Vehicle Industry

Shuqi Xuan[1], Tengye Chen[1], Wenyi Tang[1], Wenjie Wang[1], and Weilun Huang[2](✉)

[1] School of International studies, Wenzhou Business College, Wenzhou, Zhejiang, China
[2] School of Finance and Trade, Wenzhou Business College, Wenzhou, Zhejiang, China
huangwl@wzbc.edu.cn

Abstract. This study investigates the multifaceted impact of price wars on the new energy vehicle (NEV) industry chain and examines their role in industrial restructuring. The purpose of the research is to understand how aggressive pricing strategies affect critical performance metrics across the NEV supply chain, encompassing both automakers and suppliers. To achieve this, annual report data from 2015 to 2023 of four major NEV companies (BYD, Tesla, Geely Auto, and Great Wall Motor) and four key suppliers (CATL, Gotion High-Tech, EVE Energy, and Sunwoda) were analyzed. Employing a mixed-method approach that integrates descriptive statistics, correlation analysis, regression models, and game theory, the study explores the relationships between variables such as average selling price, policy support, market competition intensity, technological innovation investment, and consumer demand with key outcomes like sales volume, profitability, market share changes, and inventory turnover rate. The results reveal that technological innovation investment has a positive, albeit marginally significant, effect on NEV and battery sales ($p = 0.057$); market competition intensity significantly influences market share changes ($p < 0.001$); and policy support is significantly negatively correlated with inventory turnover rate ($p = 0.045$). Based on these findings, it is recommended that companies bolster R&D efforts and optimize supply chain management to enhance competitiveness, while policymakers should continue to provide targeted support that guides the industry towards sustainable, high value-added development.

Keywords: Price Wars · NEV Industry · Technological Innovation · Market Competition · Industrial Restructuring.

1 Introduction

The new energy vehicle (NEV) industry has emerged as a crucial direction in the global transformation of the automotive sector, propelled by significant technological advancements, robust policy support, and continuously increasing market demand (Zhang et al. 2024). This rapid growth has led to a steady rise in the market share of NEVs on an annual basis, fundamentally reshaping traditional automotive paradigms and accelerating the shift toward sustainable mobility solutions.

Y. Zhu (Ed.): ISESIE 2025, LNCE 000, pp. 173–183, 2026.
https://doi.org/10.1007/978-981-92-2389-3_15

Concurrently, intensified market competition has given rise to pervasive price wars, which have increasingly become a primary competitive strategy within the industry. These price wars not only impact the financial performance of individual companies but also have profound implications for the structure and future development of the entire NEV industry chain. In particular, in the Chinese market, numerous automakers have adopted aggressive price-cutting strategies to capture market share, thereby exerting significant pressure—and creating both challenges and opportunities—for upstream and downstream enterprises (Wu et al. 2020). Despite their importance, current research predominantly focuses on the isolated effects of price competition at the firm level, leaving a notable gap in understanding the holistic impact on various segments of the supply chain (Shao et al. 2020).

In response to this research gap, the present study is designed to thoroughly analyze the impact of price wars on different segments of the NEV industry chain and to explore their role in driving industrial restructuring. The primary objective is to employ a comprehensive data analysis and model-building approach to investigate both the short-term and long-term effects of price wars. Specifically, the study aims to assess how price wars influence the financial performance, market share, and innovation capabilities of automakers and suppliers, to predict future trends in industrial restructuring, and to develop corresponding policy recommendations and strategic measures for industry stakeholders.

Methodologically, the study utilizes a robust empirical framework based on annual report data from four leading NEV companies (BYD, Tesla, Great Wall Motor, Geely Auto) and four major suppliers (CATL, EVE Energy, Gotion High-Tech, Sunwoda) covering the period from 2015 to 2023. Key variables—such as average selling prices, levels of policy support, market competition intensity, technological innovation investments, and changes in consumer demand—are extracted and analyzed using descriptive statistics, correlation analysis, and regression analysis. Additionally, the incorporation of game theory models allows for an in-depth examination of the dynamic interactions between automakers and suppliers under competitive pressure, offering insights into the strategic interdependencies within the industry chain.

The innovative contribution of this work lies in its holistic approach to assessing the impact of price wars across the entire NEV industry chain, rather than focusing solely on individual companies. By integrating traditional quantitative methods with advanced game theory modeling, the study not only elucidates the direct effects of price wars on corporate performance but also provides theoretical support for understanding the broader industrial restructuring processes. This dual approach yields empirical evidence that is instrumental for policy-makers and corporate strategists, highlighting the necessity of continuous investment in technological innovation, adaptive cost management, and flexible supply chain strategies to navigate the challenges of intense market competition.

2 Literature Review

NEV industry chain (see Fig. 1) has witnessed rapid growth as a result of substantial technological advancements, robust policy support, and continuously increasing market demand (Zhang et al. 2024). At the heart of this transformation lies a complex industry chain that integrates upstream raw material sourcing, midstream vehicle manufacturing, and downstream charging and aftermarket services. For example, raw materials such as lithium ore, separators, electrodes, rare earth elements, and silicon wafers are supplied by key component manufacturers (e.g., CATL, BYD, EVE Energy, Gotion, Soundon), which feed into vehicle production by automakers including BYD, Tesla, Great Wall, and Geely. The end segment involves the establishment of a supportive infrastructure through charging networks and post-sales services provided by companies like Teld, Star Charge, Shouqi, Wukong, and Ping An. This tightly integrated chain highlights both collaborative structures and competitive dynamics that collectively shape the industry's overall development (Zhang et al. 2025; Xu and Mo 2024).

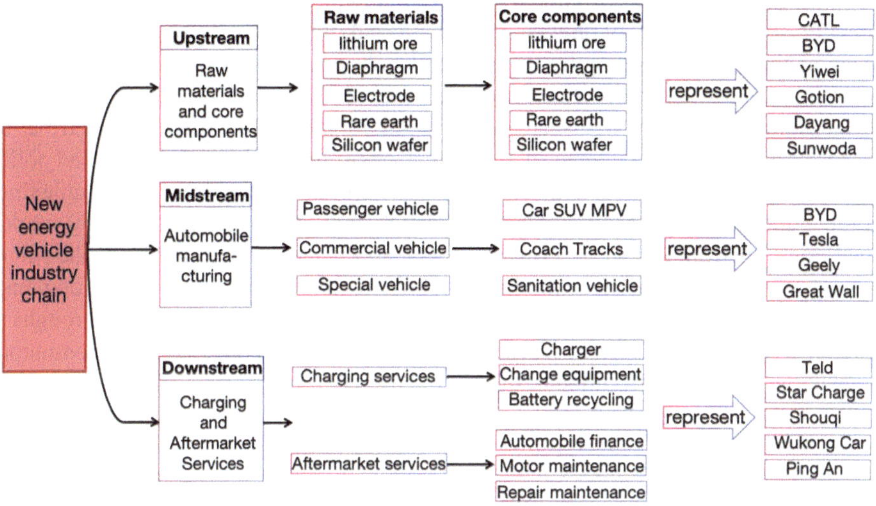

Fig. 1. Organizational structure of new energy vehicle industry chain.

Driving this rapid expansion are significant policy incentives and market forces. Government measures such as subsidies and tax breaks have proven instrumental in accelerating NEV adoption, as noted in several studies (Wang et al. 2025; Cai et al. 2025). Moreover, the high degree of specialization within the industry chain, as documented in CNKI literature, fosters a competitive yet cooperative environment among automakers, suppliers, and technology providers (Zhang et al. 2025; Xu and Mo 2024). These factors not only facilitate market penetration but also contribute to reducing urban carbon emissions, thereby reinforcing the sustainability goals underlying the NEV revolution.

However, as market competition intensifies, price wars have emerged as a prominent strategy. Grounded in theories of market competition and cost leadership (Tian and Zhao 2024), these price reductions can lead to short-term gains in market share while imposing pressure on corporate profitability. The resulting market consolidation increases industry concentration and forces companies to adopt more efficient supply chain and production models (Lu et al., 2024; Sun et al. 2025). Analyzing the strategic interactions between automakers and suppliers through game theory further reveals that coordinated responses in pricing and operational adjustments are critical to maintaining industry stability (Xing and Wang 2025).

Industrial restructuring in the NEV sector is driven by a combination of increased industry concentration, enhanced supply chain integration, and continuous technological innovation. Advances in battery technology and the development of intelligent connected vehicles are not only improving user experience but also catalyzing the reorganization of production networks (Lei 2024; Zhan et al. 2025). These innovations play a pivotal role in ensuring that the entire value chain remains competitive and sustainable over the long term, as supported by research emphasizing the importance of these factors for future industry development (Chen 2024; Deng and Xu 2024).

Academic research on the NEV industry chain has evolved significantly over the past decade. Initial studies from 2009 to 2018 maintained a steady output, but scholarly attention surged from 2019 onward, reaching a peak in 2024 (see Fig. 2). The thematic focus of this research—ranging from overall industry dynamics and vehicle production to market resilience and R&D investment—reflects the multifaceted challenges and opportunities within the NEV sector (Fig. 3). Additionally, the disciplinary distribution of these studies spans economics, management, environmental sustainability, and policy systems, underscoring the interdisciplinary nature required to address the complexities of the industry (Fig. 4). Multidimensional factors, including policy, technology, market conditions, environmental concerns, and social acceptance, further influence the stability and innovation capacity of the industry chain (Fig. 5). Finally, key developmental milestones—from early policy-driven pilot projects and technological breakthroughs in battery systems to the current phase of international competitiveness and continuous innovation—chart the evolution of the NEV industry from 2011 to 2025 (Fig. 6).

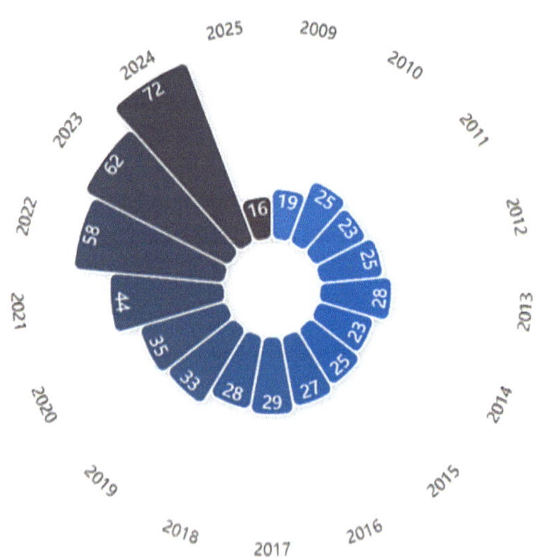

Fig. 2. Evolution of new energy vehicle industry chain research.

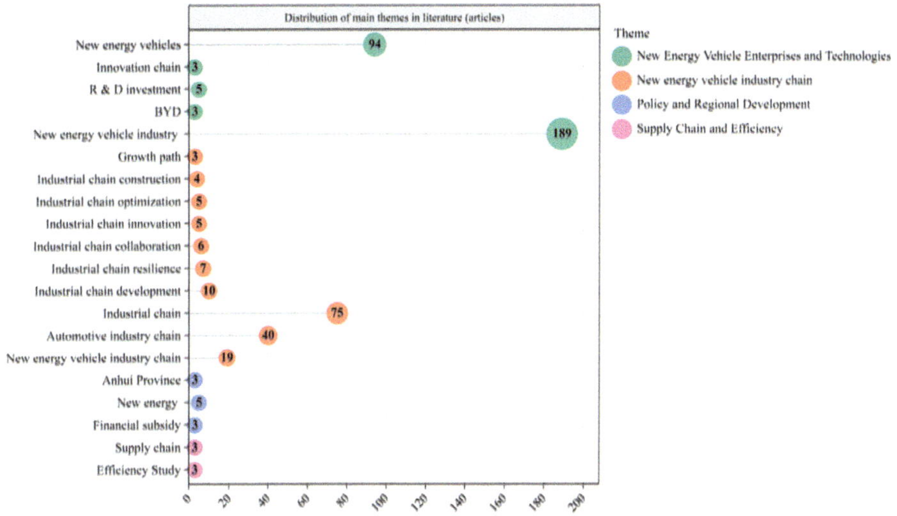

Fig. 3. Topic distribution in new energy vehicle research.

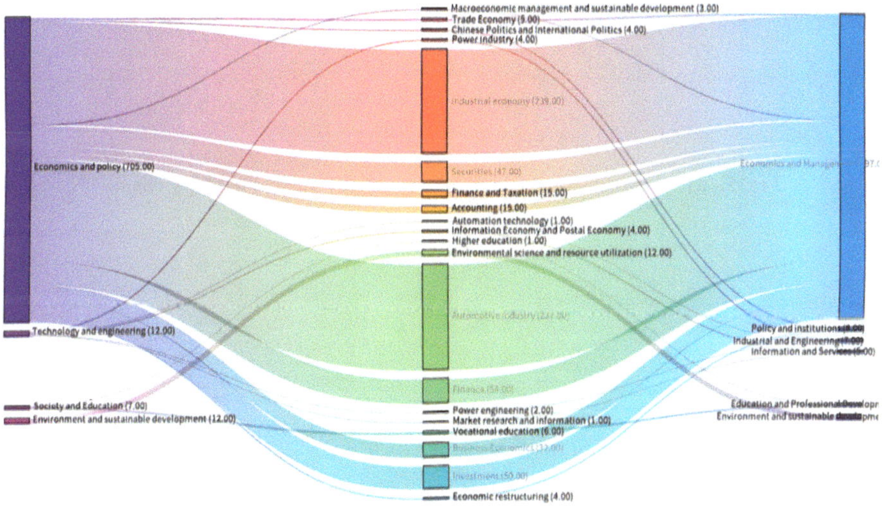

Fig. 4. Disciplinary distribution in new energy vehicle research

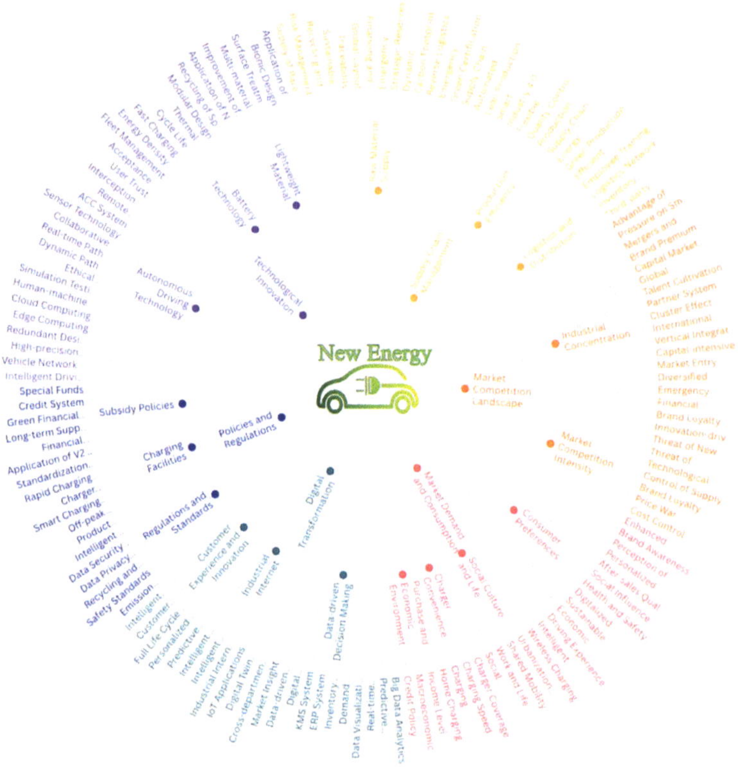

Fig. 5. Multi-dimensional factors in new energy vehicle industry chain.

Fig. 6. NEV industry chain timeline.

3 Data Analysis

This study is based on annual reports from four prominent NEV manufacturers—BYD, Tesla, Geely Auto, and Great Wall Motor—and four key suppliers—CATL, Gotion High-Tech, EVE Energy, and Sunwoda—spanning the period from 2015 to 2023. The research provides an in-depth analysis of the evolution and current status of the NEV industry chain, capturing shifts over different periods and highlighting the distinct trajectories of various market players.

Through systematic examination of the data, the study reveals that industry leaders such as BYD and Tesla have significantly propelled industry development by leveraging competitive market strategies and operational excellence. In contrast, traditional automakers like Geely Auto and Great Wall Motor have successfully transitioned into the NEV sector, demonstrating robust performance and resilience across diverse economic cycles. Moreover, the financial performance of suppliers such as CATL not only reflects the overall health of the supply chain but also underscores the pivotal role of advances in battery technology for the entire industry.

To understand these dynamics, the study employs a quantitative methodology that focuses on several key independent variables: average selling price (X1), level of policy support (X2), market competition intensity (X3), proportion of technological innovation investment (X4), and changes in consumer demand (X5). These factors are analyzed in relation to dependent variables including sales volume (Y1), profitability (Y2), market share changes (Y3), and inventory turnover rate (Y4). This approach is designed to elucidate the operational logic and development trends of the NEV industry, particularly under the influence of competitive pricing strategies.

For clarity and precision, the specific variable constructions are defined as follows: average selling price (X1) is calculated by dividing sales revenue by sales volume; policy support (X2) is measured as the ratio of government subsidy to operating revenue; market competition intensity (X3) is represented by the inverse of market share; the proportion of technological innovation investment (X4) is defined as the ratio of R&D expenses to operating revenue; and inventory turnover rate (Y4) is determined by dividing the cost of sales by the average inventory.

Table 1 presents the descriptive statistics for the main independent variables (X1–X5) and dependent variables (Y1–Y4) used in this study. These statistics reflect the numerical distribution of each variable within the sample. Specifically, X1 represents the average selling price, X2 denotes the level of policy support, X3 indicates the intensity of market competition, X4 corresponds to the proportion of technological innovation investment, and X5 reflects changes in consumer demand. The dependent variables Y1, Y2, Y3, and Y4 correspond to sales volume, profitability, market share changes, and inventory turnover rate, respectively.

Table 1. Descriptive statistical analysis (condensed version)

Variable	Maximum	Minimum	Mean	Standard deviation	Median	Variance
X1	1.85×10^7	2034	483,778	2.2×10^6	127,447	4.87×10^{12}
X2	0.21	– 0.14	0.03	0.05	0.01	0.002
X3	333	2	25	41	14	1,708
X4	1.1	– 1.04	0.04	0.28	0.05	0.08
X5	384	– 33	7	46	0.38	2,092
Y1	321	0.11	66	70	47	4,838
Y2	4.68×10^6	– 224,057	6.69×10^6	951,485	370,390	9.05×10^{10}
Y3	20	– 0.7	0.32	2	0	6
Y4	20	– 1.1	4	6	0.3	34

From Table 1, significant differences are evident across various measures, including maximum, minimum, mean, standard deviation, median, and variance. For example, although the mean value of X1 is 483,778, its high standard deviation of 2.2×10^6 indicates substantial fluctuations in price levels. In contrast, the mean of X2 is only 0.03, which suggests that the overall level of policy support is relatively low and exhibits minimal variability. Overall, these statistical results provide a solid foundation for exploring the operational logic and development trends of the NEV industry in the context of price wars.

Table 2 presents the Spearman correlation coefficients between the key independent variables—average selling price (X1), policy support level (X2), market competition intensity (X3), technological innovation investment proportion (X4), and consumer demand changes (X5)—and the dependent variables, which include sales volume (Y1), profitability (Y2), market share change (Y3), and inventory turnover rate (Y4). This analysis provides insight into how these factors interact within the NEV industry. For instance, the strong negative correlation between X1 and Y1 ($r = -0.556$) indicates that higher average selling prices are associated with lower sales volumes, while the positive correlation between Y1 and Y2 suggests that increased sales volume is linked to improved profitability. Additionally, the moderate negative correlation between market competition intensity (X3) and inventory turnover rate ($r = -0.409$) implies that intensified competition may result in lower turnover rates. By focusing on these critical relationships, the table offers valuable empirical evidence to support further strategic analysis and operational adjustments in the NEV sector.

Table 2. Spearman correlation coefficients between independent and dependent variables

Variable	Y1 (Sales volume)	Y2 (profitability)	Y3 (Market share change)	Y4 (Inventory turnover)
X1 (Avg. selling price)	−0.556	−0.089	0.147	−0.316
X2 (Policy support)	−0.100	0.130	−0.138	−0.135
X3 (Market competition)	−0.285	−0.269	0.242	−0.409
X4 (Innovation investment)	−0.061	−0.059	−0.114	−0.020
X5 (Consumer demand)	−0.233	−0.241	0.145	−0.263

Table 3 presents the key results from the stepwise regression analyses, focusing on three dependent variables: new energy vehicle/battery sales (Y1), market share change (Y3), and inventory turnover rate (Y4). In the first regression model, which examines Y1, the only predictor retained is technological innovation investment (X4). Although the overall model has low explanatory power (with R^2 of 0.09), the coefficient for X4 ($B = 73$) is marginally significant ($p = 0.057$), suggesting that increases in innovation investment tend to boost NEV/battery sales. The second model, which focuses on market share change (Y3), shows a very strong and significant effect of market competition intensity (X3). Here, the positive coefficient ($B = 0.05$, $p < 0.001$) along with a high R^2 value (0.82) indicates that more intense competition is associated with an increase in market share change. In the third regression model, investigating inventory turnover rate (Y4), policy support (X2) is the retained predictor. The analysis reveals a significant negative relationship ($B = -38$, $p = 0.045$), implying that higher levels of policy support correspond to lower inventory turnover rates. These results collectively provide valuable insights into how specific factors—innovation investment, competitive intensity, and policy support—affect different operational outcomes in the NEV industry.

Table 3. Key results from stepwise regression analyses

Dependent Variable	Predictor	B	p-value
NEV/Battery sales (Y1)	Constant	74.8	0.000
	Technology innovation investment (X4)	73	0.057
Market share change (Y3)	Constant	−0.88	0.000
	Market competition intensity (X3)	0.05	0.000
Inventory turnover rate (Y4)	Constant	5.3	0.000
	Policy support (X2)	−38	0.045

4 Conclusion and Recommendations

This study aims to explore the impact of price wars on various segments of the new energy vehicle (NEV) industry chain and to analyze their role in industrial restructuring. To achieve this objective, the researchers utilized annual report data from 2015 to 2023 from four vehicle manufacturers (BYD, Tesla, Geely Auto, and Great Wall Motor) and four suppliers (CATL, Gotion High-Tech, EVE Energy, and Sunwoda). By employing descriptive statistics, correlation analysis, regression models, and game theory models, the study comprehensively examined the effects of price wars on key indicators such as vehicle sales, market share, profitability, and inventory turnover rate over both the short and long term.

The results indicate that investment in technological innovation has a positive effect on NEV and battery sales, although this impact is only marginally significant ($p = 0.057$). Additionally, the intensity of market competition significantly influences changes in market share ($p < 0.001$), suggesting that in highly competitive environments, firms are more likely to adopt price-cutting strategies to secure market share. Furthermore, a significant negative correlation was found between policy support and inventory turnover rate ($p = 0.045$), implying that while government subsidies may not directly enhance economic performance, they can indirectly influence firms' inventory management strategies through their effect on operational decisions. Overall, price wars not only affect the financial and market performance of individual companies directly but also play a critical role in the structural adjustments and future development of the entire industry chain.

Based on these findings, it is recommended that companies facing intense price competition should increase their investment in research and development to enhance product competitiveness and optimize supply chain management to reduce costs and improve efficiency. Additionally, firms should adjust their pricing strategies flexibly in response to market fluctuations to maintain sustainable long-term growth. For the government, it is advisable to continue providing policy incentives—such as subsidies and tax breaks—while also focusing on guiding companies toward high value-added and sustainable development, thereby promoting the overall healthy development of the NEV industry chain.

References

Cai, J., Zhu, M., Wu, J., Chen, X., Xu, J., Shen, Z.: Evaluating the impact of new energy vehicle subsidies on urban carbon emissions: evidence from 261 Chinese cities. Renew. Energy. **240** (2025)

Chen, Z.: Vibration and noise control of new energy vehicle power system in vehicle engineering. Forum Res. Innov. Manag. **2**(11) (2024)

Deng, X., Xu, F.: Connectedness between international oil and China's new energy industry chain: a time-frequency analysis based on TVP-VAR model. Energy Econ. **140** (2024)

Han, J., Jiang, C., Liu, R.: Does intelligent transformation trigger technology innovation in China's NEV enterprises? Energy. **270**, 126823 (2023)

Lei, C.: New energy vehicle battery state of charge prediction based on XGBoost algorithm and RF fusion. Energy Inform. **7**(1) (2024)

Shao, L., Jin, S., Li, Y.: Formation mechanism and countermeasures of turbulence in China's new energy vehicle industry. Resour. Sci. **44**(8), 1316–1330 (2020)

Sun, H., Wang, D., Li, Q., Qin, Y.: A common DC bus circulating current suppression method for motor emulators of new energy vehicles. Machines. **13**(1) (2025)

Tian, Y., Zhao, X.: The influence of new energy vehicle enterprises' strategy on government subsidy acquisition: a configurational analysis based on dynamic QCA. Acad. J. Bus. & Manag. **6**(12) (2024)

Wang, L., Wu, T., Ren, F.: Analysis of spatial differences and influencing factors of carbon-emission reduction efficiency of new-energy vehicles in China. Energies. **18**(3) (2025)

Wu, X.B., Feng, X.Y., Li, H.B.: What influences sales market of new energy vehicles in China? Empirical study based on survey of consumers' purchase reasons. Energy Policy. **142**, 111484 (2020)

Xing, P., Wang, M.: The interplay of recycling channel selection and blockchain adoption in the new energy vehicle supply chain under the government reward-penalty scheme. J. Clean. Prod. **487** (2025)

Xu, W., Mo, W.: Institutional unlocking or technological unlocking? The logic of carbon unlocking in the new energy vehicle industry in China. Energy Policy. **195** (2024)

Zhan, W. et al.: Study of titanium/zirconium with low polymerization water-based epoxysilane composite chemical conversion coatings on multi-metals bodies of new energy vehicles. Prog. Org. Coat. **200** (2025)

Zhang, Q., Du, A.M., Lin, B.: Driving total factor productivity: the spillover effect of digitalization in the new energy supply chain. Res. Int. Bus. Financ. **75** (2025)

Zhang, Q., Feng, Y., You, L.: Research on the impact of supply chain integration on supply chain resilience in NEV manufacturing enterprises. Sustainability. **16**(19), 8546 (2024)

Research on Restoration Schemes for Reservoir Water Level Fluctuation Zone Ecosystems in the High Frigid Zone

Qi Li[1,2], Qin Yu[3,4](✉), Xiaomei Kou[1,2], Le Niu[1,2], Rui Liu[1,2], and Baicheng Niu[3,4]

[1] Power China Northwest Engineering Corporation Limited, Xi'an, Shaanxi 710065, China
[2] Shaanxi Union Research Center of University and Enterprise for River and Lake Ecosystems Protection and Restoration, Xi'an, Shaanxi 710065, China
[3] School of Geographical Science, Qinghai Normal University, Xining 810008, China
yuqinjyq@163.com
[4] Key Laboratory of Tibetan Plateau Land Surface Processes and Ecological Conservation (Ministry of Education), Qinghai Normal University, Xining 810008, China

Abstract. Longyangxia Reservoir (LYXR), located in the upper reaches of the Yellow River, is characterized by its high altitude, cold climate, and alpine, arid conditions with a short growing season. Restoring vegetation in the water level fluctuation zone (WLFZ) to intercept and impede gully runoff, alongside enhancing the reservoir's purification capacity, constitutes a fundamental approach to establishing an ecological barrier along the reservoir shoreline. This paper explores suitable ecological restoration models for reservoir WLFZs, based on an analysis of habitat characteristics and ecological remediation techniques within the LYXR WLFZ. The WLFZ of the LYXR is categorized into three distinct types: easily restorable zones (including sediment-accumulated and wind-sand-accumulated gentle slopes), difficult-to-restore zones (comprising gravel-accumulated slopes and steep slopes), and tributary water level fluctuation zones. Key technical measures proposed include both vegetative and engineering interventions. The findings provide a scientific basis for the ecological conservation and restoration of reservoir WLFZs in high-altitude cold regions, and offer valuable insights for ecological construction along the shorelines of other reservoirs across China.

Keywords: Longyangxia Reservoir · High frigid zone · Water level fluctuation zone · Ecological restoration.

1 Introduction

Water conservancy projects play an irreplaceable role in ensuring water resource allocation, flood and disaster mitigation, and energy supply; however, they also trigger ecological and environmental problems. Among these, the ecological degradation of the Water Level Fluctuation Zone (WLFZ) has become a significant challenge that restricts regional ecological security and sustainable development. Functioning as a sensitive ecotone between aquatic and terrestrial ecosystems, the WLFZ is characterized

Y. Zhu (Ed.): ISESIE 2025, LNCE 000, pp. 184–194, 2026.
https://doi.org/10.1007/978-981-92-2389-3_16

by ecological transition, environmental fluctuation, and systemic vulnerability. Frequent and drastic dry-wet alternation [1], significant habitat changes induced by water level fluctuations [2], intense erosion and scouring, and impeded vegetation succession [3] collectively render it one of the most vulnerable and difficult-to-recover areas within reservoir ecosystems [4].

The WLFZ of reservoirs in the high frigid zone is highly susceptible to severe soil erosion, soil degradation, bank collapse, sharp declines in biodiversity, and ecological function degradation. These issues not only threaten the safe operation of the reservoirs themselves (such as bank stability and reservoir capacity maintenance) but also profoundly impact the health and service functions of surrounding ecosystems (such as water purification and habitat preservation). Therefore, exploring restoration technologies and models tailored to the harsh environmental characteristics of the high frigid zone WLFZ holds significant practical importance for ensuring the sustainable operation of water conservancy projects and constructing regional ecological barriers.

The core objective of WLFZ ecological restoration is to establish a self-sustaining vegetation system, which hinges on the selection and configuration of suitable plant species and the synergistic application of engineering and biological measures [5, 6]. Given the unique constraints of the high frigid zone, successful restoration [7] depends on selecting native species with high stress tolerance (e.g., to flooding, drought, cold, and nutrient-poor conditions) and implementing zoned management strategies based on topographic features such as elevation gradient, substrate type, flooding regime, and slope. Currently, systematic research and practical applications focused on the restoration of reservoir WLFZs in high frigid zones remain relatively scarce, necessitating further in-depth exploration through detailed case studies.

This study focuses on the high-altitude, cold-temperate WLFZ of the LYXR in the upper reaches of the Yellow River. Building upon a systematic review of existing ecological restoration practices in reservoir WLFZ, it employs a combination of detailed field surveys and laboratory analyses to thoroughly diagnose the primary ecological and environmental challenges facing this WLFZ and their underlying causes. The research aims to identify key ecological degradation characteristics across different regions of the LYXR WLFZ, classified by elevation and ecological zones. For distinct sub-zones—including easily restorable sediment- or wind-deposited gentle slopes, difficult-to-restore gravel or steep slopes, and tributary water level fluctuation zones—this study explores and proposes optimized zoned restoration strategies. These strategies encompass primarily biological measures, combined bioengineering approaches, or necessary engineering interventions employed as supplementary measures. A key priority of this study is the screening of dominant plant species and their combination patterns suited to the harsh conditions of alpine-temperate WLFZs. This research aims to provide a scientific basis and technical guidance for the restoration of these unique and vulnerable ecological units in high frigid zones, thereby contributing to regional ecosystem conservation and restoration efforts.

2 Materials and Methods

2.1 Overview of the Study Area

The Longyangxia Reservoir (LYXR; 35°46′–36°13′N, 100°23′–100°50′E) is a large-scale reservoir with multi-year regulation capacity, forming a key component of the upstream cascade reservoir system on the Yellow River (Fig. 1). It is the largest reservoir on the Tibetan Plateau, with its dam situated in the eastern Gonghe Basin [8]. Its operational pattern of "winter storage and summer discharge" results in a pronounced draw-down zone along the shoreline, characterized by "winter inundation and summer exposure [9]. At the maximum operating water level of 2600 m, the reservoir has a total capacity of 2.47 2.47×10^{10} m^3 [10]. In recent years, the minimum water level has been maintained at 2580 m, resulting in a water-level-fluctuation zone with an area of approximately 56 km^2 between 2580 m and 2600 m [11]. Rapid water-level fluctuations, coupled with human activities, have contributed to a decline in vegetation diversity and overall biodiversity within these interface zones [12]. Risks such as soil erosion, bank collapse, and soil loss [13] in parts of the WLFZ may also impact the safety of the entire reservoir area and downstream regions.

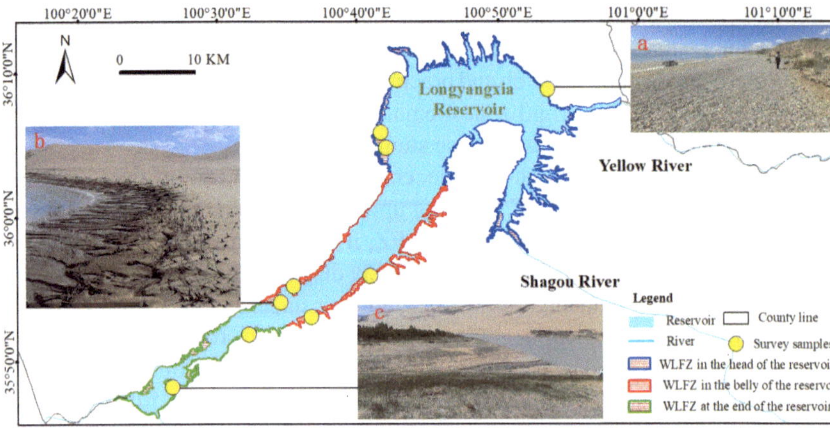

Fig. 1. Map of study area

2.2 Research Methodology

2.2.1 Theoretical Foundations

Biological communities and physical environments are the basic elements that make up an ecosystem.[14] As the last line of defense for reservoir water protection, ecosystem restoration of WLFZ is an important step in its maintenance of reservoir bank stability and prevention of soil erosion.

The unique natural environment of the high frigid zone is used to analyze the ecological status of the WLFZ fading zone of the LYXR reservoir. The reservoir faces outstanding problems such as wind-sand disaster, geological disaster, steep bank surge erosion and soil erosion. Coupled with the fact that vegetation has the effect of reducing river flow rate and promoting sedimentation, the use of plant materials and ecological materials can effectively hinder river sediment erosion, which is economical and practical [15]. Planting herbaceous plants on higher embankments can adapt to the environment of changing water levels and play a protective role. Therefore, clarifying the "characteristics of different types of WLFZ" and "existing vegetation types in the WLFZ" is the key to realize the ecological restoration of reservoir WLFZ. The preliminary understanding of the current status of WLFZ through the field survey can provide a reference for the comprehensive use of engineering and biological measures to realize ecological restoration and management.

2.2.2 Field Surveys

After determining the ecological environment of LYXR and the possibility of reaching it with reference to remote sensing images and field surveys, we set up sample plots in the typical WLFZ areas on the two sides of the reservoir, including the head, the belly and the tail of the reservoir, in order to carry out field surveys. The survey mainly included topographic characteristics of the WLFZ, existing vegetation characteristics, and recorded the composition of plant species, number, height, and cover information at the survey site, and photographed the surveyed sample plots one by one for further confirmation in the laboratory.

3 Results

3.1 Types of Longyangxia Reservoir Water Level Fluctuation Zone

The topography and geomorphology of the Longyangxia Reservoir (LYXR) are complex. The soil texture within its Water Level Fluctuation Zone (WLFZ) exhibits significant spatial heterogeneity, with distinct surface layer characteristics including mudflats, sediment accumulation, and gravel deposits. Based on the slope classification standards established by the International Geographical Union's Commission on Geomorphological Survey and Mapping, the slopes within the WLFZ were categorized into seven types: plain, micro-slope, gentle slope, slope, steep slope, cliff, and vertical wall (Table 1). Integrating slope and soil conditions, the WLFZ was classified into three primary types: easily restorable (including sediment-accumulated and wind-sand-accumulated gentle slopes), difficult-to-restore (comprising gravel-accumulated slopes and steep slopes), and tributary inflow zones (Table 2).

Table 1. Slope of water level fluctuation zone of Longyangxia reservoir

Terrain type	Slope	Classification	Area	%	Site conditions
Gentle slope zone	0–0.5	Flat land	0.48	0.86	The area features gentle slopes with extensive exposed surfaces, providing favourable conditions for plant growth and exhibiting moderate erosion intensity
	0.5–2	Gentle slope	5.5	9.83	
	2–5	Moderate slope	18.4	32.87	
	5–15	Steep slope	24.44	43.65	
Steep slope zone	15–35	Very steep slope	6.62	11.83	The steep gradient of this area renders the soil susceptible to erosion, adversely affecting plant growth. Where necessary, engineering measures must be implemented to stabilise the slopes and retain the soil. The lower sections of the steep slopes are becoming exposed, indicating severe erosion
Collapse zone	35–55	Perpendicular wall	0.52	0.92	This area features steep gradients where reservoir banks are susceptible to scouring and erosion, hindering vegetation growth; bedrock is exposed, resulting in severe erosion intensity
	55–90		0.02	0.04	

Table 2. Types of water level fluctuation zones in Longyangxia Reservoir

Type of WLFZ	Surface conditions	Classification	Distribution
Easily restorable WLFZ	This type of WLFZ is primarily composed of loose material deposits, with the surface characterised by tidal flats and sediment accumulation	Sediment-accumulating gentle slope type of WLFZ; Wind-sand accumulation gentle slope type of WLFZ	Gentle slope area; Sand dunes on the northern shore of the reservoir
Difficult-to-restore WLFZ	This area exhibits a high gravel content, forming exposed gravel beds with minimal soil content, tending towards barrenness and poor soil fertility	Gravel-accumulation type WLFZ; Steep slope-type WLFZ	Steep slope zone; Collapse Zone.
River mouth WLFZ	The river mouth is susceptible to sedimentation and erosion from tributaries. The complex terrain and sparse vegetation in this area make pollution reduction difficult and result in landscape degradation	The area where the river enters the reservoir	Shagou River

3.2 Restoration of Different Types of Water Level Fluctuation Zone

Easily Restorable Water level fluctuation zone. Easily restorable WLFZs are characterized by gentle slopes formed through the accumulation of soft materials. They primarily comprise two subtypes: sediment-deposited zones and gently sloping zones formed by wind-blown sand accumulation. The sediment-accumulation subtype is most extensively distributed along the southern shore of the reservoir. This subtype manifests as a sediment layer overlying the soil, often adjacent to agricultural land or slopes, and features a gentle topographic gradient with an overall mildly sloping terrain. During the emergent period of the WLFZ, numerous annual herbaceous species thrive, exhibiting rapid growth rates and substantial biomass accumulation. Plant families such as Fabaceae, Poaceae, and Amaranthaceae are well-represented. For instance, the "white thorn beach" area within the reservoir's WLFZ (Fig. 1c) is a peninsula, surrounded by water on three sides. The upper section of this WLFZ features artificial ecosystems, including dryland cultivated vegetation and commercial forests. These planted forests serve as a crucial ecological buffer zone, despite being subject to intense human activity and agricultural practices. Local villagers frequently utilize this area, reflecting strong demands for agricultural development and water use for domestic and agricultural purposes. This demonstrates that both water-level fluctuations and human activities significantly alter the ecosystems within the reservoir's WLFZ. The gently sloping sand-accumulation type of the WLFZ is primarily found in the sand dune areas along the northern shore of LYXR. Sand burial restricts plant germination and growth, leading to a significant reduction in the germination rate of vegetation within this zone. Surveys reveal that this type of WLFZ harbors few plant species, with dominant species being annual herbaceous plants such as *Agriophyllum pungens (Vahl) Link.*, *Salsola collina Pall.*, and *Persicaria lapathifolia (L.) Delarbre.*

Plant selection. Given the formation of the WLFZ and the short growing season in the high frigid zone, the selection of suitable plants for WLFZ restoration should adhere to the following principles: (1) Suitability of plant species; (2) Primarily native plants; (3) When selecting vegetation for WLFZ in high frigid zone reservoirs, consider both flood tolerance and drought resistance, while also prioritizing cold hardiness characteristics. Given this, certain plant species discovered during field surveys of the LYXR WLFZ resemble those listed in the Appendix of Suitable Plant Species for Desertified Lands in High Frigid Zones (Table 3). These species may be considered for ecological restoration in WLFZ of reservoirs located in the high frigid zone.

Table 3. Suitable plant species for sandy land in alpine zone

Plant type	Name
Shrub	*Hippophae rhamnoides, Salix oritrepha, Salix psammophila, Dasiphora fruticose, Salix linearistipularis, Lonicera japonica, Hippophae tibetana, Myricaria squamosa, Artemisia stelleriana, Caragana korshinskii, Salix cheilophila*
Herbaceous	*Poa alpigena, Avena sativa, Elymus nutans, Festuca rubra, Leymus secalinus, Carex brunnescens, Carex praeclara, Kobresia robusta, Agropyron desertorum*

Quoted from: Technical Standards for Comprehensive Management of Desertified Land in High Frigid Zone LY/T 2997–2018

Based on the existing plant species in the WLFZ of LYXR, restoration efforts should prioritize Fabaceae, Poaceae, and Amaranthaceae. The nitrogen-fixing ability unique to Fabaceae plants not only supports their own growth, reducing reliance on chemical fertilizers, but also releases nitrogen into the soil through root exudates and the decomposition of plant debris, thereby enhancing soil fertility. Common Fabaceae species in the WLFZ of LYXR include *Medicago lupulina L.*, *Melilotus officinalis (L.) Lam.*, *Sophora alopecuroides L.*, *Vicia cracca L.*, and *Oxytropis ochrantha Turcz.* Poaceae plants such as *Setaria viridis (L.) P. Beauv.*, *Phragmites australis (Cav.) Trin. ex Steud.*, *Calamagrostis pseudophragmites (Haller f.) Koeler*, *Leymus secalinus (Georgi) Tzvelev*, *annual Poa annua L.*, *Avena sativa L.*, *Elymus dahuricus Turcz.*, *Leymus secalinus (Georgi) Tzvelev*, and *Agropyron desertorum (Fisch.) Schult.* Plants of the Amaranthaceae family are predominantly mesophytic species that fill ecological niches between Poaceae and

Fabaceae plants, forming multi-layered vegetation structures. Examples include dominant species in the WLFZ such as *Salsola collina Pall.* Herbaceous plants like *Caragana korshinskii* and *Artemisia ordosica Krasch.*, found in the upper, non-flooded zones of the WLFZ, are also common shrub species. These can be trial-planted in the upper zones to mitigate wave impact.

Species allocation. Considering that even amphibious plants struggle to survive in fully submerged environments, implementing spatial vegetation configurations across different water level zones within easily restored WLFZ can enhance plant survival rates.[16] In lightly flooded zones with lower inundation stress, consider three-dimensional planting of trees, shrubs, and grasses with strong flood tolerance, well-developed root systems, and high soil stabilization capacity. Simultaneously, for sediment-accumulating gentle slopes with complex habitat conditions and low inundation stress, implement shrub-grass mixed planting to enhance sediment trapping, pollution interception, and water quality protection benefits.

Difficult to restore WLFZ. The surface layer of the difficult to restore WLFZ consists primarily of gravel deposits with steep slopes. This area requires a combined approach of engineering protection and vegetated slope stabilization to establish a three-dimensional ecological restoration space. Measures such as slope clearing, manual compaction, excavation of planting pits, and construction of plant barriers should be implemented synergistically to restore the damaged ecosystem (Fig. 2).

Among these, gravel-accumulation-type WLFZ are commonly found in the upstream reservoir WLFZ area. Under intense, high-frequency periodic scouring, severe soil erosion occurs, forming extensive exposed gravel bars with minimal soil content, poor fertility, and extremely low vegetation survival rates and coverage. Therefore, planting vegetation in gravel bar areas to enhance coverage and restore ecological functions presents significant challenges. Steep slope and landslide-prone zones typically occur near the dam with extremely steep gradients. Due to the steep reservoir banks, significant topographic variation, narrow width, and thin terrestrial soil layers, these areas are highly susceptible to geological hazards like landslides and debris flows, resulting in fragile ecosystems. Furthermore, in cliff-type WLFZ, the water body directly abuts the retaining wall without a buffer zone. Special attention must be given to reinforcing the retaining wall to prevent collapse.

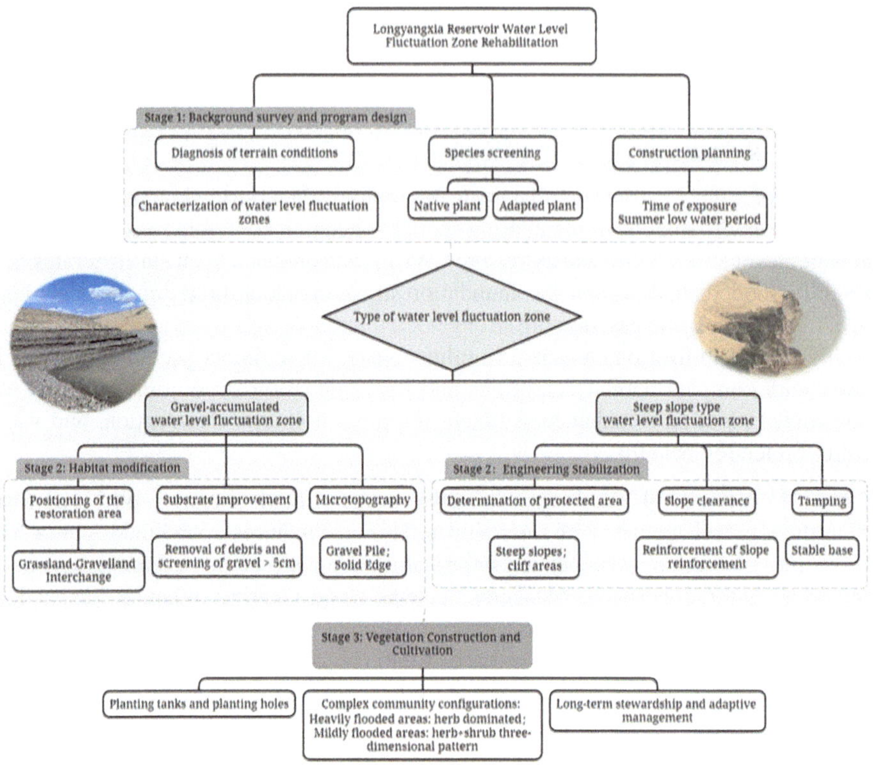

Fig. 2. Schematic diagram of ecological construction in the water level fluctuation zone

Tributary water level fluctuation zone. The Shagou River is a significant tributary of LYXR. Beyond sharing common issues with the main channel, the WLFZ at the tributary's reservoir inlet exhibits predominantly open and flat topography due to the river's scouring and deposition processes. During the dry season, persistent river scouring forms a narrow, deep main channel within the WLFZ. Water flow concentrates within this channel, while low groundwater levels in the WLFZ result in poor vegetation conditions. Consequently, the tributary WLFZ fails to intercept or degrade pollutants entering the reservoir, causing significant damage to the overall landscape. Building upon existing research findings, installing multi-tiered weirs at appropriate locations within WLFZ—with crest elevations matching the overall floodplain level at each section—can intercept surface runoff and groundwater seepage, thereby enhancing regional vegetation growth. By altering local dam configurations to create topographical diversity, regional biodiversity can be enhanced. This approach simultaneously improves ecological conditions for vegetation and microorganisms while increasing the hydraulic retention time of tributary water. Consequently, it strengthens the purification capacity of the WLFZ for tributary water, holding significant importance for ecological restoration in river-to-reservoir WLFZ and for safeguarding and purifying reservoir water quality. This approach represents a currently underutilized engineering technique that addresses the ecological challenges and constraints inherent in tributary WLFZ. It effectively balances dual objectives of ecological restoration and water purification, offering a long operational lifespan with proven safety and efficacy. Further research should prioritize its deeper understanding and broader adoption.

4 Conclusion and Outlook

4.1 Conclusion

Through field investigation and analysis, this study preliminarily categorizes the WLFZ of LYXR into three types: easily recoverable WLFZ, difficult-to-restore WLFZ, and river-inflow WLFZ. Specifically, easily recoverable WLFZ comprise sediment-accumulation gentle-slope types and wind-sand-accumulation gentle-slope types; difficult-to-restore WLFZ consist of gravel-accumulation types and steep-slope types. In the ecological restoration of easily recoverable WLFZ, species selection should prioritize existing Fabaceae, Poaceae, and Amaranthaceae. The primary plant species include annual herbs such as *Agriophyllum pungens*, *Salsola collina*, and *Setaria viridis*, which thrives on sand dunes or shifting sands. This is complemented by a "herbaceous + shrub" model, featuring a multi-layered arrangement of native plants with strong flood tolerance and well-developed root systems. Meanwhile, for the difficult-to-restore WLFZ characterized by gravel accumulation on the surface and steep slopes, a multi-pronged approach centered on protection is required. This involves integrating engineering measures with biological measures through a synergistic model, employing techniques such as slope clearing, retaining wall stabilization, artificial compaction, and excavation of planting pits to achieve vegetation restoration and regional conservation. In addition to sharing common issues with the main channel's WLFZ, the WLFZ at the river's confluence with the reservoir exhibits a more complex ecological environment. Measures such as installing multi-level weirs to promote vegetation growth and altering dam types to enhance biodiversity and water purification capacity could be considered. Restoration strategies require further research in the future.

4.2 Prospect

While existing ecological restoration techniques for WLFZ in China can improve vegetation conditions, they fail to fundamentally optimize plant growth conditions. Combining near-natural restoration techniques with engineering measures offers greater sustainability. Due to the significant heterogeneity of site conditions in the study area's high-altitude cold regions and WLFZ, management efforts must prioritize the alignment between WLFZ typology and vegetation configuration. This includes integrating measures such as combining grass grid sand barriers for sand fixation with vegetation restoration, and pairing earthen retaining walls for geological hazard prevention with bank protection works. Planting drought-tolerant shrubs and herbaceous plants such as *Salix cheilophila*, *Caragana korshinskii*, and *Artemisia ordosica* can enhance vegetation coverage and improve sand fixation.

To address the issue of windblown sand entering the upper and middle reaches of LYXR, grass-grid sand barriers can be implemented in the upper part of the WLFZ, supplemented by planting drought-tolerant shrubs and herbaceous plants suited to the local environment. For potential geological hazards along the reservoir shoreline, retaining walls can be constructed at critical points. Concurrently, adhering to the principle of "the right tree in the right place," native tree species with strong adaptability should be selected and planted based on site-specific factors. These approaches warrant further exploration and application in subsequent research.

References

1. Diao, C.T., Huang, J.H.: A preliminary study on land resources of the water-level-fluctuating zone in the three-Gorces reservior. Resour. Environ. Yangtze Basin. (01), 75–80 (1999)
2. Battaglia, L.L., Collins, B.S.: Linking Hydroperiod and vegetation response in Carolina Bay wetlands. Plant Ecol. **184**(1), 173–185 (2006)
3. Yuan, X.Z., Zhang, Y.W., Liu, H., et al.: The littoral zone in the three gorges reservoir, China: challenges and opportunities. Environ. Sci. Pollut. Res. Int. **20**(10), 7092–7102 (2013)
4. Lin, S.B., Fu, Q.F., Cheng, T.F., et al.: Study of the ecological slope along the water fluctuation belt in Songzikeng reservoir in Shenzhen. Pearl River. (03), 20–21+38 (2007)
5. Diamond, J.S., Epstein, J.M., Cohen, M.J., et al.: A little relief: ecological functions and autogenesis of wetland microtopography. Wiley Interdiscip. Rev. Water. **8**(1), e1493 (2021)
6. Bruland, G.L., Richardson, C.J.: Hydrologic, edaphic, and vegetative responses to microtopographic reestablishment in a restored wetland. Restor. Ecol. **13**(3), 515–523 (2005)
7. Singh, R., Tiwari, A., Singh, G.: Managing riparian zones for river health improvement: an integrated approach. Landsc. Ecol. Eng. **17**(2), 195–223 (2021)
8. Xie, Y.Y., Wang, Y.M., Huang, Q.: Year-end water level control and water discharge in flood season for Longyangxia reservoir. J. Northwest & F Univ. (Nat. Sci. Ed.). **42**(01), 223–227+234 (2014)
9. Hou, S.Z., Wang, P., Guo, Y., et al.: Analysis of water and sediment regulation of Longyangxia reservoir. Journal of hydroelectric Engineering. **32**(06), 151–156 (2013)
10. Shang, K.Z., Yang, D.B., Wang, S.G., et al.: Influfnce of water conservancy and power projects on local climate at upper reache of the yellow river. Arid. Land Geogr. **01**, 57–64 (1997)
11. Yu, Q., Li, Q., Fang, H.Y., et al.: Evaluating the ecological sensitivity of water level fluctuation zone in Longyangxia reservoir using coefficient of variation method. Bull. Soil Water Conserv., 1–10 (2025)
12. Qi, D.H., He, L., Zhou, X., et al.: Species composition and species diversity of Plant Community in the Water-lever-fluctuation Zone of three gorges reservoir. Acta Agrestia Sin. **22**(05), 966–970 (2014)
13. Zhou, W.C.: Impact of land nature slope and sea level elevation on the economic development in the three Gorges' area. Resour. Environ. Yangtze Basin. (01), 15–21 (2001)
14. Dang, C.X., Guo, R., Song, W.N., et al.: Study on the ecological restoration Technology of Soil and Wter in the drawdown zone of reservoir in Shenzhen. Resrarch Soil Water Conserv. **22**(01), 324–328 (2015)
15. Zierholz, C., Prosser, I., Fogarty, P., et al.: In-stream wetlands and their significance for channel filling and the catchment sediment budget, Jugiong Creek, New South Wales. Geomorphology. **38**(3–4), 221–235 (2001)
16. Rao, J., Tang, Q., Feng, Y., et al.: Habitat condition and vegetation restoration patterns in the water level fluctuation zone of the three gorges reservoir. J. Soil Water Conserv. **38**(01), 310–318 (2024)

Research on Carbon Accounting for a Water Supply Plant: A Case Study of the S Plant in Urumqi

Wanying Xie[1(✉)], Xiao Yang[1], Jinhui Li[1], Hongxin Zhao[1], Xinran Li[1], Shangye Yang[1], and Xiangyu Cheng[2]

[1] School of Civil Engineering and Architecture, Xinjiang University, Urumqi 830017, China
xwy@xju.edu.cn

[2] Urumqi Water Industry Investment and Development Group Co., Ltd., Shidunzishan Water Plant, Urumqi 830017, China

Abstract. In response to the "dual carbon" goals, this study focuses on carbon emissions during the operational phase of the S Water Supply Plant in Urumqi. The plant is equipped with two treatment lines. Line 1 adopts the process route of raw water processed by pretreatment and high-density clarifier, then filtration and disinfection. The Line 2 replaces the high-density clarification process with conventional coagulation-sedimentation, with all other treatment units remaining identical. Using the emission factor method, this study accounted for the plant's carbon emissions from 2024 to 2025, and systematically quantified indirect emissions (from electricity consumption) and extended responsibility carbon emissions (from chemical consumption and transportation). Results showed that Line 1 not only had higher production capacity but also exhibited a significantly lower carbon emission intensity (0.0598 $kgCO_2$-eq/m^3), when compared to Line 2 (0.0813 $kgCO_2$-eq/m^3). This is primarily attributed to its higher equipment efficiency and integrated control technology, which minimizes energy waste. To address the high carbon emissions of Line 2, specific measures are proposed, including optimizing high-energy-consuming equipment, implementing intelligent upgrades, and constructing energy recovery devices. Water supply plants can achieve both carbon reduction and efficiency improvement through process optimization and technological innovation, providing empirical support for the green and low-carbon development of the water industry.

Keywords: Water Supply Plant · Carbon Accounting · Carbon Emissions · Carbon Emission Intensity · Carbon Reduction Measures

1 Introduction

Greenhouse gas emissions have led to increasingly severe global climate issues, making the reduction of such emissions a common goal of worldwide concern [1]. Carbon accounting has consequently become a mandatory task for various industries [2, 3], driven by multiple macro-trends that relate not only to national strategic transformation but also to international industrial competitiveness [4]. In 2015, the Fifth Plenary

Y. Zhu (Ed.): ISESIE 2025, LNCE 000, pp. 195–210, 2026.
https://doi.org/10.1007/978-981-92-2389-3_17

Session of the 18th Central Committee of the Communist Party of China first proposed the dual control of energy consumption mechanism, aiming to enhance China's energy efficiency and achieve high-quality development [5]. In 2023, the Central Commission for Comprehensively Deepening Reform adopted the Opinions on Promoting the Transition from Dual Control of Energy Consumption to Dual Control of Carbon Emissions, explicitly advocating for this strategic shift [6]. In 2024, the State Council issued the Work Plan for Accelerating the Construction of a Dual Carbon Control System (State Council Document [2024] No. 39) [7]. The essence of the dual carbon control policy is to promote a comprehensive green and low-carbon socioeconomic transition in China by constraining both the total amount and intensity of carbon emissions. Carbon accounting serves as the fundamental prerequisite for transitioning from "dual control of energy consumption" to "dual control of carbon emissions".

China's Dual Carbon Control policy is closely related to the water treatment industry, which has been incorporated into the nationwide whole-chain carbon management system. Newly constructed or expanded water treatment projects are required to conduct carbon emission assessments, and large water utilities are designated as key energy-consuming and carbon-emitting units. These entities must regularly report carbon emission data and participate in carbon quota allocation. Conducting carbon accounting in the water supply industry is not only a mandatory requirement under China's Dual Carbon Control policy but also a core tool for the industry to reduce costs, improve efficiency, and adapt to the green and low-carbon transition [8]. Carbon accounting helps identify carbon emission sources in processes such as water transmission and distribution [9], chemical dosing in water treatment [10], and electricity consumption [11], thereby accurately locating hidden carbon cost hotspots. It provides guidance and a basis for formulating carbon reduction measures, achieving cost savings and efficiency gains, and facilitating the low-carbon transformation of the water treatment industry.

In response to the Dual Carbon goals, this study takes the S Water Supply Plant in Urumqi as its research object. Based on operational data from the plant at 2024, the emission factor approach was applied to quantify carbon emission indicators from two different process lines. The study determined the plant's water treatment capacity and carbon accounting system boundaries, assessed the annual carbon emissions and carbon emission intensity of the two process lines, and conducted a comparative analysis of indirect carbon emissions and scope 3 carbon emissions. Furthermore, targeted carbon reduction measures were proposed to address the identified emission issues in the S Plant.

2 Methodology

2.1 System Boundary for Carbon Accounting in the Water Supply Plant

Operational Overview of the Water Supply Plant. The S Water Supply Plant in Urumqi utilizes two parallel drinking water treatment process lines (Fig. 1), with a total designed capacity of 240,000 m^3/d. Process Line 1 has a treatment capacity of 170,000 m^3/d, while Process Line 2 handles 70,000 m^3/d. Both lines include core units for coagulation, sedimentation, filtration, and disinfection, but differ in specific structure selection and process details.

Process Line 1 is summarized as follows: (1) Pretreatment. Raw water first passes through an automatic bar screen (medium-sized) to remove large impurities, while sodium hypochlorite (NaClO) is dosed for pre-disinfection.(2) High-Density Clarification. The water flows through an underground tubular static mixer, where coagulant (polyaluminum chloride, PAC) and coagulant aid (polyacrylamide, PAM) are added for thorough mixing. The water then enters a plate-type flocculation tank for flocculation reaction to form flocs, it flows into multiple inclined tube settlers for high-speed solid-liquid separation subsequently. (3) Filtration. The clarified water enters a V-type filter for deep filtration. Backwashing of filter medias combines water washing (supplied by water pumps) and air scouring (supplied by air compressors and air dryers). (4) Disinfection and Clear Water Storage. NaClO is added to the filtered water for final disinfection to ensure water safety, after which it is conveyed to clear water tanks for storage.

Process Line 2 is outlined as follows. (1) Integrated Pretreatment and Coagulation-Sedimentation. Raw water similarly undergoes pre-disinfection (NaClO) and PAC dosing via an underground tubular static mixer. The difference lies in the flocculation and sedimentation processes, which are completed within a mechanically agitated clarifier. This tank integrates mechanical mixing (to promote flocculation) and sludge scraping (for sludge discharge) functions, achieving flocculation reaction and mud-water separation. (2) Filtration. The effluent from the clarifier enters a backwashing Filter for filtration. Its backwashing method is similar to Line 1, also employing combined water and air scouring. (3) Disinfection and Clear Water Storage. The filtered water is similarly disinfected with NaClO before entering clear water tanks for storage.

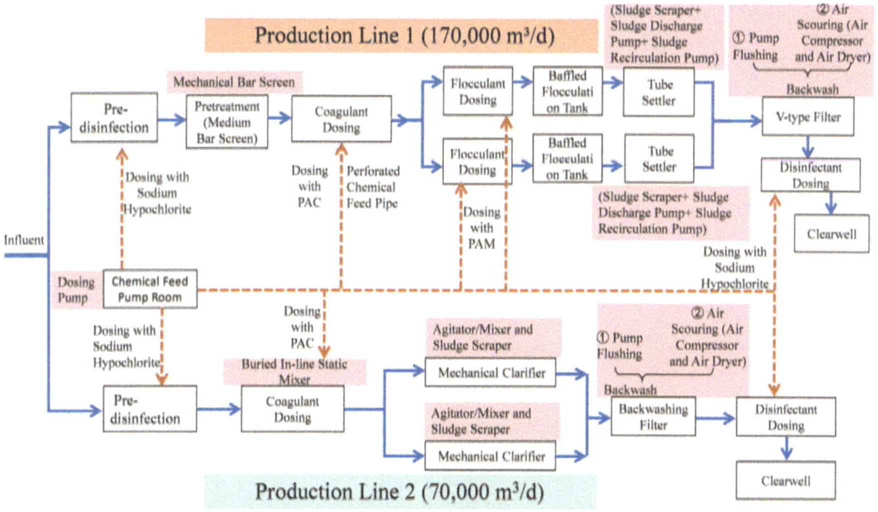

Fig. 1. Schematic diagram of the two process lines at the S water supply plant.

System Boundary for Carbon Accounting. Water supply plants serve as the fundamental operational units and the smallest entities for conducting carbon accounting within municipal water supply systems. The lifecycle of a water plant is divided into three stages: design and construction, operation, and asset renewal/decommissioning [12]. The design and construction phase is relatively short-term, and its associated carbon emissions are comparatively low. In contrast, the operational phase typically spans several decades. This study focuses on quantifying the annual carbon emission intensity during the operational phase of the water plant.

The operational boundary for carbon accounting is categorized into three scopes: Scope 1, Scope 2, and Scope 3 [13]. Scope 1 encompasses direct emissions, primarily from fossil fuel combustion and chemical reactions in certain processes. Scope 2 refers to indirect emissions from purchased energy, such as electricity, heat, and steam consumed during operations. Scope 3 includes other indirect emissions, such as those associated with chemical usage and energy transportation. Since neither process line in this plant utilizes fossil fuels or facilities such as boiler rooms during operation, the direct carbon emissions (Scope 1) are considered negligible and are excluded from the accounting boundary. Therefore, this study defines the system boundary to include Scope 2 (indirect emissions from purchased electricity) and Scope 3 (extended responsibility emissions from chemical consumption and transportation). The accounting methodology follows the Greenhouse Gas Protocol Corporate Accounting and Reporting Standard [13].

Identification of Carbon Emission Sources in the Water Supply Plant. Based on the operational overview of the two process lines at the S Plant, neither process involves fossil fuel combustion. Therefore, this carbon accounting study only includes indirect carbon emissions from electricity consumption and scope 3 emissions from chemical usage and transportation [14].

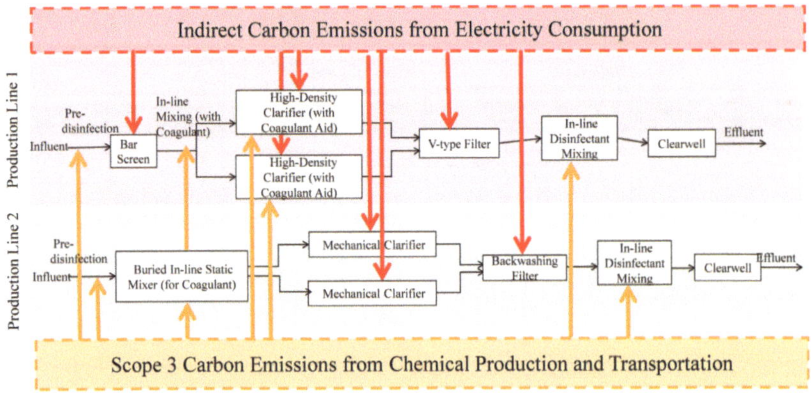

Fig. 2. Schematic diagram of carbon emission sources identification at the S water supply plant.

The carbon emission sources of the two lines at Plant S are illustrated in Fig. 2. For Line 1, the pretreatment bar screen, the flocculation-clarification units within the High-Density Clarifier, and the filtration units in the V-type filter are all equipped with numerous high-energy-consuming devices, and these are the primary sources of electricity-related carbon emissions. For Line 2, the mechanically agitated clarifier, backwashing filter, and chemical dosing room are the direct sources of carbon emissions from electricity consumption.

In both process lines, the use of water treatment chemicals such as NaClO, PAC, and PAM contributes to scope 3 emissions from chemical consumption. Additionally, the transportation of chemicals used in both process lines represents another source of scope 3 emissions.

2.2 Carbon Accounting Methodology

Emission Factor Approach. For this carbon emission accounting of the water treatment plant, the Emission Factor Method (EFM) was selected. Its fundamental operational framework involves multiplying data on human activity levels by a coefficient that quantifies the emissions produced per unit of activity [15] (known as the emission factor [16]). It is a standardized methodology for estimating emissions based on the multiplication of activity data by emission factors. Compared to other carbon accounting methods, the Emission Factor Method offers advantages of simplicity, ease of implementation, readily available data, and standardization.

Carbon Emission Accounting. From the perspective of carbon emission equivalents, no fossil fuels are used in any process units of this water treatment plant. Therefore, the direct carbon emissions from the plant are zero in this accounting. The carbon emission accounting herein includes the following two types: (1) Indirect carbon emissions, which originate from the consumption of externally supplied electricity by equipment operation across various process units. (2) Extended responsibility carbon emissions, which encompass downstream emissions from the product lifecycle. This specifically includes emissions generated during the dosing and application of chemicals and their transportation process.

Accounting for Indirect Carbon Emissions. All indirect emission activities in this plant are associated with electricity consumption, e.g., CO_2 is produced by the operation of equipment such as pumps and motors. The carbon emission intensity (CES_d) resulting from electricity consumption for operating equipment is calculated using Eq. (1) [17]:

$$CES_d = \frac{(E_d \times EF_d)}{Q} \tag{1}$$

where, E_d represents the total annual electricity consumption for the operation and maintenance of the water supply plant facilities (kW·h/yr); EF_d denotes the regional electricity emission factor ($kgCO_2$-eq/kW·h); Q is the total annual treated water volume of the process line being assessed (m^3/yr).

Accounting for Extended Responsibility Carbon Emissions. The extended responsibility carbon emissions of this water plant specifically include two types. Firstly, emissions generated from the use of chemicals such as coagulants and disinfectants; and secondly, emissions resulting from the external transportation of these chemicals during operation. Carbon emission intensity (CES_{cl}) due to chemical consumption is calculated using Eq. (2) [17], and the carbon emission intensity (CES_{sys}) due to chemical transportation is calculated using Eq. (3) [17]:

$$CES_{cl} = \sum_{i=1}^{n} \frac{\left(M_{cl,i} \times EF_{cl,i}\right)}{Q} \quad (2)$$

where, $M_{cl,\,i}$ represents the total annual consumption of the i-th chemical or material (kg/yr); $EF_{cl,\,i}$ is the emission factor of the i-th chemical or material ($kgCO_2$-eq/kg); n denotes the total number of chemical types used; Q refers to the total annual treated water volume of the process line being assessed (m^3/yr).

$$CES_{sys} = \sum_{i=1,j=1}^{n,l} \frac{\left(M_{ys,i,j} \times L_{ys,i,j} \times EF_{ys,j}\right)}{Q} \quad (3)$$

where, $M_{ys,\,i,\,j}$ represents the total quantity of materials transported by the j-th mode during the i-th shipment in the accounting year (t/yr); $L_{ys,\,i,\,j}$ is the transportation distance of the j-th mode during the i-th shipment in the accounting year (km); $EF_{ys,\,j}$ denotes the emission factor of the transportation vehicle used by the water plant ($kgCO_2$-eq/(t·km)); n indicates the total number of shipments conducted during the accounting year; Q refers to the total annual treated water volume of the target process line (m^3/yr).

2.3 Activity Data and Emission Factor

The activity data for this carbon emission accounting were obtained from the nameplate information of equipment in the water plant and relevant reliable literature. The selected carbon emission factors for various categories are detailed in Table 1.

The carbon emission factor for electricity consumption in the water plant was adopted from the 2023 National Electricity Carbon Emission Factor published by the Ministry of Ecology and Environment of China in 2025 [18]. The carbon emission factors for various chemicals were determined based on the Technical Guide for Carbon Accounting and Emission Reduction Pathways in Urban Water Systems [19]. The emission factor for the heavy-duty diesel truck (18 t) used in transportation was sourced from the Carbon Emission Factors for Various Modes of Building Materials Transportation [20].

3 Carbon Accounting for a Water Supply Plant

3.1 Activity Data Collection for Emission Sources.

For this carbon accounting study, a systematic collection of activity data related to carbon emissions from major equipment operation and chemical usage in the water treatment process was conducted. Regarding equipment operation, as shown in Tables 2 and 3, the indirect carbon emissions resulting from electricity consumption by the two process lines cover the energy consumption of key equipment such as sludge discharge pumps, electric motors, and mixers. Simultaneously, NaClO, PAC, and PAM were selected as the main water treatment chemicals. Data on indirect carbon emissions generated by the transportation of these chemicals are detailed in Table 4. In terms of filter media

Table 1. Selection of carbon emission factors for the case study water plant.

Emission source	Emission factor	Data source
Electricity	0.6205 ($kgCO_2$-eq/kW·h)	2023 National Grid Average Emission Factor (Ministry of Ecology and Environment of China) [18]
PAC	0.53 ($kgCO_2$-eq/kg)	Technical Guide for Carbon Accounting and Emission Reduction Pathways in Urban Water Systems [19]
NaClO	0.99 ($kgCO_2$-eq/kg)	Technical Guide for Carbon Accounting and Emission Reduction Pathways in Urban Water Systems [19]
PAM	1.48 ($kgCO_2$-eq/kg)	Technical Guide for Carbon Accounting and Emission Reduction Pathways in Urban Water Systems [19]
Quartz Sand	0.0025 ($kgCO_2$-eq/kg)	Unit energy consumption data provided by the process technology manufacturer
Heavy-duty diesel truck (18 t payload)	0.1211 ($kgCO_2$eq/t·km)	Carbon Emission Factors for Various Modes of Building Materials Transportation [20]

selection, both process lines use quartz sand with a particle size range of 1.35–1.37 mm as the filtration medium. For the transportation of materials, heavy-duty diesel trucks were used as the transport vehicles, and the carbon emissions resulting from diesel combustion have also been included within the scope of this accounting.

Table 2. Activity data of indirect carbon emissions from electricity use in process line 1.

Process stage	Equipment	Function	Quantity	Power (kW)
Bar screen	Sludge discharge pump	To evacuate sludge generated from the tube settler	2	22
	Circulation pump	To maintain a high-density "sludge blanket," enhance "contact flocculation," and improve sedimentation efficiency	4	22

(*continued*)

Table 2. (*continued*)

Process stage	Equipment	Function	Quantity	Power (kW)
	Mechanical bar screen	To remove screenings collected by the medium bar screen	1	2.2
	Three-phase asynchronous motor	Serves as the prime mover (or drive unit) for the mechanical bar screen	1	2.2
High-density sedimentation (HDS) basin	Three-phase asynchronous motor	Acts as the central power source for all moving components within the high-density clarifier	1	5.5
	SEW gear reducer	Functions as the critical link (interface) and speed governor between the three-phase asynchronous motor and the working mechanism	1	0.087
	Gear reducer	Serves as an intermediary for speed reduction and torque increase	1	5.5
	Agitator	To achieve rapid mixing and facilitate flocculation	1	5.5
	Electric motor	Serves as the power source for the entire system	1	7.5
	Agitator	To accomplish mixing and flocculation	1	7.5
	Sludge scraper	To collect and concentrate sludge	1	0.75
V-type filter	Electric motor	Provides the primary motive power for the blower	3	55

(*continued*)

Table 2. (*continued*)

Process stage	Equipment	Function	Quantity	Power (kW)
	Blower	Supplies the air source for air scouring, which is the initial step and the core driving force in the backwashing process	3	45.4
	Air compressor	Generates the instrument air source and the power gas for pneumatic actuators	2	15
	Air dryer	To remove moisture and ensure the dryness of the compressed air	1	1
	Backwash water centrifugal pump	Provides the water source and motive force for water backwashing	3	37
	Three-phase asynchronous motor	Delivers the primary power to the backwash water centrifugal pump	3	22
	In-line circulation pump	Employed for the recirculation of filtered water to stabilize and regulate hydraulic conditions	3	22
	Vertical multi-stage centrifugal pump	To supply seal water, lubrication, and auxiliary water for the backwash system	2	5.5

Table 3. Activity data of indirect carbon emissions from electricity use in process line 2.

Process stage	Equipment	Function	Quantity	Power (kW)
Mechanically agitated Clarifier	Variable-frequency drive (VFD) three-phase asynchronous motor	To drive the agitator and steplessly regulate its rotational speed via a Variable Frequency Drive (VFD)	1	7.5

(*continued*)

Table 3. (*continued*)

Process stage	Equipment	Function	Quantity	Power (kW)
	Sludge scraper	To continuously and uniformly scrape the sludge settled at the tank bottom towards the central sludge hopper for evacuation	1	7.5
	Wide-Frequency Three-Phase asynchronous motor	It is typically employed to drive the sludge scraper, providing stable and reliable motive power.	1	1.5
Backwashing filter	Three-Phase asynchronous motor	To serve as the prime mover for the double-suction centrifugal pump	2	37
	Double-Suction centrifugal pump	To function as the backwash water pump, delivering the high-intensity water flow required for backwashing	2	37
	Three-Phase Asynchronous motor	To drive the tri-lobe Roots blower, acting as its core power source	1	37
	Air compressor	To generate instrument-grade compressed air	1	7.5
	Tri-Lobe roots blower roots blower	To supply the air source necessary for air scouring	1	37
Chemical dosing room	Electric motor	To drive the metering pump, providing its primary power	3	1.5
	PAC dosing pump	To accurately meter and transport Polyaluminum Chloride (PAC) to the dosing points	3	0.75

(*continued*)

Table 3. (*continued*)

Process stage	Equipment	Function	Quantity	Power (kW)
	Chlorine dosing pump	To precisely and continuously meter and dose disinfectant into the water to be treated	3	1.5

Table 4. Activity data of indirect carbon emissions from chemical usage.

Process line	Chemical	Consumption (kg)	Water flow rate (m^3/d)	Transport distance (km)
Process line 1	PAC	186150	170000	20
	NaClO	496400	170000	40
	PAM	3102.5	170000	20
	Quartz Sand	196500	170000	30
Process line 2	PAC	102200	70000	20
	NaClO	229950	70000	40
	PAM	1788.5	70000	20
	Quartz Sand	47600	70000	30

3.2 Carbon Accounting Results

Carbon emissions refer to the total amount of CO_2 and equivalent greenhouse gases released directly or indirectly into the atmosphere due to human activities, serving as a core indicator for quantifying the human impact on climate. Carbon emission intensity, on the other hand, denotes the amount of CO_2 emissions produced per unit of economic output [21]. In this study, carbon emission intensity is used to evaluate the efficiency relationship between water production activities and carbon emissions at the water plant. The carbon emission indicators from the operation and maintenance of the two process lines were calculated, which presented in Figs. 3 and 4, where EPR denotes extended responsibility carbon emissions. According to the accounting results, the total annual carbon emissions of the water plant are 5,784,660 kgCO_2-eq, and the annual carbon emission intensity is 0.1411 kgCO_2-eq/m^3.

Among these, Process Line 1 accounts for 3,108,787 kgCO_2-eq, while Process Line 2 accounts for 2,075,873 kgCO_2-eq. As shown in Fig. 3a, in Process Line 1, carbon emissions from electricity consumption amount to 3,110,033 kgCO_2-eq, and extended responsibility carbon emissions total 598,752 kgCO_2-eq. Among the three stages in Process Line 1, the backwash process of the V-type filter involves the largest number of devices (20 units) and relatively high total power consumption. Consequently, this stage contributes the highest proportion of emissions, accounting for 82.04% of the carbon emissions from electricity consumption and 68.8% of the total carbon emissions of Process Line 1 (Fig. 3b).

As indicated in Fig. 4a, in Process Line 2, carbon emissions from electricity consumption reach 1,789,754 kgCO_2-eq, and extended responsibility carbon emissions amount to 286,117 kgCO_2-eq. Among its three stages, the reverse filtration process exhibits relatively high energy consumption. This process contributes the largest share of carbon emissions, representing 69.26% of the electricity-related emissions and 59.7% of the total emissions in Process Line 2 (Fig. 4b). The carbon emission intensity during the operation and maintenance of Water Plant S is illustrated in Figs. 3a and 4a. The carbon emission intensity of Process Line 1 is 0.0598 kgCO_2-eq/m^3, while that of Process Line 2 is 0.0813 kgCO_2-eq/m^3.

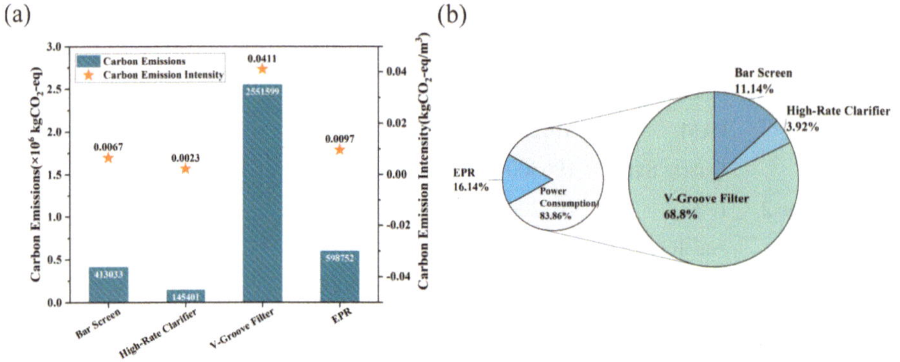

Fig. 3. Carbon emissions from process line 1 in plant S: (a) annual carbon emissions and carbon emission intensity, and (b) carbon emission proportion by process unit.

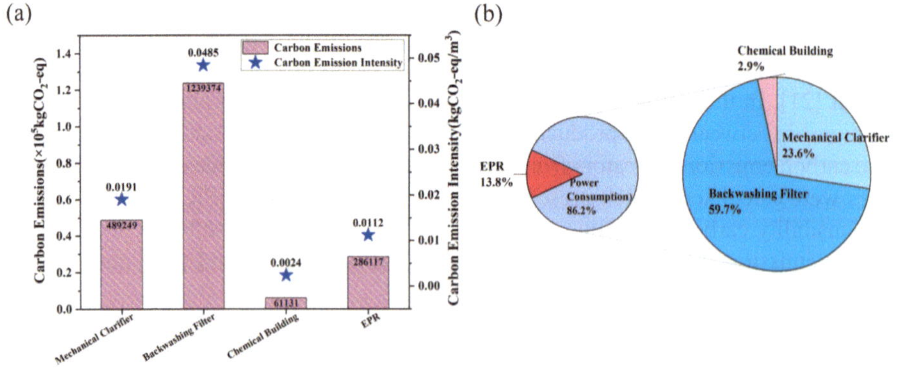

Fig. 4. Carbon emissions from process line 2 in plant S: (a) annual carbon emissions and carbon emission intensity, and (b) carbon emission proportion by process unit.

As shown in Fig. 5, the annual carbon emissions from the two process lines account for different proportions of the total carbon emissions of Plant S. The shares of Process Line 1 and Process Line 2 are 64.10% and 36.00%, respectively. The significantly higher proportion of emissions from Process Line 1 is primarily attributed to its larger water production capacity and correspondingly higher annual carbon emissions. Specifically, the carbon emissions from electricity consumption in Process Line 1 and Process Line 2 are the main contributors, accounting for 53.76% and 30.94% of the total plant emissions, respectively. This is followed by the extended responsibility carbon emissions of Process Line 1 and Process Line 2, which represent 10.35% and 4.95% of the total, respectively.

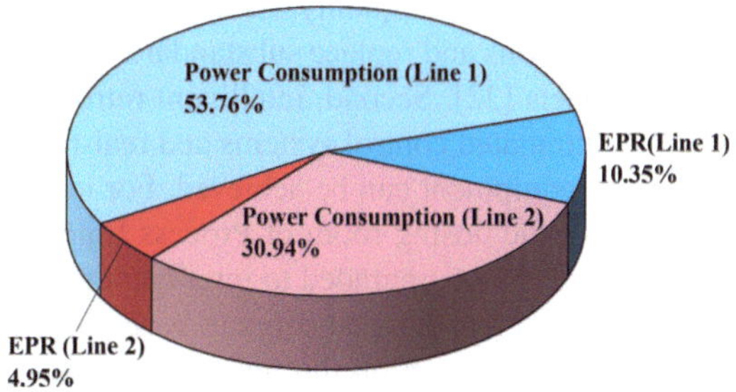

Fig. 5. Proportion of carbon emission accounting.

Compared to Process Line 2, Process Line 1 has a larger water production capacity and incorporates integrated control across multiple facilities. While ensuring water production efficiency, this effectively avoids unnecessary energy consumption caused by equipment idling or excessive functional capacity, thereby enhancing the overall operational efficiency of the process. Correspondingly, more equipment and electrical facilities are used in the water treatment process. For instance, there is a greater quantity of certain core equipment (e.g., blowers, circulation pumps, and centrifugal wash-water pumps) with higher configured power ratings. The higher operational load results in greater annual carbon emissions from Process Line 1. A unique advantage of Process Line 2 is the use of variable frequency drive (VFD) technology in the mechanical mixing clarifier, specifically, like variable frequency three-phase asynchronous motors. However, Process Line 2 has a smaller water production scale, and the power configuration of some facilities does not match the production volume. Moreover, in the subsequent reverse filtration process, the double-suction centrifugal pumps and three-phase asynchronous motors still operate at a fixed speed, leading to energy waste.

3.3 Considerations for Carbon Reduction Pathways

Analysis of the carbon emission accounting results indicates that electricity consumption serves as the primary source of carbon emissions in both process lines, contributing to as much as 84.7% of the plant's total emissions. Hence, optimizing electrical equipment and reducing the energy consumption of plant facilities represent critical pathways for achieving energy conservation and carbon reduction. Additionally, systematic employee training should be strengthened to standardize the operation of high energy-consuming electrical devices, enhance overall energy-saving awareness and operational management competence, and ensure the effective implementation and sustainable practice of energy-saving and emission-reduction measures.

In Process Line 1, equipment such as motors, blowers, and centrifugal wash-water pumps are configured with excessively high power. Process Line 2 also faces high energy consumption issues due to its double-suction centrifugal pumps and three-phase asynchronous motors. These devices are major contributors to the high carbon emissions. To reduce electricity consumption, improvements can be made in three aspects. First, outdated equipment should be upgraded. Some aging devices have shown significantly increased failure rates due to prolonged operation, which not only compromises power supply continuity but also reduces water quality assurance capability. The plant should conduct comprehensive evaluations and replace substandard equipment with new, high-efficiency energy-saving models [22]. Second, intelligent retrofitting should be implemented [23]. By installing automated control systems and real-time monitoring devices [24], optimized operation of equipment can be achieved. For instance, the PAC pumps and chlorine dosing pumps in the dosing room of Process Line 2 currently operate as fixed-rate metering pumps and can be upgraded to intelligent precision dosing systems that allow online monitoring and real-time adjustment based on water quality, thereby reducing non-essential energy consumption through autonomous regulation and optimized operation. Third, energy recovery devices should be installed [25]. Waste heat generated during the treatment process can be recovered and converted into reusable electricity through relevant technologies, further enhancing energy efficiency.

Employee management and training are vital components for ensuring efficient, stable, and safe operation of the water plant. This also plays a key role in effectively promoting energy conservation and consumption reduction initiatives. First, systematic training should be provided to operators on energy-efficient operational procedures and methods for critical equipment such as blowers, motors, and pumps [26], along with enhancing their understanding and compliance with environmental regulations and the plant's specific environmental requirements. Second, specialized training programs on new technologies and equipment should be introduced, such as big data analysis [27] and IoT technologies [28]. Meanwhile, an intelligent water plant management system [29] can be established in accordance with the Technical Guide for Smart Urban Water Management [30]. The adoption of these new technologies will help optimize equipment operation and reduce energy consumption, while also improving the level of automation, alleviating the operational burden on staff, and reducing human errors. Finally, a regular employee feedback mechanism should be established. By systematically collecting and analyzing operational opinions and suggestions, continuous improvement and optimization of plant performance can be achieved.

3.4 Conclusions and Prospects

The carbon emission accounting data indicate that during the operational period, the annual carbon emissions of Process Line 1 and Process Line 2 were 3,108,787 $kgCO_2$-eq and 2,075,873 $kgCO_2$-eq, respectively. However, the carbon emission intensity of Process Line 1 (0.0598 $kgCO_2\text{-eq/m}^3$) is significantly lower than that of Process Line 2 (0.0813 $kgCO_2\text{-eq/m}^3$). By comparison, for the same volume of water produced, Process

Line 2 generates more carbon emissions. A comparative analysis reveals that although Process Line 1 uses more water treatment equipment, it employs advanced integrated control technology with higher electrical energy conversion efficiency, resulting in lower carbon emission intensity. Process Line 2 has a smaller treatment capacity and accordingly uses fewer pieces of equipment. However, most of the equipment exhibits low energy conversion efficiency, and the power configuration of some devices does not match the water flow rate, leading to energy waste. It is recommended to replace unsuitable water treatment equipment in Plant S to optimize electricity consumption and align energy use with output demand without unnecessary waste. Furthermore, implementing intelligent upgrades and constructing energy recovery facilities is advised to leverage the plant's inherent advantages and move toward energy self-sufficiency.

References

1. Huang, J., Cui, J., Dong, F.: Impact of e-governments on carbon emission reduction targets: evidence from global 51 countries. Clean Techn. Environ. Policy. **26**, 2357–2377 (2024)
2. Geng, W., Liu, X., Liao, X.: Mechanism analysis of the influence of intelligent manufacturing on carbon emission intensity: evidence from cross country and industry. Environ. Dev. Sustain. **26**, 15777–15801 (2024)
3. Page, N., Gholami, A., Zhang, Q.: Demystifying the landscape of carbon quantification and reporting standards: a practical note for the financial sector. Environ. Res. Commun. **7**(5), 052003 (2025)
4. Ren, X., Li, W., Cheng, X., Zheng, X.: Economic freedom and corporate carbon emissions: international evidence. Bus. Strateg. Environ. **33**(8), 8388–8412 (2024)
5. Xia, L., Liu, Y.: Electricity generation and china's GDP growth under dual control policy. Energy Explor. Exploit. **42**(4), 1422–1431 (2024)
6. The State Council of the People's Republic of China: Opinions of the State Council on Comprehensively Promoting the Construction of a Beautiful China (State Council Document No. 4 (2024))
7. General Office of the State Council, People's Republic of China: Work plan for accelerating the construction of a dual-control system for carbon emissions (Guo Ban Fa No. XX). http://www.gov.cn/zhengce/content/20240111/content_6448346.htm (2024). Last accessed 02 Aug 2024
8. Lam, K.L., Liu, G., Motelica-Wagenaar, A.M., van der Hoek, J.P.: Toward carbon-neutral water systems: insights from global cities. Engineering. **14**, 77–85 (2022)
9. Dandy, G., Wu, W., Simpson, A., Leonard, M.: A review of sources of uncertainty in optimization objectives of water distribution systems. Water. **15**(1), 136 (2023)
10. Yateh, M. et al.: Understanding the influence of energy and chemical use on water treatment plants carbon emissions accounting. J. Water Process Eng. **69**, 106669 (2025)
11. Ang, B.W., Zhou, P., Tay, L.P.: Potential for reducing global carbon emissions from electricity production - a benchmarking analysis. Energy Policy. **39**(5), 2482–2489 (2011)
12. Li, C., Ma, S.J., He, X., Zhang, H.J., Xu, B., Tang, Y.L.: Carbon emission calculation and carbon emission reduction path of typical water works operation. Water & Wastewater Engineering. **49**(7), 1–7 (2023)
13. World Resources Institute (WRI) and World Business Council for Sustainable Development (WBCSD): The Greenhouse Gas Protocol: A Corporate Accounting and Reporting Standard (Revised Edition). World Resources Institute, Washington, DC (2015). https://ghgprotocol.org/corporate-standard. Last accessed 01 Jan 2015

14. Zhang, X.Y. et al.: Carbon emission characteristics and influencing factors of typical processes in drinking water treatment plant. Environ. Sci. **45**(1), 123–130 (2024)
15. Jiang, X., Wang, D., Yang, Q.: Carbon Emission Accounting Methodology. China Social Sciences Press, Beijing (2021)
16. Nikolaisen, M., Cornulier, T., Hillier, J., Smith, P., Albanito, F., Nayak, D.: Methane emissions from rice paddies globally: a quantitative statistical review of controlling variables and modelling of emission factors. J. Clean. Prod. **409**, 137245 (2023)
17. Cheng, Z.Q., Fang, H.M., Dong, X.L., Chen, Z.W., Peng, X.H.: Carbon emission accounting and carbon emission reduction strategies of urban water supply system operation. China Water Wastewater. **40**(14), 13–18 (2024)
18. Ministry of Ecology and Environment (MEE): 2023 National Electricity Carbon Footprint Factors (Announcement No. 3) (2025). https://www.mee.gov.cn/xxgk2018/xxgk/xxgk01/202501/t20250123_1101226.html. Last accessed 23 Jan 2025
19. Ministry of Housing and Urban-Rural Development, Science and Technology & Industrialization Development Center.: Technical Guide for Carbon Accounting and Emission Reduction Pathways of Urban Water Service Systems. China Architecture & Building Press, Beijing (2023)
20. Ministry of Housing and Urban-Rural Development of the People's Republic of China, State Administration for Market Regulation.: Standard for Building Carbon Emission Calculation (GB/T 51366–2019). China Architecture & Building Press, Beijing (2019)
21. Wang, J., Li, Y., Zhao, Z.: Analysis of Energy Carbon Emission Systems. Science Press, Beijing (202X)
22. Kwon, M.-J., Jung, B.-G., Choi, Y.: A study on energy diagnosis and efficiency improvement for public sewage treatment facilities. J. Korean Soc. Environ. Technol. **25**(3), 145–153 (2024)
23. Shen, Y. et al.: An artificial neural network-based data filling approach for smart operation of digital wastewater treatment plants. Environ. Res. **224**, 115549 (2023)
24. Seo, H.J., Shin, H.-J.: THC emission management at workplaces using real-time measurements. J. Odor Indoor Environ. **23**(4), 287–291 (2024)
25. Li, K.L. et al.: Balancing energy recovery and direct greenhouse gas emissions in wastewater treatment. Water Res. X. **28**, 100328 (2025)
26. Xie, W.Y. et al.: Study on carbon emission accounting and carbon reduction measures for a secondary Pumping Station in Jiashi County, Xinjiang. In: Proceedings of the 4th International Symposium in Environmental Science and Industrial Ecology (ISESIE 2024), pp. 93–111 (2025)
27. Müller-Czygan, G., Tarasyuk, V., Wagner, C., Wimmer, M.: How does digitization succeed in the municipal water sector? The WaterExe4.0 meta-study identifies barriers as well as success factors, and reveals expectations for the future. Energies. **14**(22), 7709 (2021)
28. Okoli, N.J., Kabaso, B.: Building a smart water city: IoT smart water technologies, applications, and future directions. Water. **16**(4), 5577 (2024)
29. Jian, D. et al.: Interpretation of guidelines for urban smart water. China Water Wastewater. **40**(2), 1–9 (2024)
30. Ministry of Housing and Urban-Rural Development (MOHURD) of the People's Republic of China.: Technical Guide for Smart Urban Water Management. China Architecture & Building Press, Beijing (2023).

Hydrogeochemical Classification and a Comprehensive Analysis for Water Quality Modelling Towards Drinking Water Assessment in Taladanda Canal, Cuttack City, Odisha

Abhijeet Das(✉)

Department of Civil Engineering, C.V. Raman Global University (CGU), Bhubaneswar, Odisha, India
das.abhijeetlaltu1999@gmail.com

Abstract. Surface water (SW) contamination is a significant challenge in Taladanda Canal, located near Jobra, Cuttack, Odisha, which demands a thorough assessment of water quality. For the analysis, 12 surface water samples, collected from five sampling locations between 2020 and 2025 are used, focusing mainly on pre-monsoon season. This research aims to develop a new WQI model by integrating objective weighting methods, such as entropy (E) and CRITIC (C), with weighted arithmetic (WA) WQI. The goal is to evaluate its predictive performance to facilitate effective local water resource management, aligning with the objectives of SDG 6 (Clean Water and Sanitation). Relying on physicochemical outputs, elevated BOD, F^-, and Cl^- were detected, indicating a significant risk of microbial and anthropogenic contamination. According to WAWQI observation, nearly 45% of the collected water samples were categorized under very poor–unsuitable drinking water quality. The entropy theory revealed that BOD, F^-, Cl^-, and Fe^{2+}, were the most influencing parameters. Again, the result of EWQI classification offers more realistic assessment, depicting a value of 77–210, signifying good to extremely poor water quality. The CWQI (30–254) revealed 33.33% as unsuitable water quality. In contrast, irrigation water quality was generally suitable, with classifications of C1-S2, C2-S3 and C2-S4 on the Wilcox diagram, reflecting low to moderate salinity and sodium hazards. Distinct variations were observed due to magnesium-rich rocks and minimal human disturbance. As a result, these findings underscore the urgent need for continuous monitoring and adaptive water management to safeguard drinking water quality and irrigation suitability.

Keywords: Surface Water · Objective Weighting Methods · Clean Water and Sanitation · Entropy · CRITIC · Irrigation

1 Introduction

Surface water is a vital subsurface resource that plays a crucial role in fulfilling water needs for households, agriculture, and industry worldwide [1]. It significantly contributes to socio-economic development in both developed and remote areas. Surface water is

Y. Zhu (Ed.): ISESIE 2025, LNCE 000, pp. 211–234, 2026.
https://doi.org/10.1007/978-981-92-2389-3_18

often located close to where it is needed, typically provides high-quality water, and can be developed more economically than other water sources [2]. However, the decline in surface water quality due to direct and indirect human activities reduces the availability of this essential resource and poses serious risks to human health. Numerous research efforts have been undertaken to monitor the physicochemical properties of water and evaluate its quality, with the goal of supporting effective water resource management [3]. Various methods have been developed to assess water quality and identify the main factors influencing its variation. Among these methods, the Water Quality Index (WQI) is widely recognized and used globally as a simple yet effective tool for evaluating the quality of water resources [4]. The WQI plays an increasingly vital role in managing water resources. Notably, heavily weighted parameters can greatly influence the sensitivity of the WQI, making it essential to assign weights according to the specific purpose of water use. As a result, even slight variations in weight distribution can lead to considerable shifts in the overall evaluation of water quality status.

The Weighted Arithmetic (WA) WQI is an extensively adopted approach and has been utilized by various researchers, including [5–7]. This method assumes that the weight of each water quality parameter is inversely related to its prescribed standard. In the WAWQI method, when two water quality parameters are considerably equally important, the one with a lower reference standard is assigned a higher weight and sub-index value, making it appear more influential than a parameter with a higher guideline limit. An important portion of research on surface water quality has primarily concentrated on the application of the five-scale weighting WQI method [8]. The assignment of weights from 1 to 5 to drinking water constituents represents their relative impact, typically informed by expert insights or regulatory guidelines. However, utilizing a five-point weighting scale may introduce errors in accurately evaluating drinking water quality.

To eliminate the subjectivity of conventional weight assignment, the entropy-weighted WQI (EWQI) was developed. This method calculates weights objectively by analyzing the informational content present in the given dataset [9]. The EWQI has been employed in several studies to assess surface water suitability for human consumption [10]. To minimize the subjectivity involved in conventional weight assignment, the criteria importance through inter-criteria correlation (CRITIC) method has also been utilized to derive the relative weights of the parameters. The concept relies on two main principles: the standard deviation of the parameters and the degree of conflict or interdependence among them. Several studies [11, 12] have applied the CRITIC-based WQI (CWQI) to evaluate water quality for potable use. Hence, incorporating entropy and CRITIC weighting methods into WQI analysis is a sensible approach, as it allows for leveraging the strengths of both WAWQI and five-point weighting approaches. Following a thorough literature review and analysis of globally acknowledged methodologies, this study has selected the EWQI and CWQI techniques for evaluating surface water quality.

Remote sensing and Geographical Information System (GIS) work together to monitor and analyze water quality by detecting indicators such as chlorophyll, turbidity, and temperature [13]. Remote sensing provides the capability to collect spatial data on water quality, while GIS enables the integration of these data with other sources, such as field measurements and ancillary data, to create detailed maps and models [14]. This combination allows for the identification of pollution sources, trend analysis, and impact assessment, ultimately aiding in the development of effective management strategies.

The suitability of river water for irrigation has become an increasingly critical concern for ensuring agricultural sustainability in the Cuttack region. Spatiotemporal variations in key irrigation water quality indicators, such as Electrical Conductivity (EC), sodium percentage (Na%), sodium adsorption ratio (SAR), Permeability Index (PI), magnesium adsorption ratio (MAR), Kelly's Index (KI) and the cation ratio of soil structural stability (CROSS), play a pivotal role in determining the appropriateness of river water for irrigation [15]. These fluctuations are driven by both natural factors, including weathering and seasonal shifts, as well as anthropogenic influences such as agricultural runoff and industrial effluents [16]. Understanding these dynamics is essential for developing region-specific water management strategies. Regular monitoring of these parameters can help identify potential risks to soil health and crop productivity. Consequently, sustainable irrigation planning must integrate water quality assessments to ensure long-term agricultural viability in the region [17].

Surface water contamination is frequently observed in the Taladanda Canal of Cuttack district, comprising of 5 sampling sites, located within the Mahanadi Basin of Odisha. The vulnerability of surface water and the poor water quality significantly contribute to health hazards, putting residents at risk of various life-threatening conditions. The Twin Cities of Cuttack and Bhubaneswar in Odisha have a long history when it comes to water quality issues. Both cities are known for their rich cultural heritage and historical significance, but in recent years, the water quality in the region has been a major cause for concern [18]. identified the water quality issues in the Twin Cities as the rapid urbanization and industrialization that has taken place in the area. As more and more people migrate to these cities in search of better opportunities, the water demand has increased significantly. This has put a strain on the existing water resources, leading to contamination and pollution of water bodies [15]. Reference [18] contributed to the water quality issues in the Twin Cities is the lack of proper infrastructure for sewage treatment and waste management. The existing sewage treatment plants are outdated and insufficient to handle the increasing volume of wastewater generated by the growing population. As a result, untreated sewage is often discharged directly into water bodies, further worsening the water quality. The use of pesticides and fertilizers in agriculture has also contributed to water pollution in the region [19].

Only a few studies on surface water quality have been conducted in this district, mostly focusing on specific issues such as salinity [20], saltwater intrusion [21], nitrate, BOD [22], WQI based on WAWQI across blocks [18], fluoride levels [15], and arsenic and fluoride contamination [19]. A thorough and well-organized assessment of surface water conditions, along with reliable forecasting of its quality, is essential for developing an effective strategic framework for regional water resource management. Notably, no research has been conducted in this area to evaluate overall surface water quality using an objective weight-based WQI approach. Additionally, there is currently no specific statistical model available for monitoring surface water quality. This remains a crucial yet unexplored topic in the region, highlighting an urgent need to adopt this approach to protect human health in the future.

Based on the factors above, the Taladanda Canal, situated nearby Jobra, Cuttack, has been selected as the focus of this study, and this research represents a structured effort to fill the existing gaps in surface water quality studies within this region. Hence, it can serve as a reference point for future researchers to benchmark their results against the present study in this specific region. Moreover, this study aims to develop a new surface water WQI model that is based on the objective weighting (like entropy and CRITIC) WQI approaches, offering a novel contribution that can aid in the effective management of existing water resources for local communities.

The main objectives of this research are:

(a) Implementing an assessment of the concentration and distribution of major ions and trace elements in the surface water of the study area;
(b) To evaluate the surface water quality using both the EWQI and CWQI;
(c) To conduct a comparative analysis between EWQI and CWQI;
(d) Analyze sources of surface water pollutants from natural and anthropogenic activities;
(e) To evaluate the seasonal and spatial suitability of the quality of water for drinking and irrigation.

Thus, the study aims to generate evidence-based insights that support local water management strategies and contribute to the achievement of SDG-6 targets. The resulting water quality map is expected to serve as a valuable resource for researchers, policy-makers, and local communities, facilitating effective decision-making for sustainable groundwater management.

2 Material and Methods

2.1 Study Area

Taladanda Canal originates from the Jobra Barrage in Cuttack, Odisha, drawing water from the Mahanadi River. Constructed in the early twentieth century, it plays a vital role in the irrigation and water management system of coastal Odisha [21]. The canal stretches over 85 km and flows southeast toward the Paradip Port, supporting a significant agricultural and industrial belt. The climate in the Cuttack region is tropical with a significant monsoon season. The region experiences a tropical monsoon climate, with average annual rainfall of approximately 1500–1800 mm, mostly between June and September [18]. It is noticed that this high rainfall supports intensive agriculture but also increases the risk of flooding and runoff into the canal. Temperature varies widely with the seasons. Summers (March to June) are hot, with maximum temperatures soaring up to 42 °C, while winters (December to February) are relatively cooler, with minimum temperatures dropping to around 10 °C. This climatic diversity influences cropping patterns and water requirements in the region [17]. Geologically, the area comprises alluvial deposits from the Mahanadi River system. The soils around the Taladanda Canal are mainly alluvial and loamy, suitable for intensive agriculture. These fertile soils support multiple cropping systems [22]. The canal irrigates vast tracts of farmland where paddy is the dominant crop, followed by pulses, vegetables, and oilseeds. The availability of canal water ensures both Kharif and Rabi cropping seasons are productive. As a result, studying the water quality of Taladanda Canal is crucial due to increasing pollution from industrial discharges, domestic sewage, and agricultural runoff. As the canal supports drinking water, irrigation, and industrial use, poor water quality can affect human health, reduce crop yield, and damage aquatic ecosystems. According to [20], regular monitoring of the canal at multiple locations is necessary to assess pollutants and ensure water safety. Rapid urbanization and industrialization in Cuttack and nearby areas further necessitate the implementation of water quality management and pollution control strategies to preserve this critical water resource. Figure 1 illustrates the distribution of the sampling locations.

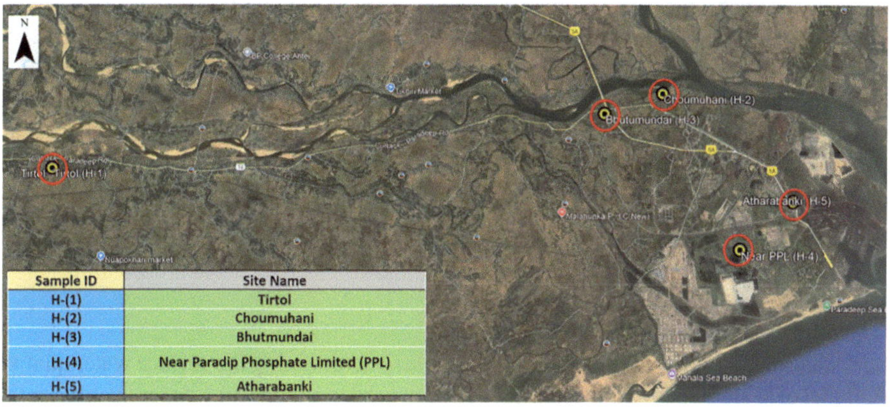

Fig. 1. Location map of the study area with plotted sampling locations

2.2 Surface Water Sampling and Analysis

This study utilizes a dataset consisting of 12 water quality parameters, collected from five sampling points, across the canal. A hand-held Global Positioning System (GPS) was used to record the geographic position of the sampling point. Figure 1 illustrates the distribution of the sampling locations. The selection of these stations was guided by the need to represent hydrological, topographic, and land-use heterogeneity across the basin, with particular attention to capturing the orographic effects, climatic variability, and geological differences that influence water quality [23]. The data sourced from the State Pollution Control Board, Government of Odisha. To capture spatial variation influenced by land use, river morphology and confluences, samples were collected at 5–10-km intervals along the main river, while the samples from the tributaries provided insight into localized hydro-chemical inputs. All collected water samples were stored in small refrigerated boxes, with ice packs maintaining a constant internal temperature of approximately 4 °C to ensure sample stability. The sampling cycle was set at three days, ensuring that samples were transported to the laboratory on the fourth day.

During transportation, shockproof measures were implemented, and samples were protected from direct sunlight [1]. The dataset comprises concentrations of key parameters, including temperature, pH, total dissolved solids (TDS), total hardness (TH), dissolved oxygen (DO), biochemical oxygen demand (BOD), sulphate (SO_4^{2-}), nitrate (NO_3^-), phosphate (PO_4^{3-}), Chloride (Cl^-), Fluoride (F^-), Iron (Fe^{2+}), and Chromium (Cr^{2+}), measured across different examined sites, during the 5-year period from 2020 to 2025 in pre-monsoon season. The temperature of the water samples was recorded using a standard laboratory thermometer. In situ parameters such as pH and BOD were assessed using a pH meter and a BOD meter, respectively. TH and Cl^- levels were analyzed via volumetric titration methods. DO was measured by employing the Winkler's titration method, while TDS were determined using a TDS meter. The concentrations of NO_3^- and F^- were determined using ion-selective electrode (ISE) methods. Sulphate (SO_4^{2-}) was estimated through gravimetric analysis. The concentration of PO_4^{3-} was measured using the ascorbic acid colorimetric method [24]. Finally, a double-beam UV-Visible spectrophotometer was used to quantify cations such as Fe^{2+} and Cr^{2+} ions. The process of sample collection, transportation, storage, and subsequent analysis were conducted by experts from pollution board, following standard analytical procedures and adhering to quality assurance and quality control protocols as per APHA [25] guidelines. The experiments were performed in triplicate and the average results are reported with a standard deviation of $\pm$ 2.4%. On this background, this analysis offers a general overview of surface water quality and facilitates comparison of each water quality parameter's concentration with the drinking water standards outlined in WHO [26].

2.3 Water Quality Assessment Method

Surface water chemistry plays an important role in determining the surface water quality. A model of Water Quality Index (WQI) has been used to evaluate the water quality for the purpose of human consumption. In this context, WQI represents the bearing of discrete parameters on overall water quality. The flowchart (Fig. 2) outlines the comprehensive methodology used for assessing surface water quality of the Taladanda Canal from 2020 to 2025, based on five sampling locations. The process begins with data collection and proceeds through summary statistics and objective weighting. Water Quality Indices such as WAWQI, EWQI, and CWQI are applied to provide detailed analysis, including surface water quality overview and classification [2]. A comparison by ambiguity and eclipsing identifies the best WQI, followed by input selection via correlation and dataset partitioning. Finally, a combined CWQI model integrates all indices to determine surface water suitability, aiding in accurate and informed water quality management.

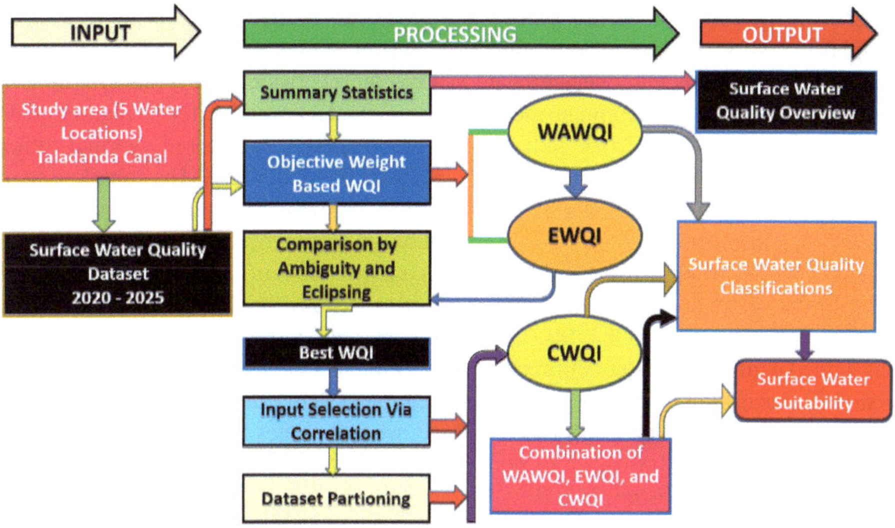

Fig. 2. Schematic representation of the study process

2.3.1 Analysis of Water Quality Parameters Using Water Quality Indices (WQIs)

The Weighted Arithmetic Water Quality Index (WAWQI) method is a widely used approach for evaluating the overall quality of surface water based on multiple physicochemical parameters [27]. It assesses the combined effect of different water quality indicators by assigning weights to each parameter according to its relative importance for drinking water suitability. The index is then calculated using a mathematical formula that integrates both the weight and the measured concentration of each parameter, producing a single value that reflects the overall water quality. The entire procedure for evaluating the WQI model was carried out following the methodology outlined by [6]. The advantages of the WAWQI method include its simplicity, ease of interpretation, and ability to consolidate complex data into a single, understandable index. It requires fewer criteria to make comparisons across a wide range of water uses, making it especially useful for regions with limited monitoring infrastructure [8]. Additionally, it allows for quick identification of pollution sources and trends over time. The importance of WAWQI in water quality investigation lies in its effectiveness in ranking water bodies based on their suitability for potable use [7]. This aids in the early detection of water quality degradation, supports regulatory decision-making, and helps prioritize water treatment and conservation efforts. As such, WAWQI serves as a practical and accessible tool for sustainable water resource management and public health protection. The WAWQI classification scheme defines surface water quality as value < 50 (Excellent); 26–50 (Good); 51–75 (Poor); 76–100 (Very Poor); and > 100 (Unsuitable for drinking activities).

The entropy theory is grounded in the concept of measuring uncertainty within a system. The Entropy Weight Water Quality Index (EWQI) method quantifies the randomness or arbitrariness of water quality parameters using a mathematical framework. This approach effectively addresses and overcomes the technical limitations often encountered in other evaluation techniques for surface water quality assessment [4]. This model assigns weights to assessment parameters using their entropy values, combining multiple water quality measurements into one aggregated value that represents the overall condition of surface water quality. To eliminate the effects of varying units and differing quality grade scales among the characteristic indices, data preprocessing is required. Normalization functions for benefit type and cost-type indicators are computed [28]. The complete step-by-step approach is elaborated as per [8]. By assigning objective weights to parameters based on their informational entropy, EWQI reduces human bias in decision-making. It enables a more comprehensive and data-driven interpretation of water quality conditions [11]. In the present investigation, EWQI has been utilized to characterize the surface water quality due to its reliability, high precision, and simplicity. As a result, EWQI has gained significance in environmental monitoring, resource management, and policy planning. According to [8], the EWQI classification system categorizes water quality as excellent, when the EWQI is less than 50; good, when it ranges from 50 to 100; medium, when it falls between 100 and 150; poor for values between 150 to 200, and extremely poor, when the EWQI exceeds 200.

The CRITIC (Criteria Importance Through Intercriteria Correlation/C)-WQI method, introduced by [9], is a robust approach for evaluating water quality based on objective weighting of parameters. It determines the significance of each water quality indicator by considering both its standard deviation, indicating as a measure of contrast intensity and the correlation with other indicators, which represents conflict or redundancy. Prior to weight calculation, the raw data is normalized to eliminate the influence of differing measurement units, ensuring comparability across parameters [10]. One key advantage of the CWQI method is its objectivity, as it does not rely on expert judgment or subjective input. Instead, it uses intrinsic data characteristics to determine weights, making it especially useful in large or complex datasets. It also accounts for redundancy by reducing the influence of highly correlated parameters, thereby improving accuracy. The complete framework for calculating this mechanism is suggested by [29]. The importance of the CWQI in water quality investigations lies in its ability to provide a clear, quantitative assessment of surface or groundwater quality. It helps policymakers, researchers, and environmental managers identify critical pollution factors and prioritize remediation efforts. By offering a more refined analysis, CWQI supports sustainable water resource management and informed decision-making. The CWQI classification scheme defines surface water quality as value < 50 (Excellent); 50–100 (Good); 100–150 (Medium); 150–200 (Poor); and > 200 (Extremely Poor).

As a result, WQI values for each sample are computed using both EWQI and CWQI methods, after which a comparative evaluation is performed to determine the most suitable objectively weighted WQI. The chosen index is subsequently used for sustainable water resource model development [11].

2.3.2 Irrigation Water Quality

The suitability of water for irrigation purposes was evaluated by EC, % Na, SAR, MAR, KR, PI and CROSS (Table 1) and the Wilcox diagram.

Sodium Percentage (%Na) It measures the amount of sodium relative to total cations (Na^+, K^+, Ca^{2+} and Mg^{2+}) in the water. Higher %Na levels in irrigation water can lead to sodium build-up in the soil, which can lower permeability and harm soil structure and crop yield [30].

Sodium Adsorption Ratio (SAR) SAR was used as an indicator of the ability of sodium to displace calcium and magnesium ions and subsequently affect soil permeability. Elevated SAR can diminish infiltration and soil structure, thereby making water unsuitable for irrigation [17].

Magnesium Adsorption Rate (MAR)/Magnesium Hazard (MH) MAR, denotes the dominance of magnesium's and calcium's concentrations in water. The more magnesium in water, the more alkaline it is, which negatively affects crop yield, especially when MAR exceeds 50% [3].

Permeability Index (PI) PI was used as a measure of the impact of sodium, calcium, magnesium, and bicarbonates on soil permeability. The more desirable, the PI the better quality of irrigation water, that allows for stable soil structural stability and therefore hydraulic conductivity [30].

Cation Ratio of Soil Structural Stability (CROSS) CROSS is a more recent field indicator used to reflect the risk of soil dispersal related to the relative abundance of monovalent and divalent cations. Values of CROSS < 10 generally occur when soil has good structural integrity and low sodicity risk [31].

Kelly's Ratio (KR) /Kelly's Index (KI) KR was used to look at the sodium status in relation to calcium and magnesium. A KR value greater than 1 relates to sodium-rich water, which has a potential for reducing soil quality and the water may not be accepted for irrigation [32].

The classification of these irrigational water quality parameters according to their suitability for irrigation is presented in Table 1.

Table 1. Equations used irrigational water quality parameters

Indices	Equations
Sodium percentage (%Na)	$Na\% = \frac{Na^{+}+K^{+}}{Na^{+}+Ca^{2+}+Mg^{2+}+K^{+}} \times 100$
Sodium adsorption ratio (SAR)	$SAR = \frac{Na^{+}}{\sqrt{\{(Ca^{2+}+Mg^{2+})/2\}}}$
Magnesium hazard (MH)	$MAR = \frac{Mg^{2+}}{Ca^{2+}+Mg^{2+}} \times 100$

(continued)

Table 1. *(continued)*

Indices	Equations
Kelly's ratio (KR)	$KR = \frac{Na^+}{Ca^{2+}+Mg^{2+}}$
Permeability index (PI)	$PI = \frac{Na^+ + \sqrt{HCO_3^-}}{Na^+ + Ca^{2+} + Mg^{2+}} \times 100$
Cation ratio of soil structural stability (CROSS)	$CROSS = \frac{Na^+ + 0.56K^+}{\sqrt{\{(Ca^{2+}+Mg^{2+})/2\}}}$

3 Result and Discussion

3.1 Surface Water Quality Derived from Descriptive Statistics

The summary statistics of individual water quality parameters collected from the study area are presented here. The variations in water quality parameters across the study area are shown through box plots (Fig. 3a-l). These basic statistical analyses were performed on the complete surface water (SW) dataset to provide a general understanding and to detect variations within the data. In this context, water quality parameters are fundamental indicators for determining the nature, classification, and variability of surface water.

Sometimes, in this investigation, higher temperatures in the downstream can be linked to lower elevations and warmer climatic conditions, whereas the colder upstream temperatures align with the high-elevation, glacial-fed origin [23]. The computed temperature for five water samples records between 23 and 29 °C. The corresponding average temperature is around 26 °C. The highest temperature (29 °C) was observed at H-(5), a downstream site, while the lowest (23 °C) was recorded at H-(2), an upstream location in the coldest region of the canal.

The pH levels of the surface water samples varied between 7.04 and 7.15 (Fig. 3a), with a computed average of 7.098, indicating a slightly above-neutral to alkaline nature. This confirms that all samples fall within the recommended pH range for drinking water (6.5–8.5). Slightly alkaline to alkaline pH of surface water amplifies owing to loss of CO_2 and precipitation of mineral salts [2].

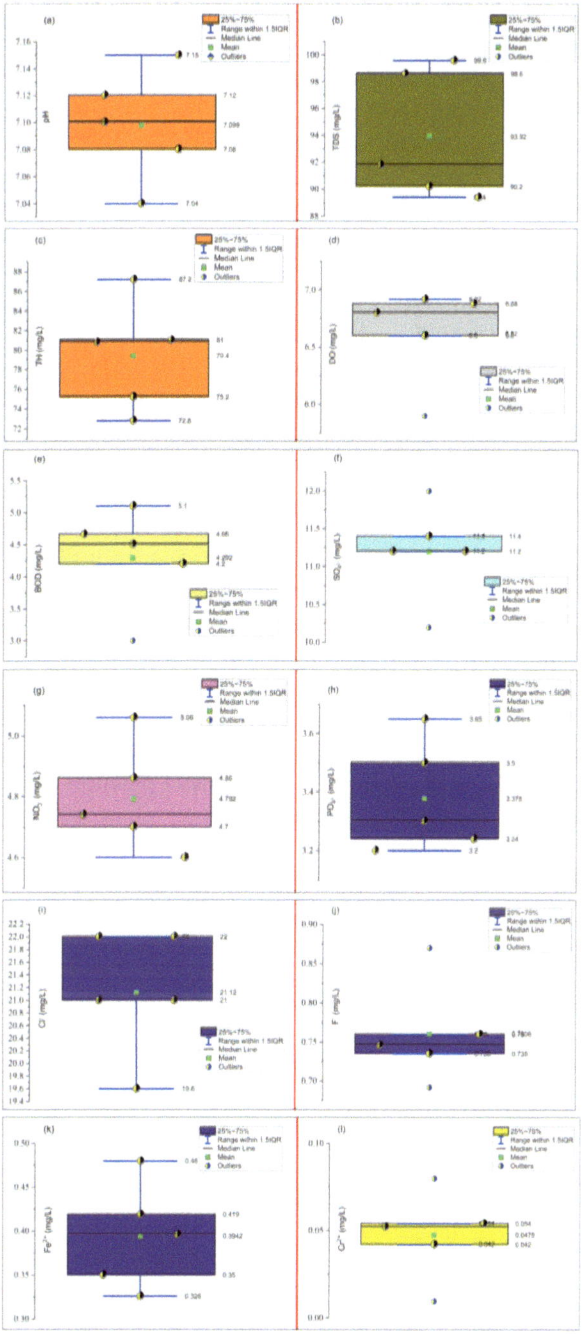

Fig. 3. Concentration of surface water quality parameters through box plots (a–l)

TDS indicates the concentration of total dissolved salts present in surface water, serving as a measure of salinity and helping in adjustments regarding its use in daily household activities [7]. TDS levels ranged from 89.4 to 99.6 mg/L (Fig. 3b), having a mean of 93.92 mg/L. About 100% of the sampling locations are considered to have fresh water, as their TDS concentrations are below 100 mg/L.

TH represents the combined concentration of Ca^{2+} and Mg^{2+} in surface water, expressed as mg/l of $CaCO_3$, spanning a concentration range of 72.8–87.2 mg/L, with a corresponding average of 79.4 mg/L (Fig. 3c). As per [12], water with TH values between 150 and 300 mg/L is rated as hard, while values above 300 mg/L are designated as very hard. Across the study site, nearly all sampling locations fall within the threshold of WHO standards (< 100 mg/L), indicating water suitability for human consumption.

DO levels in the surface water samples fell within the range of 5.9–6.92 mg/L (Fig. 3d), recording an average of 6.62, satisfying well with the permissible limit set by WHO (> 5 mg/L). It is noticed that the high DO levels indicate a well-oxidized river system with minimal pollution. The highest DO values were recorded at the site H-(2). It could be due to increased turbulence and dilution from rainfall [10]. The river's turbulent flow, photosynthetic activity in the river, cooler temperatures in certain regions and atmospheric diffusion also contribute to the high DO levels [33].

BOD concentrations fluctuated between 3 and 5.1 mg/L, yielding an average concentration of 4.29 mg/L (Fig. 3e). The lowest BOD values at H-(1) is recorded, likely due to dilution from rainfall, meltwater, and spring sources while the highest values at H-(2), (3), (4), and (5) were observed, particularly near human settlements, indicating organic pollution from domestic and agricultural sources. However, localized places increases in BOD near settlements, suggesting the influence of untreated domestic and organic waste, highlighting the need for improved waste management practices to mitigate potential organic and chemical pollution [11]. Based on the observation, approximately, 100% of the samples showed BOD levels above the allowable threshold (3 mg/L).

Sulphate ($SO_4{}^{2-}$) concentrations in surface water influence water taste and, at very high levels, may have a laxative effect. Contaminated water and wastewater frequently exhibit high concentrations of sulphates, primarily due to the dissolution of sulphate-bearing minerals and the discharge of industrial effluents [34]. When consumed, such elevated levels of sulphate can cause gastrointestinal disturbances, including diarrhoea, abdominal discomfort, and dehydration, particularly in sensitive populations such as children and the elderly. In the study area, sulphate concentrations range from 10.2 to 12 mg/L (Fig. 3f), with an average of 11.20 mg/L, and all measured values comply with WHO standards (< 100). The current assessment shows a slight upward trend in sulphate concentrations across the samples (from H-2 to H-1), likely due to dissolution processes. The highest sulphate levels are observed primarily in the north-west, south-east, and south-west regions during the pre-monsoon season.

From a health perspective, nitrogen-containing compounds pose significant impacts on human health. The computed $NO_3{}^-$ concentration ranged from 4.6 to 5.06 mg/L (Fig. 3g). For nitrate ($NO_3{}^-$), the national standard is 45 mg/L; however, in this study, 5 water sources were found to have concentrations lower than 10 mg/L, with the highest even as 5.06 mg/L. The high nitrate levels at H-(4) is closely linked to land use practices in the channel, which is referred as a major grain-producing base, it relies heavily on high fertilizer application to boost yields, which contributes to the elevated surface water nitrate concentrations [18].

Throughout the study, PO_4^{3-} concentrations ranged from 3.2 to 3.65 mg/L, demonstrating a mean value of 3.378 mg/L (Fig. 3h). It is noticed that, variations in surface water PO_4^{3-} levels across different sampling sites are closely related to local differences in fertilizer application, soil geology, and climate change. Quantifying the impact of agricultural activities on groundwater quality remains challenging, primarily due to uncertainties in agricultural production processes—fertilizer types and dosages are heavily influenced by subjective human experience, and these uncertainties are exacerbated by annual climate variability [8]. In this study, PO_4^{3-} readings were below the recommended limit, as suggested by WHO regulations (5 mg/L).

Cl^- plays an important role in human physiology; however, elevated levels can lead to high blood pressure. In general, the atmospheric precipitation, dissolution of salt deposits, and weathering of halite's and evaporites are considered as the major lithogenic sources of Cl^- in the surface water [19]. The soluble salts emplaced in sediments by the evaporation of surface water in discharge or shallow water level areas also raises the Cl^- concentration in surface water. However, excessive chloride levels can impart a salty taste and may have a laxative effect in unaccustomed individuals. Throughout this study, Cl^- concentrations ranged from 19.6 to 22 mg/L, yielding an average concentration of 21.12 mg/L (Fig. 3i). Around 100% of the surface water samples did not surpass the recommended limit of Cl^-.

F^- is usually present at elevated levels in surface water than in other type of water resources, primarily due to natural geological process. In view of F^- contents, its main origin is fluoride-bearing minerals and chemical fertilizers [35]. Consuming water with high fluoride is known to result in dental and skeletal fluorosis [14]. Within this research, F^- ranged between 0.6 and 0.87 mg/L (Fig. 3j). Approximately all five tested locations recorded concentrations exceeding the recommended limit of 0.3 mg/L for F^- in drinking water. Though it suggests a geogenic source of the solutes, the carbonate source (dissolution of limestones and dolomites) of the solutes seems to be high, as indicated by high loading at all investigation locations.

Consumption of Fe^{2+} rich water can lead to harmful health outcomes, including a condition known as red hot disease. Fe^{2+} concentrations spanned between 0.3 and 0.48 mg/L, with a corresponding average of 0.39 mg/L (Fig. 3k). This confirms that all samples fall above the recommended Fe^{2+} range (> 2 as per WHO criteria). The poor water quality at all locations, may impact the yield production, the physical state of the soil, and the number of dissolved particles in irrigation water, which results in inhibited vegetable growth and yield production [31]. The surface water quality also differs with the complexity of water, seasonal variations, trickled dissolved salts, and sub-surface environment.

Cr^{2+} is naturally present in surface water, particularly under aerobic conditions. Its level depends on factors like rainfall, aquifer composition, geochemistry, and surface water flow. Chromium (Cr^{2+}) exposure through drinking water can result in severe and irreversible conditions affecting the skin, lungs, and bladder [3]. In this work, the estimated Cr^{2+} value for all surface water samples fluctuated as 0.0095–0.08 mg/L (Fig. 3l). The average score recorded as 0.0475 mg/L. As a result, this outcome confirms that 3 water sites exceeding the permissible limit set by WHO standards (< 0.05). In this scenario, the concentrations in some areas exceeded the drinking water standards; its sources are mainly animal manure and fertilizer use, which contribute to the elevated concentrations.

Water quality parameters primarily reflect the contamination degree of water resources. Figure 3a-l summarizes the statistical results using box-and-whisker plots, revealing significant variations in the compositional concentrations among different sampling sites. Particularly, Fe^{2+}, BOD, and F^- exhibit substantial spatial distribution differences. The plots indicate that the cation concentrations follow the order: $Fe^{2+} > Cr^{2+}$ (mg/L), while the anion concentrations follow: $Cl^- > SO_4{}^{2-} > NO_3{}^- > PO_4{}^{3-} > F^-$. Elevated concentrations of these ions can render water unfit for human consumption and biological survival [5]. Given that the primary land use in the selected places, the high concentrations of these ions (likely referring to Cl^- and Fe^{2+}) are mainly attributed to fertilizer application, which in turn may affect food production. Irrigation return flow, leachates from landfill, waste dumps, septic tanks, industrial and animal wastes are some of the possible anthropogenic sources of high BOD in surface water [29]. Again, F^- considered as essential nutrient for human health and crops, has two main sources: natural occurrence and anthropogenic inputs, primarily from fertilizer application. Therefore, farmers should be advised to adopt sustainable practices, such as meticulous farming, to minimize nitrate-based fertilizers. These assessments are essential for evaluating fluoride, BOD, and Fe^{2+} contamination in drinking water, supporting health interventions, infrastructure development, and sustainable resource management.

3.2 Surface Water Quality Based on WAWQI

The Weighted—Arithmetic (WA) Water Quality Index (WQI) provides a comprehensive assessment of surface water suitability for drinking purposes. Radar graphs provides valuable insights into the spatial variability of surface water quality and facilitated the identification of areas requiring contamination control and remedial measures. Figure 4 presents the WQI values for all 5 sampling locations. The WAWQI measurements vary between 43.16 and 147.08, yielding an average score of 101.14, reflecting considerable variability in water quality across the region. The analysis indicates that most of the sample falls within the "Very Poor" or "Unsuitable" water quality categories, confirming that nearly 45% of the surface water in the study area is unfit for human consumption. The radar distribution of WQI values is illustrated in Fig. 5a. The map reveals that the highest water quality ("Unsuitable") is concentrated in the northern and south-eastern regions of the study area. In contrast, surface water quality gradually decreases from "Good" to "Very Poor" towards the north-east. This declining trend is primarily influenced by the hydraulic gradient and the direction of water flow, which converges towards the north-eastern boundary of the aquifer. Urbanization and industrial development inevitably generate large volumes of sewage, and improper treatment can lead to substantial pollutants entering the groundwater system, thereby contaminating surface water sources. Overall, the water quality variation of WQI highlights that while surface water in the study area is generally suitable for drinking at one location (H-1); remaining locations renders localized variations in water quality, that reflect the natural hydrogeological controls on water chemistry.

3.3 Surface Water Quality Based on EWQI and CWQI

The EWQI represents a unified, thorough, and consistent approach to surface water quality assessment. Elevated entropy-derived weights signify a higher significance of the respective parameters in influencing water chemistry variations, particularly for drinking water applications. The study reveals that BOD, and pH exert the highest and lowest influence, respectively, on surface water quality in the area. Among the water quality parameters, BOD exhibits the highest entropy weight (0.34), whereas pH shows the lowest entropy weight (0.032). Figure 4 represents the surface water quality classification pattern across the studied region.

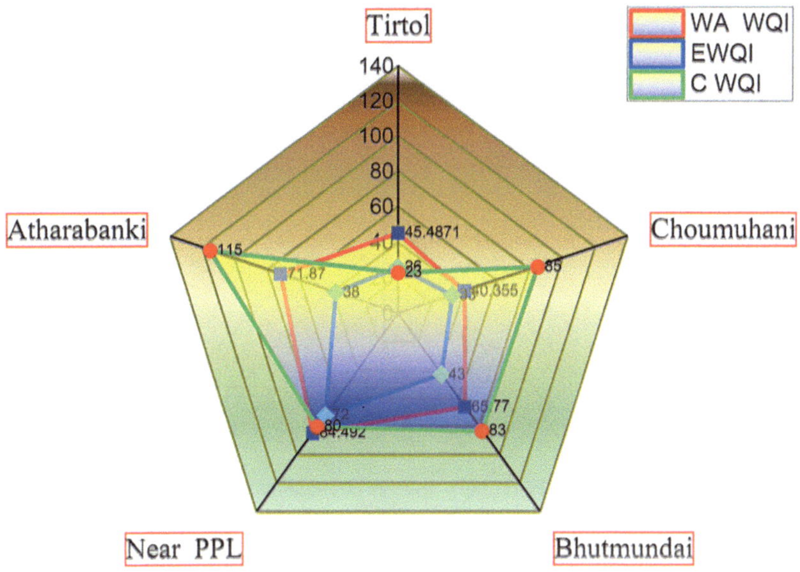

Fig. 4. WQI scores of WAWQI, EWQI, and CWQI

The EWQI values derived from the analysis of surface water specimens gathered across the study region span a range from 77 to 210, with a computed mean value of 157, reflecting considerable variability in water quality across the region. Based on these values, 2 samples are identified as having extremely poor water quality, making them entirely unsuitable for drinking (Fig. 5b). Additionally, one samples are identified as having poor quality, which is also considered unfit for consumption. About 11.11% of the samples are classified as medium quality, suggesting that their quality is borderline acceptable for drinking and should undergo treatment before domestic use. In this context, at 3 sites (H-3, 4, 5), surface water quality is influenced by multiple factors, including recharge sources, geological formations, hydrological conditions, mineral composition of the basin and aquifer, and water-rock interactions. Meanwhile, 11.11% of the samples are assessed as good category, and none of the samples were rated as excellent.

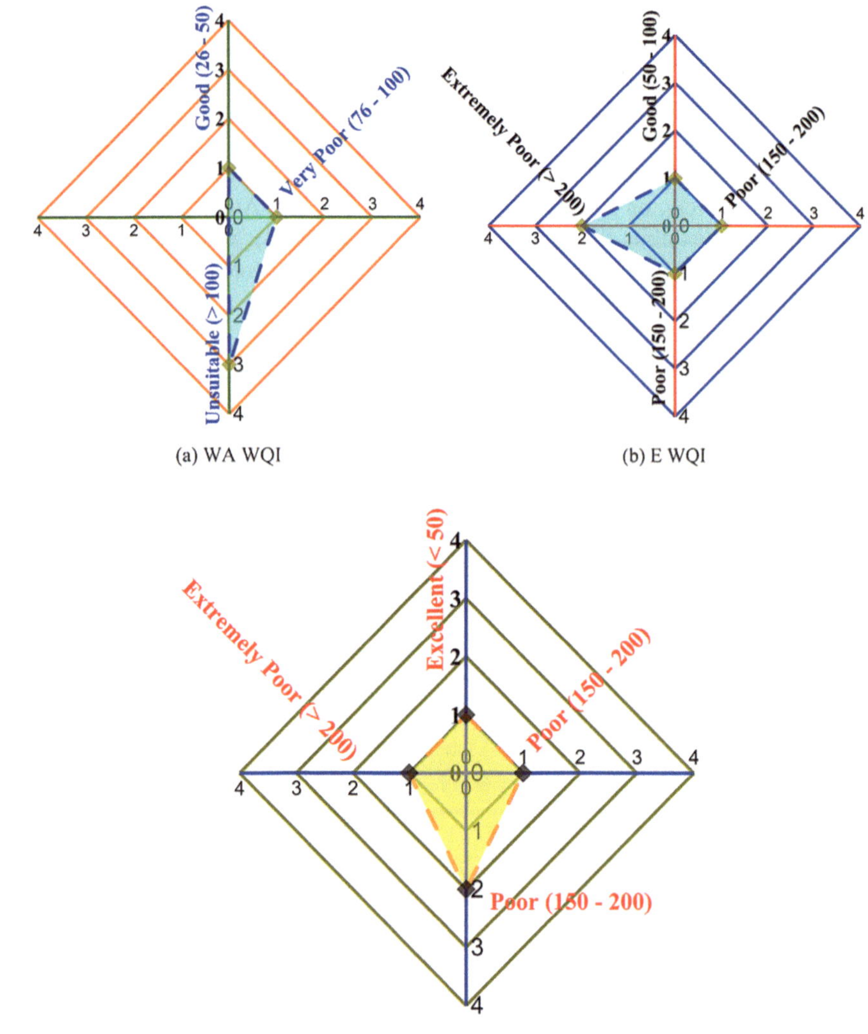

(c) CRITIC (C) WQI

Fig. 5. Representation of radar plots rendering Water quality index classification (a–c)

Figure 4 illustrates the C-WQI values for the five collected surface water samples. The C-WQI measurements vary between 30 and 254, with a mean value of 142.4. Among these samples, 33.33% sample is identified as exhibiting poor to extremely poor water quality, whereas 22.22% are grouped under the excellent to medium category, indicating their appropriateness for drinking purposes (Fig. 5c). The excellent WQI values across the H-(1) can be attributed to the high flow rate of the river, which aids in the dispersion and dilution of contaminants. Furthermore, the homogeneity of water quality throughout the basin, with minimal differences between the main river and tributaries, is likely due to this high flow rate, which ensures consistent dilution and dispersion of pollutants.

Pollution stemming from agricultural and industrial activities, combined with large-scale historical over-extraction of surface water, has resulted in widespread contamination and declining of water quality at locations H-(4). Furthermore, in agricultural areas, large amounts of chemical fertilizers are applied to produce food essential for human subsistence, which can elevate F^- and BOD concentrations in surface water. Therefore, Effective waste management and water treatment solutions are essential to mitigate these risks and ensure safe drinking water.

The WQI outcomes for the study area, derived using both the EWQI and C-WQI approaches, are illustrated in Figs. 4 and 5b–c. A comparative evaluation of the two methods reveals that they predominantly classify the surface water quality within the full range of categories, spanning from excellent, good, to extremely poor. As anticipated, both the EWQI and C-WQI methods classify the samples as excellent to good, when all water quality parameters remained within permissible limits, demonstrating the reliability and rationality of both approaches in representing water quality status. Moreover, the number of samples identified under the poor and extremely poor classifications is quietly similar in both the models.

Ambiguity and eclipsing represent key sources of uncertainty in WQI modelling. As noted by [14, 16], ambiguity arises when the overall WQI value exceeds permissible limits even though individual parameters remain within acceptable ranges. Conversely, eclipsing occurs when certain parameters surpass critical thresholds, yet the composite index fails to capture or reflect this exceedance. Therefore, it is essential to determine which WQI method demonstrates a lower susceptibility to ambiguity and eclipsing, thereby providing a more accurate representation of water quality. In this study, five sampling locations are identified where nine water quality parameters, out of twelve, meet the acceptable standards for potable use. The corresponding final WQI scores for these points, as determined by both methods, demonstrate compliance with drinking water standards. This consistency confirms that both WQI approaches are free from ambiguity and eclipsing in these cases. As a result, the EWQI and C-WQI method accurately classifies the water quality status of 5 sampling locations, matching exactly with the expected status based on number of guideline value breaches.

3.4 Irrigation Water Quality

Suitability for irrigation is mainly determined by examining the dissolved salts in the water. In too much-salinized water, Na^+ might displace Ca^{2+} and Mg^{2+}. As a result of the displacement soil permeability is decreased which in turn restrict air and water movement [36]. Surface water suitability for irrigation (Table 2) was assessed by calculating the EC, Na%, SAR, PI, MAR, KR and CROSS values for the pre-monsoon season for the examined study area.

Table 2. Classification of surface water quality indices for irrigation purpose

Parameters	Threshold	Class	Number of samples	% of samples
EC	> 3000	Fair	0	0

(continued)

Table 2. *(continued)*

Parameters	Threshold	Class	Number of samples	% of samples
	700–3000	Good	0	0.00
	< 700	Excellent	5	55.56
Sodium percentage (%Na)	> 80	Unsuitable	2	22.22
	60–80	Doubtful	2	22.22
	40–60	Permissible	0	0
	20–40	Good	1	11.11
	< 20	Excellent	0	0.00
SAR	> 26	Poor	1	11.11
	18–26	Fair	2	22.22
	10–18	Good	2	22.22222222
	< 10	Excellent	0	0.00
MAR	< 50	Suitable	1	11.11
	> 50	Unsuitable	4	44.44
PI	> 75%	I (Suitable/excellent))	3	33.33
	25–75%	II (Good)	1	11.11111111
	< 25%	III (Unsuitable)	1	11.11
CROSS	> 26	Poor	1	11.11
	18–26	Fair	2	22.22
	10–18	Good	2	22.22
	< 10	Excellent	0	0
KR	< 1	Acceptable	3	33.33
	> 1	Unsuitable	2	22.22

In the present investigation, the value of EC derived from the analysis of surface water specimens gathered across the study region span a range from 143 to 566 μS/cm, with a computed mean value of 358.6 μS/cm (Fig. 6a). It is reflected that the EC values for all five locations were below 700 signifying water falls in excellent category for irrigational purpose. The low EC value in tributaries may be due to their smaller size and steeper slope with less land-water interactions, they may have quicker water velocities than the main river channel, which limits the dissolving and accumulation of dissolved particles and lowers the EC.

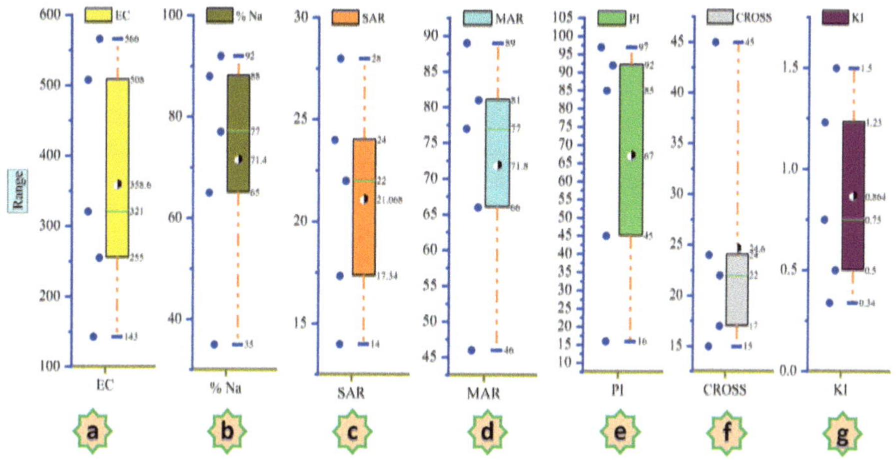

Fig. 6. Presentation of box plots to detect the concentration values and outliers in the irrigation water quality dataset (a–g)

The % Na levels of the surface water samples varied between 35 and 92, recording an average % Na of 71.4, respectively (Fig. 6b). From the examined outcomes, the %Na value falls into the good, doubtful, and unsuitable categories and is termed suitable for irrigation purposes for 3 places, such as H-(1), (2), and (3). Higher %Na levels were observed at H-(4) and (5) during the pre-monsoon season, that could lead to soil structural deterioration and reduced infiltration rates, with a noticeable decline during the H-(1–3), likely due to the dilution effect from increased rainfall and runoff [36].

SAR levels ranged from 14 to 28, having a mean of 21.068. From the obtained results, the value of SAR at H-(1) and (3), was recorded between 10 and 18, which indicates excellent water quality suitable for irrigation (Fig. 6c), thus indicating that the underlying geology has a significant impact on the solute chemistry which has dominant silicate mineral concentrations have a distinct ionic composition, with lower sodium concentrations [24]. In other way, 22.22% (H-2 and 5) are identified as fair water quality, whereas 11.11% (H-4) is grouped under poor water category. Although within acceptable ranges for irrigation, prolonged use of the tributary water with higher sodium content could lead to soil structural deterioration and reduced infiltration rates.

The obtained MAR observations vary between 46 and 89, with a mean value of 71.8, in the ongoing assessment of the studied area (Fig. 6d). The highest value of MAR is noticed at H-(4), whereas its value at H-(1) is recorded lowest. Only one sample (H-1) has MAR, that remained below the threshold value of 50, indicating suitability for irrigation. The highest value of MAR at 4 sites (H-2 to 5), which is the upstream downstream portion of the river basin might be due to extended interaction between the water and the nearby rocks that contain magnesium and can result in the leaching and dissolution of magnesium in water [17]. In the case of H-(4), the possible reason for the high MAR value could be due to inputs from anthropogenic activities like agricultural input and domestic wastewater.

The PI values can indicate how the soils permeability is influenced by Na^{+}, Mg^{2+}, Ca^{2+}, HCO_3^{-}, and the type of soil (Doneen 1954). The computed value of PI in the current research fluctuated as 16–97 (Fig. 6e). The average value of PI estimated as 67. The PI values indicated water quality suitable for irrigation, with all 4 sites (H-1 to 3, and 5), falling into Class I and II of the PI value classification (Doneen 1954). The influx of rainwater and anthropogenic inputs from increased agricultural activities during the monsoon season tend to increase the dissolved salt content in surface water sources, leading to a decrease in permeability index compared to the pre-monsoon period [10].

Figure represents the box plot of surface water quality scores across the studied region. The CROSS value for 5 water specimens recorded between 15 and 45. The average value of CROSS (Fig. 6f) was 24.60 in the pre-monsoon season, respectively. The CROSS values for two samples such as H-(1) and (2), were among (10–18), which indicates its suitability for irrigation. Specifically, sparse agriculture and low population density, coupled with its unique geological composition, lead to minimal fertilizer use and sodium input into the river. The difference in CROSS value between the sites H-(3), (4), and (5), indicates that the tributaries may have a much higher tendency of mineral precipitation and could result in damage to the irrigation system and soil structure [22].

KR concentrations (Fig. 6g) fluctuated among 0.34 and 1.5, yielding an average concentration of 0.864. The KR value also indicates that water is suitable for irrigation purposes for three sites such as H-(1), (2), and (3), accounting 33.33% of the total water samples. At locations H-(4), and (5), the KR value is greater than 1, which might be primarily due to anthropogenic influence in the sapling site [34]. The observed variance in KR values between the districts can be ascribed to the varied rock formations and complicated terrain of the basin, land use practices and hydrological regime in these places. Also, the intensity of distinct land use pattern directly influences the chemical composition of the water, ultimately resulting in contrasting water quality in both the leeward and windward sides of the Himalayas [5].

3.5 Wilcox Diagram

After interpreting the Wilcox diagram, two major classes are observed, i.e., C1–S2, C2–S3, and C2–S4 across the Pre-monsoon season (Fig. 7). The C1–S1 category (low salinity and low sodium hazard) exhibits the best quality for irrigation needs that are productive in the agricultural sector. Similarly, the C2–S1 category (medium salinity and low sodium hazard) exhibits the best quality for irrigation needs that are productive in the agricultural sector. C1–S2, represents Brackish water, approximately perfect for agriculture, contains only one sample. The Wilcox diagram analysis reveals that during the pre-monsoon season, a greater number of water samples fall within the C2–S3 (2 samples) and C2–S4 (2 samples) zone, that denotes harmful to agriculture. This shift can be attributed to the increased rainfall during the monsoon, which significantly enhances surface runoff. As this runoff flows over degraded soils, it mobilizes and transports dissolved salts into rivers, lakes, and aquifers [7]. Consequently, despite the abundance of water, the salinity levels in these water bodies rise due to the influx of salts carried by the runoff [12]. Overall, the research indicates that the river's water quality is suitable for agricultural irrigation and satisfies the requirements for effective usage in the agricultural sectors.

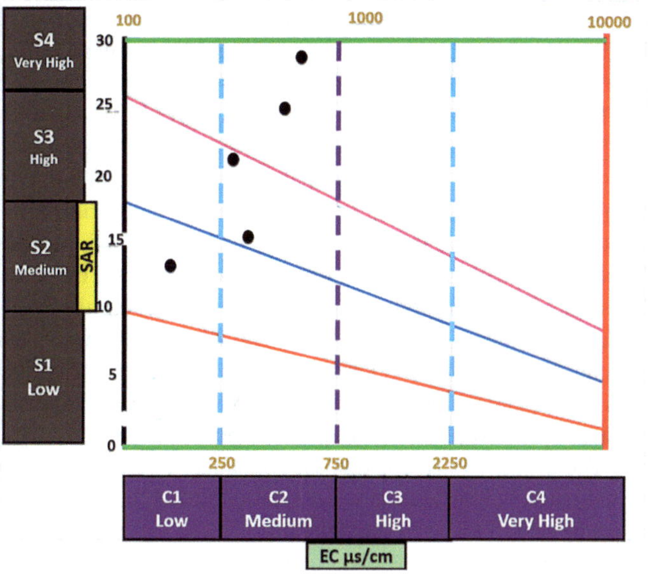

Fig. 7. Wilcox chart for the classification of irrigation water based on EC and SAR

Overall, the results of analyzing several irrigational water quality parameters are comparable to other similar studies conducted in the Taladanda Canal, Odisha. For example, [19] found that monsoon dilution lowered salinity on the Koshi River while increasing sediment and nutrient transport, as seen in chosen region. Likewise, [34] found low sodium hazards in the Bhagirathi River but raised concerns about agricultural practices and the hydropower sector, as in the selected channel, where localized sodium and magnesium spikes were noted in tributaries. These studies highlight the dominance of calcium and magnesium ions because of carbonate weathering, monsoon-driven seasonal changes, and growing anthropogenic pressures such as agricultural runoff and untreated sewage. Although previous studies have enhanced our understanding of regional hydrochemistry and geochemical processes, this study extends that foundation by integrating a broader suite of irrigation-specific indices with fine-scale spatial and seasonal analyses across multiple districts. By incorporating parameters such as %Na, SAR, MAR, CROSS, KR and WQI, and evaluating both irrigation and drinking water suitability, the study adopts a more ecologically grounded and context-specific approach.

4 Conclusion

Surface water is a vital source of water in the examined study region. This study utilized the WAWQI, EWQI, and CWQI methodologies to assess surface water quality and developed an objective-weighted WQI model, highlighting seasonal variations and their implications for Sustainable Development Goal 6 (Clean Water and Sanitation) and long-term ecosystem sustainability. The resulting integration showed great promise in accurately evaluating surface water quality and serves as a useful tool for effective management of local water resources. The key findings of the study are summarized below: Out of 12 analyzed surface water samples, 75% met acceptable limits for all parameters. While pH levels remained within permissible limits in all samples indicating slightly alkaline condition, BOD, F^-, and Fe^{2+} surpassed the recommended threshold of WHO (2012). The WQI method: specifically, the WAWQI (43.16–147.08), EWQI (77–210), and CWQI (30–254) effectively categorized surface water quality across a wide range, from good–unsuitable, Good–Extremely Poor, and Excellent–Extremely Poor drinking water quality. When all water quality parameters were within permissible limits, both methods consistently rated the samples as excellent to good, demonstrating their reliability. However, the EWQI method outperformed the CWQI by better minimizing eclipsing effects, making it the preferred choice for developing the best-WQI model. For irrigation, key parameters such as SAR, %Na, PI, and EC were well within safe limits and are considered suitable for agricultural use. However, agricultural indicator's variation across the detection locations indicates the influence of underlying geology and land-use practice. At few places, substantial higher values of MAR, KI, and CROSS were observed. The primary reason is that the Mahanadi River and its tributaries show different seasonal patterns in irrigation water quality during the pre-monsoon, the main river experiences increased levels of most parameters due to heavy runoff and sediment input, while tributaries tend to show lower concentrations, reflecting their unique local water flow patterns. Hence, this study provides clear and practical contributions to the assessment of surface water quality for drinking water in specific areas. Additionally, the findings serve as a valuable reference for future research, aiding the development of more effective and reliable strategies for surface water monitoring and management.

Acknowledgements. The present research work is supported under the C.V. Raman Global University (CGU), Bhubaneswar, Odisha, India.

References

1. Dharmani, A.B. et al.: Unravelling groundwater contamination and health-related implications in semi-arid and cold regions of India. J. Contaminant Hydrol. **261**, 104303 (2024)
2. Gad, M. et al.: Combining water quality indices and multivariate modeling to assess surface water quality in the northern Nile Delta, Egypt. Water. **12**(8), 2142 (2020)
3. Hussein, E.E. et al.: Groundwater quality assessment and irrigation water quality index prediction using machine learning algorithms. Water. **16**(2), 264 (2024)
4. Gupta, D., Mishra, V.K.: Development of entropy-river water quality index for predicting water quality classification through machine learning approach. Stoch. Env. Res. Risk A. **37**(11), 4249–4271 (2023)
5. Sajib, A.M., Mahammad Diganta, M.T., Rahman, A., Dabrowski, T., Olbert, A.I., Galal Uddin, M.: developing a novel tool for assessing the groundwater incorporating water quality index and machine learning approach. Groundwater Sustain. Dev. **23**, 101049 (2023)
6. Patel, D.D., Mehta, D.J., Azamathulla, H.M., Shaikh, M.M., Jha, S., Rathnayake, U.: Application of the weighted arithmetic water quality index in assessing groundwater quality: A case study of the South Gujarat region. Water. **15**(19), 3512 (2023)
7. Ferdous, M.N., Hoque, M., Popy, S., Hossain, M.I., Munna, R.I.: Assessment of the groundwater quality of the Jessore municipality by using weighted arithmetic water quality index. Next Res. **2**(1), 100187 (2025)

8. Ukah, B.U., Ameh, P.D., Egbueri, J.C., Unigwe, C.O., Ubido, O.E.: Impact of effluent-derived heavy metals on the groundwater quality in Ajao industrial area, Nigeria: an assessment using entropy water quality index (EWQI). Int. J. Energy Water Resour. **4**(3), 231–244 (2020)
9. Khan, I., Umar, R.: Machine learning-driven optimization of water quality index: a synergistic ENTROPY-CRITIC approach using Spatio-temporal data. Earth Syst. Environ. **8**(4), 1453–1475 (2024)
10. Jahan, A., Rai, N., Khan, M.U., Dar, T.A.: Hydrogeochemical evolution, natural background level quantification, quality assessment of groundwater using integrated entropy and CRITIC weighing methods based water quality index (IWQI), and health risk assessment in the middle Gangetic Plain, India. Acta Geochimica, 1–42 (2025)
11. Han, L., Wang, Y., Li, S., Li, W., Chen, X.: Evaluation of water resource carrying capacity and analysis of driving factors in the Dadu River basin based on the entropy weight method and CRITIC comprehensive evaluation method. Water. **17**(16), 2360 (2025)
12. Zhang, Y., Xiao, Y., Yang, H., Wang, L., Wang, J., Hu, W., Wang, N., et al.: Hydrogeochemical features, genesis, and quality appraisal of confined groundwater in a typical large sedimentary plain. Water Environ. Res. **96**(8), e11088 (2024)
13. Patrick, M., Hennink, M., Dasmohapatra, M., Rout, M., Clasen, T., Caruso, B.A.: Influences on water source selection and use among women in rural Odisha, India. Frontiers in Water. **5**, 1040981 (2023)
14. Khan, S.N., Anjum, L., Arshad, A., Ali, S., Aleem, M., Nasir, A.: Nature-based solution for wastewater treatment and reuse using phytoremediation with floating plants. Water. **17**(7), 1080 (2025)
15. Lenka, A.K.: Urban poor and water supply issues: a study of a planned city in India. In: One Health Approach-Advancing Global Health Security with the Sustainable Development Goals. IntechOpen (2024)
16. Wijayaweera, N. et al.: Use of machine learning and indexing techniques for identifying industrial pollutant sources: a case study of the lower Kelani River basin, Sri Lanka. Water. **16**(19), 2766 (2024)
17. Hagage, M., Hewaidy, A.G.A., Abdulaziz, A.M.: Groundwater quality assessment for drinking, irrigation, aquaculture, and industrial uses in the waterlogged northeastern Nile Delta, Egypt: a multivariate statistical approach and water quality indices. Model. Earth Syst. Environ. **11**(1), 59 (2025)
18. Sulaiman, M.A., Divya, Zafar, M.M., Anjum, S., Kumari, A.: groundwater contamination by fluoride and mitigation measures for sustainable management of groundwater in the Indo-Gangetic Plains of India. In: Groundwater in Arid and Semi-arid Areas: Monitoring, Assessment, Modelling, and Management, pp. 289–314. Springer Nature, Cham (2023)
19. Karki, B.K. et al.: Risk assessment of heavy metals in the major surface water system of Nepal with potential remediation technologies. Environ. Challenges. **14**, 100865 (2024)
20. Mangueina, D., Awuah, E., Fonteh, M.F., Antwi-Agyei, P., Nadji, E.T.: Assessment of current water, sanitation, and hygiene (WASH) practices in the third and ninth districts of N'Djamena, Chad. J. Water Health. **22**(2), 414–435 (2024)
21. Prasad, D., Kumar, J., Rai, P.K., Biswas, B., Singh, A., Ranjan, M.: Assessment of seasonal variation in water quality of Gomti River, Jaunpur City, India. In: Rivers of India: Past, Present and Future, pp. 153–164. Springer International Publishing, Cham (2024)
22. Jena, M.R., Tripathy, J.K., Sahoo, D., Sahu, P.: Geochemical evaluation of groundwater quality in the coastal aquifers of Kujang block, Jagatsinghpur District, eastern Odisha, India. Water Air Soil Pollut. **235**(6), 349 (2024)
23. Abba, S.I. et al.: trace element pollution tracking in the complex multi-aquifer groundwater system of Al-Hassa oasis (Saudi Arabia) using spatial, chemometric and index-based techniques. Environ. Res. **249**, 118320 (2024)

24. Banda, T.D., Kumarasamy, M.: Application of multivariate statistical analysis in the development of a surrogate water quality index (WQI) for south African watersheds. Water. **12**(6), 1584 (2020)
25. APHA: Standard Methods for the Examination of Water and Wastewater, 23rd edn. American Public Health Association, Washington, DC (2017)
26. WHO: Guidelines for Drinking-Water Quality. World Health Organization (2017)
27. Mohammadpour, A. et al.: Assessment of drinking water quality and identifying pollution sources in a chromite mining region. J. Hazard. Mater. **480**, 136050 (2024)
28. Zhang, Y., Xiao, Y., Yang, H., Wang, L., Wang, J., Hu, W., Wang, N., et al.: Hydrogeochemical features, genesis, and quality appraisal of confined groundwater in a typical large sedimentary plain. Water Environ. Res. **96**(8), e11088 (2024)
29. Ye, H., Chen, L., Zou, K., Wu, W., Dan, R., Wang, Y.: Coupling Monte Carlo simulation with CRITIC-enhanced water quality assessment for the Weishui reservoir. Water Supply. **24**(2), 601–614 (2024)
30. Li, J., Wang, Z.: Emerging irrigation and drainage technologies: A bibliometric review of the journal irrigation and drainage from 2010 to 2024. Irrigation Drainage. **74**(1), 3–26 (2025)
31. Altés, V., Pascual, M., Escorihuela, M.J., Villar, J.M.: Assessing irrigation impact on water quality conditions: a case study in the river Noguera Ribagorçana (NE Spain). Agric. Water Manag. **296**, 108809 (2024)
32. Gharaibeh, M.A., Albalasmeh, A.A., Obeidat, M.M.: Assessment of water quality of key dams in Jordan for irrigation purposes with insights on parameter thresholds. Water. **16**(12), 1726 (2024)
33. Das, A.: Multivariate statistical approach for the assessment of water quality of Mahanadi basin, Odisha. Mater. Today Proc. **65**, A1–1 (2022)
34. Gupta, D., Shukla, R., Srivastava, P.K., Mishra, V.K.: Assessment of heavy metal pollution level, ecological and human health risks in surface water of Narmada River, India. Sustain. Water Resour. Manage. **10**(4), 154 (2024)
35. Das, A.: An optimization based framework for water quality assessment and pollution source apportionment employing GIS and machine learning techniques for smart surface water governance. Discover Environ. **3**(1), 1–35 (2025)
36. Laaraj, M., Benaabidate, L., Mesnage, V., Lahmidi, I.: Assessment and modeling of surface water quality for drinking and irrigation purposes using water quality indices and GIS techniques in the Inaouene watershed, Morocco. Model. Earth Syst. Environ. **10**(2), 2349–2374 (2024)

Integrating Industrial Safety and Circular Economy Principles in Wastewater Treatment: A Nexus Approach for Sustainable Resource Management

Abhijeet Das[1](✉) and Satchidananda Mishra[2]

[1] Department of Civil Engineering, C.V. Raman Global University (CGU), Bhubaneswar, Odisha, India
das.abhijeetlaltu1999@gmail.com

[2] Department of Civil Engineering, Bhubaneswar Institute of Technology, Bhubaneswar, Odisha, India

Abstract. Wastewater treatment plants (WWTPs) are critical nodes in urban-industrial systems, influencing water security, energy use, and environmental sustainability. Although circular economy (CE) practices, such as energy recovery, nutrient recycling, and waste valorization, offer substantial potential to enhance resource efficiency, their integration with industrial safety remains underexplored. This paper presents a conceptual framework that couples CE strategies with industrial safety protocols in WWTP operations, adopting a nexus perspective to capture interdependencies across water, energy, and climate systems. Through a literature analysis encompassing green supply chain management, CE-enabled waste management, and nexus approaches, the study identifies operational trade-offs and synergies that shape both process safety and sustainability outcomes. The framework emphasizes hazard identification, risk mitigation, and occupational safety alongside energy recovery and emissions reduction, providing a holistic pathway to decouple environmental impacts from economic growth. Findings indicate that a nexus–CE–safety approach optimizes resource use while enhancing resilience against operational hazards, regulatory non-compliance, and environmental disruptions. This integrated perspective offers actionable insights for policymakers, industrial managers, and researchers seeking to design safe, circular, and sustainable WWTPs aligned with global sustainability goals.

Keywords: Circular Economy · Waste Water Treatment Plants · Occupational Safety and Health · Resource Recovery · Energy Recovery · Nexus Approach

1 Introduction

Water is one of the most vital resources for human development, yet urbanisation, industrialisation, and population growth have increased demand while reducing availability. Wastewater generation has risen sharply from domestic, agricultural, and industrial sources, often carrying organic matter, nutrients, suspended solids, heavy metals, and

Y. Zhu (Ed.): ISESIE 2025, LNCE 000, pp. 235–254, 2026.
https://doi.org/10.1007/978-981-92-2389-3_19

pathogens. Combined with climate change–induced rainfall variability, these pressures have intensified water scarcity, resulting in economic, human, and environmental instability [1]. Wastewater management therefore represents both a challenge and an opportunity for advancing sustainable development. Properly collected, treated, reused, and disposed of, wastewater can alleviate scarcity while contributing to multiple Sustainable Development Goals (SDGs), particularly SDG 6 (Clean Water and Sanitation), SDG 7 (Affordable and Clean Energy), SDG 11 (Sustainable Cities and Communities), and SDG 13 (Climate Action) [2]. Conversely, poor management results in serious health and ecological risks.

Urban Wastewater Treatment Plants (WWTPs) are central to achieving this balance. They are no longer mere end-of-pipe facilities but evolving eco-technological systems that integrate water purification with energy generation and resource recovery. By enabling nutrient recycling, energy self-sufficiency, and safe water reuse, WWTPs advance the principles of the circular economy while supporting resilient urban development. WWTPs operate through three fundamental stages:

(i) *Influent Reception*—collection and handling of domestic sewage, industrial effluents, and stormwater. Efficient management at this stage is critical, as influent variability directly impacts operational efficiency and opportunities for resource recovery.
(ii) *Treatment Processes*—a sequence of physical, chemical, and biological operations, including screening, sedimentation, activated sludge, anaerobic digestion, and tertiary polishing. Worker and process safety are integral here, reducing risks of chemical exposure, accidents, or uncontrolled releases.
(iii) *Effluent and Resource Outputs*—safe discharge or reuse of treated water, along with valorisation of sludge through composting, nutrient recovery, or biogas generation. This stage embodies the shift toward circularity by transforming waste streams into value-added products and renewable energy.

Recent advancements demonstrate the potential for extracting high-value materials such as cellulose, bioplastics, phosphate, and alginate-like exopolymers from sludge, alongside biogas and biochar production. Such innovations not only reduce environmental burdens but also create economic opportunities. At the same time, occupational safety and process safety remain critical enablers, ensuring that circularity-driven operations are sustainable for both workers and communities. The growing recognition of WWTPs as hubs for circular economy innovation and sustainable urban infrastructure has prompted extensive scholarly attention. Recent studies have examined wastewater within broader nexus frameworks, linking water management with energy, food, and climate systems, while also highlighting the integration of industrial safety and occupational health into circular practices. A critical review of this body of literature is therefore essential to understand how circular economy principles, nexus approaches, and safety considerations converge in wastewater management, providing the foundation for the present study.

2 Literature Review

The literature on wastewater management increasingly emphasizes the integration of circular economy principles, nexus frameworks, and occupational safety considerations. Studies highlight how WWTPs are evolving from pollution-control utilities into resource recovery hubs that contribute to water reuse, energy generation, and nutrient recycling. At the same time, nexus approaches—particularly the water–energy–food–climate linkages—offer system-level perspectives that align wastewater treatment with sustainable development objectives. Parallel to these themes, a growing body of work underscores the importance of embedding occupational safety and health (OSH) practices into wastewater operations to ensure both worker protection and resilient, sustainable processes. This section critically reviews the state of knowledge across these interconnected areas, beginning with circular economy applications (Sect. 2.1), followed by nexus approaches (Sect. 2.2), and industrial safety integration (Sect. 2.3).

2.1 Circular Economy in Wastewater Treatment Plant

According to the circular economy (CE) philosophy, waste should be viewed as a secondary raw resource that can be recycled for processing and repurposing, while goods, materials, and raw inputs remain in circulation for as long as possible. This sets CE apart from the linear economy, which follows the "take–make–use–dispose" model where waste is typically the final stage of a product's life cycle [3]. By reducing waste production and reintroducing residuals as secondary materials, the CE approach promotes sustainable material and energy management. Despite its potential, wastewater management has historically received less societal and political attention compared to other water resource issues. Yet neglecting wastewater management negatively influences the long-term sustainability of water supplies, human health, economic growth, and ecosystems. The United Nations reported in 2017 that more than 80% of wastewater worldwide is discharged without adequate treatment [4]. Treatment coverage strongly correlates with income level: high-income countries treat over 70% of their wastewater, compared with 38% in upper middle-income, 28% in lower middle-income, and only 8% in low-income countries [5].

A paradigm shift is therefore required, positioning wastewater as not merely a waste stream but as a carrier of valuable resources including nutrients, organic matter, and embedded energy. Designing treatment systems with an emphasis on energy recovery and treatment has garnered attention in recent years as global energy consumption rises by nearly 50% between 2010 and 2040 [6]. When effectively recovered, these can be transformed into marketable products such as biofertilizers, biogas, or reclaimed water, thus contributing to both economic viability and environmental sustainability. For instance, the Strass WWTP in Austria has demonstrated the feasibility of energy-positive operations by optimizing anaerobic digestion for biogas recovery, while Singapore's NEWater program illustrates how advanced treatment enables potable reuse at scale. These examples highlight the dual role of WWTPs as environmental protectors and resource recovery hubs [7]. The adoption of a CE framework ensures that wastewater treatment extends beyond contaminant removal to actively closing resource loops. Within this framework, safety and quality assurance in water reuse are critical: treatment processes must meet quality standards appropriate to the intended application, whether for agricultural irrigation, industrial processes, or potable supply. This necessitates rigorous risk management and monitoring protocols to safeguard both workers and end-users. Advanced technologies, including energy-efficient treatment systems, nutrient recovery modules, and smart digital monitoring, can enhance performance while lowering long-term costs and regulatory risks.

Generally, CE-driven WWTP strategies must be designed with system-level integration in mind, connecting urban water cycles, energy systems, and agricultural resource flows through industrial symbiosis. This systemic perspective enhances resilience while aligning with the Sustainable Development Goals. However, the shift toward circularity also introduces complex trade-offs between water, energy, safety, and environmental outcomes. Addressing these interdependencies requires moving beyond CE principles alone and adopting a nexus-based approach, which is explored in the Sect. 2.2.

2.2 Analytical Foundation—CE Strategies, Nexus Framing, Safety Integration

CE principles in wastewater treatment aim to minimize virgin resource extraction, extend material life cycles, and close water–energy–nutrient loops. Unlike conventional treatment systems focused primarily on pollutant removal, CE emphasizes resource recovery and valorization, positioning wastewater as a carrier of secondary raw materials. Recent applications demonstrate the breadth of CE strategies in this sector:

(i) Nutrient Recovery—Techniques such as struvite crystallization, ammonia stripping, and algae-based assimilation recover phosphorus, nitrogen, and potassium from municipal and industrial effluents, reducing reliance on finite fertilizers and closing nutrient cycles [8–10]
(ii) Energy Recovery—Anaerobic digestion (AD) of sewage sludge produces biogas (~ 65% methane) for combined heat and power (CHP) generation. Process enhancements (e.g., thermal hydrolysis, co-digestion) improve yields and promote WWTP energy self-sufficiency [11, 12].
(iii) Water Reuse—Treated effluent is increasingly repurposed for agricultural irrigation, industrial operations, urban non-potable use, and even potable applications, reducing freshwater withdrawals and enhancing water security [13, 14]
(iv) Other Resource Pathways—Emerging technologies enable recovery of metals from sludge ash, bioplastic (PHA) production, and direct electricity generation via bio electrochemical systems [15, 16].

While these CE pathways provide economic and environmental gains, their integration into existing WWTPs requires balancing process efficiency, treatment reliability, and regulatory compliance.

Nexus Approaches in Wastewater Management: The nexus perspective, linking water, energy, food, and industrial systems, offers a structured framework to address interdependencies in WWTP operations. In practice, nexus thinking enhances CE implementation by ensuring resource recovery does not compromise other system objectives:

(i) Water–Energy–Nutrient Nexus—Trade-offs emerge when optimizing one stream. For example, maximizing nutrient recovery may increase energy demand, while energy-efficient processes can reduce recovery efficiency [17, 18].
(ii) CE–Nexus Integration—Embedding CE pathways within a nexus framework enables holistic planning of WWTPs, ensuring balanced outcomes for reuse, recovery, and environmental protection. Such integration is increasingly central to Smart City and urban resilience agendas [19, 20].
(iii) Multi-Scale Applications—Nexus-informed planning operates at plant, city, and regional levels, linking treatment optimization with broader energy grids, food systems, and water distribution networks [21, 22].

Safety Integration within CE–Nexus Approaches: Embedding safety into CE–nexus frameworks ensures that recovery processes do not introduce new occupational, environmental, or health risks. Biogas handling, chemical recovery, and effluent reuse require robust risk assessments and safety management systems to safeguard workers, communities, and ecosystems. Thus, the analytical foundation for sustainable WWTPs lies in the triad of CE strategies, nexus framing, and safety integration, which collectively enable resilient, efficient, and safe circular systems.

By framing wastewater treatment within a nexus approach, researchers and practitioners can maximize the benefits of CE strategies while identifying trade-offs and interdependencies among water, energy, nutrients, and other resources. However, the integration of these circular and nexus-driven innovations also introduces new layers of industrial safety challenges. From chemical handling in nutrient recovery to explosion risks in biogas systems, advancing sustainability goals requires simultaneous attention to occupational and process safety. This sets the stage for a critical examination of industrial safety integration in WWTPs.

2.3 Industrial Safety Integration

The transition of WWTPs toward circular and nexus-driven models introduces new layers of industrial safety challenges that must be addressed in parallel with sustainability objectives. As resource recovery pathways expand, so too do risks across chemical, biological, energy, and mechanical domains:

(i) Chemical Hazards—Processes such as struvite precipitation, ammonia stripping, and biosolid stabilization involve the use of acids, alkalis, flocculants, and disinfectants [23]. These create risks of chemical burns, inhalation exposure, and accidental releases, necessitating robust handling protocols and containment systems [24].
(ii) Biological Risks—Operators are routinely exposed to pathogens, bioaerosols, and residual contaminants during nutrient and water recovery operations. Effective biohazard management, protective equipment, and occupational health monitoring remain essential safeguards [25].
(iii) Energy and Process Safety—Biogas generation, storage, and utilization carry explosion and fire hazards, while thermal processes (e.g., sludge drying, thermal hydrolysis, combined heat and power [CHP] integration) add risks of burns and high-pressure failures [26]. Process safety management systems are critical in mitigating these risks [27].

(iv) Mechanical and Electrical Hazards—Pumps, mixers, conveyors, and automated control units pose entanglement, crushing, and electrocution risks. As CE technologies are integrated into WWTPs, safety protocols must evolve to address increasingly complex electromechanical interactions [28].

Recent studies underscore that industrial safety cannot remain an afterthought in circular wastewater systems. Instead, safety-by-design principles, systematic risk assessments, and continuous operator training must be embedded into CE implementation strategies. This integration ensures that efforts to recover resources and optimize system performance do not compromise worker health, community safety, or environmental integrity [29]. Embedding safety as a core pillar of the CE–nexus framework transforms it from a compliance requirement into an enabler of sustainable and resilient WWTP operations.

2.4 Synthesis and Research Gaps

The literature clearly establishes that circular economy (CE) strategies and nexus frameworks can transform wastewater treatment plants (WWTPs) from end-of-pipe facilities into resource recovery hubs. Nutrient recycling, energy generation, and water reuse are increasingly proven, while multi-resource nexus thinking supports integrated planning. At the same time, industrial safety integration has emerged as a non-negotiable condition for sustainable operations. Yet, despite these advances, the field remains fragmented and underdeveloped in several critical respects:

(i) Holistic Risk–Sustainability Integration: Research has not sufficiently embedded occupational and process safety into CE implementation. Most studies address safety as a regulatory add-on rather than as a core enabler of sustainable and resilient WWTP transitions.
(ii) Operational Nexus Modelling: While conceptual nexus frameworks are widely discussed, empirical modelling at the plant scale—capturing trade-offs between energy use, nutrient recovery, emissions, and safety performance—remains sparse. Without process-level data, optimization across resources and risks cannot be fully achieved.
(iii) Socio-technical Acceptance and Governance: Worker safety culture, operator training, and public trust in reclaimed resources are consistently underexplored. Social resistance to water reuse and recovered fertilizers undermines circularity, while governance frameworks have yet to fully integrate safety and circular economy objectives.

Addressing these gaps requires a **genuinely interdisciplinary research agenda**, linking process engineering with occupational health, environmental management, and urban policy. Advancing CE–nexus–safety integration in WWTPs will not only enhance sustainability outcomes but also secure operational reliability, regulatory compliance, and social legitimacy, foundations without which circular transitions cannot succeed. Section 2.5 develops a conceptual framework that bridges these gaps by explicitly integrating CE pathways, nexus approaches, and industrial safety considerations into a unified model for sustainable wastewater management.

2.5 Circular Economy, Nexus, and Industrial Safety Framework

Figure 1 illustrates the convergence of three dimensions within wastewater treatment plant (WWTP) operations: circular economy (CE) pathways, nexus-based system optimization, and industrial safety management. This integration highlights that CE initiatives in WWTPs cannot be pursued in isolation; they must be designed with explicit attention to safety and evaluated through multi-resource trade-off analyses. In this framework, the WWTP functions as the central node, enabling multiple recovery pathways for energy, nutrients, and water, while simultaneously serving as a point of risk governance and sustainability planning.

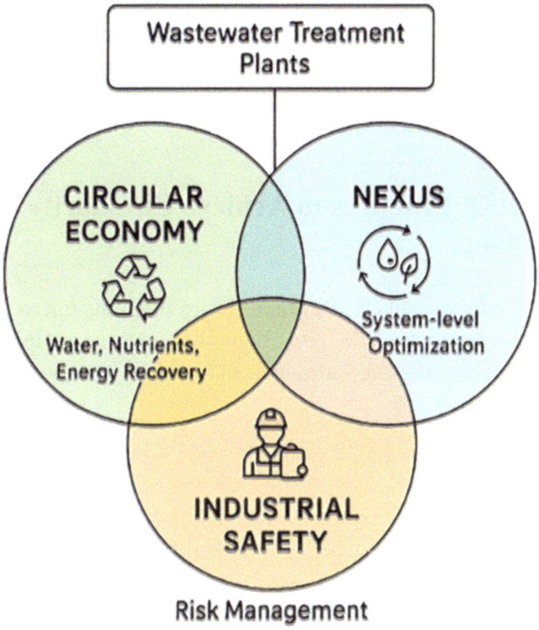

Fig. 1. Conceptual framework for integrating circular economy, nexus, and industrial safety in Wastewater Treatment Plants

(i) Circular Economy Dimension: The CE loop emphasizes continuous recovery and reintegration of resources from wastewater streams. Examples include biogas production from sludge digestion, nutrient extraction from biosolids, and treated water reuse for irrigation or industrial applications. These closed-loop strategies reduce dependence on virgin resources, minimize ecological impacts, and enhance the resilience of urban and industrial systems.

(ii) Industrial Safety Dimension: Safety is embedded as the **core** safeguard of the CE framework. Every recovery process, nutrient extraction, energy generation, or water reuse, introduces chemical, biological, mechanical, or process-related risks. Integrating safety-by-design, robust hazard management, and continuous operator training ensures that circularity does not compromise human health, plant reliability, or community well-being.

(iii) **Nexus Dimension**: The nexus perspective captures the interdependencies among water, energy, and climate systems. By applying nexus-based optimization, WWTPs can achieve multiple co-benefits: reducing greenhouse gas emissions, conserving freshwater resources, and producing renewable energy. This approach allows decision-makers to balance competing priorities and achieve sustainable outcomes across multiple domains simultaneously.

Together, these dimensions form a holistic framework where CE initiatives are technically feasible, economically viable, environmentally sustainable, and socially acceptable. Embedding nexus trade-offs and safety imperatives into circular wastewater planning provides the systemic resilience required for long-term urban water management in line with global sustainability goals.

3 Integrating HSE Practices to Achieve Circularity in Wastewater Treatment Plants

Health, Safety, and Environment (HSE) practices are fundamental to achieving circularity and sustainable operations in WWTPs. Beyond regulatory compliance, they act as a strategic enabler, ensuring the safe, efficient, and socially accepted implementation of circular economy initiatives.

(i) Safe Resource Recovery and Reuse: Circular processes in WWTPs frequently involve handling materials that pose health risks. Land application of biosolids, water reuse, and nutrient recovery can expose workers and communities to pathogens, odours, or contaminants if not carefully managed. Rigorous safety protocols mitigate these risks, ensure safe handling of by-products, and build societal trust in resource recovery initiatives. High-quality effluent and properly managed biosolids reinforce confidence among regulators and end-users, supporting the long-term sustainability of reuse practices.

(ii) Operational Efficiency and Reliability: Robust HSE management prevents disruptions, enhances reliability, and safeguards workforce performance. By reducing workplace incidents and environmental hazards, WWTPs maintain uninterrupted energy production, nutrient recovery, and water treatment—key elements for closing material and energy loops. Digital tools, such as AI and real-time monitoring, strengthen both safety and operational efficiency by predicting hazards, optimizing energy use, and streamlining compliance reporting.

(iii) Corporate Governance and Reputation: Strong HSE practices reinforce corporate governance, enhance reputation, and ensure regulatory compliance. They attract investment, build stakeholder confidence, and sustain the social license to operate, factors critical for public acceptance of circular economy projects. By integrating HSE into strategic governance, WWTPs align with ESG objectives and strengthen long-term resilience.

These interconnections between resource recovery, operational performance, and governance in enabling circularity are summarized in Fig. 2.

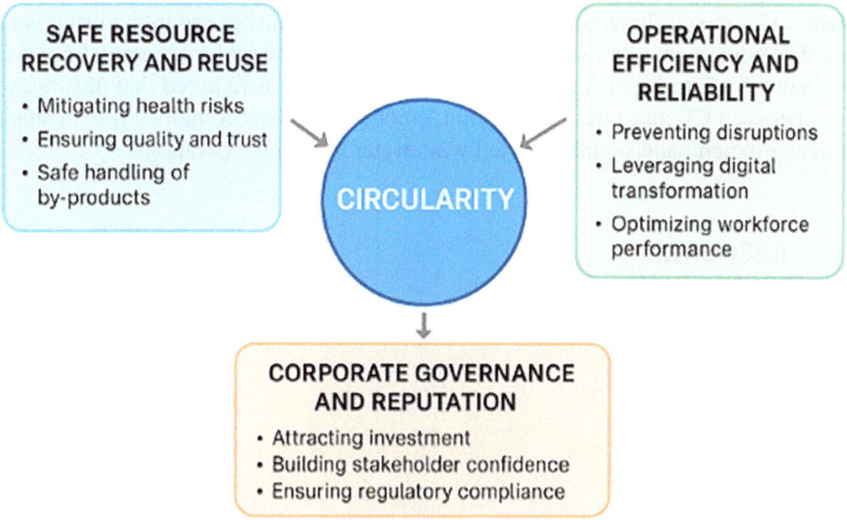

Fig. 2. Conceptual linkages between HSE practices and circularity outcomes in wastewater treatment plants.

4 Resource Circularity and Operational Safety (Process Level)

The WWTPs are increasingly recognized not only as essential infrastructure for water sanitation but also as hubs for resource recovery and circular economy implementation. Achieving circularity in these systems requires more than technological upgrades; it demands careful integration of operational safety, process reliability, and environmental protection. The quality and variability of influent, biological treatment efficiency, and energy recovery processes are deeply interdependent, with fluctuations in one component potentially affecting others.

This section examines how CE principles and safety considerations intersect in WWTP operations. Section 4.1 presents the conceptual and practical nexus of CE and safety integration, highlighting interconnections between operational processes and risk mitigation. Section 4.2 focuses on safe resource recovery and reuse, detailing strategies for sludge handling, nutrient reclamation, and energy recovery while protecting workers and the environment. Section 4.3 explores operational efficiency and reliability, emphasizing process optimization and monitoring as enablers of both circularity and safety. Together, these subsections illustrate how WWTPs can operationalize circular economy objectives without compromising human or ecological health.

4.1 Nexus of CE and Safety Integration in WWTP Operations

The preceding section highlighted how HSE practices reinforce circularity in WWTPs by ensuring safe resource recovery, operational reliability, and enhanced corporate governance. However, achieving sustainability requires moving beyond individual practices toward a systemic integration of safety within the broader circular economy framework. To capture this interdependence, a conceptual framework is required that illustrates the nexus between CE and safety integration, emphasizing their combined role in shaping resilient, efficient, and socially trusted wastewater treatment operations.

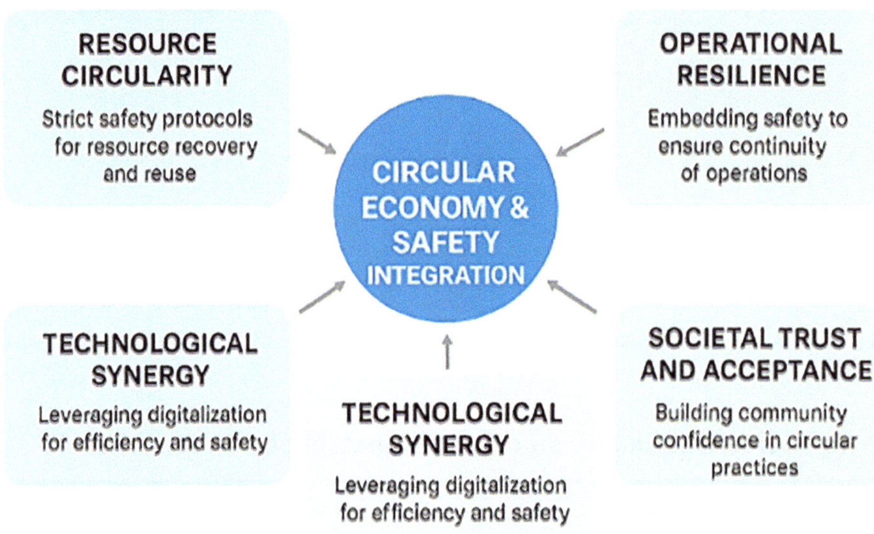

Fig. 3. Conceptual framework—Nexus of CE and safety integration in WWTP operations

The transition toward CE in WWTPs requires not only the recovery of resources and energy but also the systematic integration of occupational safety and environmental safeguards. Traditional models of WWTPs emphasize linear "treat-and-dispose" approaches, often overlooking the interplay between circularity and safety. In contrast, the proposed conceptual framework (Fig. 3) positions safety integration as both an enabler and outcome of circular practices, ensuring that resource recovery, reuse, and process optimization remain sustainable, resilient, and socially acceptable. The Core Nexus Dimensions are

(i) Resource Circularity—Recovery of water, energy, nutrients, and materials must follow strict occupational and environmental safety protocols. For example, phosphorus recovery, sludge-to-energy conversion, or effluent reuse are contingent on safe handling and monitoring to avoid health hazards.
(ii) Operational Resilience—Embedding safety in daily operations minimizes accidents, prevents contamination events, and ensures reliable system performance. This resilience strengthens the continuity of circular flows.

(iii) Regulatory and Governance Alignment—Integrating HSE with CE allows WWTPs to align with ESG standards, SDG targets, and evolving safety regulations, thereby creating a supportive governance environment.
(iv) Societal Trust and Acceptance—Circular solutions such as potable reuse or biosolid application are only viable when communities perceive them as safe. Transparent safety integration enhances social acceptance and stakeholder trust.
(v) Technological Synergy—Digitalization, real-time monitoring, and predictive analytics form a shared platform for both CE efficiency and safety assurance, enabling proactive risk management while optimizing circular loops.

Framework Contribution: The framework illustrates that CE adoption in WWTPs cannot be separated from safety considerations. Instead, safety acts as a cross-cutting enabler that links technological innovation, governance mechanisms, and societal acceptance. By adopting this integrated model, WWTPs can achieve not only environmental sustainability but also operational continuity, worker protection, and long-term socio-economic value creation.

WWTPs are no longer just sanitation facilities but emerging hubs for resource recovery and CE practices. While reclaiming water, energy, and nutrients creates value, it also introduces risks that can threaten safety, reliability, and public trust. Embedding OSH within these processes is therefore essential. Safety and circularity must advance together, ensuring that recovery and reuse deliver sustainable, resilient, and socially accepted outcomes. This interdependence is captured in Fig. 3, which illustrates the dynamic interactions between wastewater flows, resource recovery, and operational safety within a circular economy framework

4.2 Safe Resource Recovery and Reuse

Building on the principles of safe resource recovery and reuse, this interdependence is captured in Fig. 3, which illustrates the dynamic interactions between wastewater flows, resource recovery, and operational safety within a circular economy framework. Resource recovery is central to circular economy practices in WWTPs but carries inherent health and operational risks. Influent variability and high organic loads can trigger odour, volatile compound release, and pathogen exposure, particularly during biological treatment where insufficient aeration leads to incomplete degradation and anaerobic pathogen proliferation. Again, Sludge recirculation stabilizes microbial processes but introduces risks from concentrated solids and pathogens. Anaerobic digestion enables energy recovery but requires strict monitoring and handling of methane, hydrogen sulfide, and other hazardous gases to prevent explosions and protect personnel. Consequently, Robust OSH measures, including PPE, pathogen monitoring, controlled sludge handling, and careful process oversight—allow safe recovery of nutrients and energy while maintaining effluent quality. Figure Y illustrates the interdependence of operational safety, resource recovery, and environmental performance, highlighting how circular economy principles can be applied without compromising human or ecological health.

4.3 Operational Efficiency and Reliability

Building on safe recovery practices, operational efficiency and reliability are equally critical for sustaining circular processes in WWTPs. Effective Health, Safety, and Environment (HSE) management not only prevents accidents and environmental hazards but also minimizes unplanned downtime, ensuring continuity in water treatment, nutrient recovery, and energy generation—the essential loops of a circular economy framework [30]. Digital innovations now play a transformative role in coupling safety with efficiency. Artificial intelligence (AI)-driven hazard prediction, real-time effluent monitoring, and automated compliance systems enable proactive control of treatment operations. AI-based adaptive control allows operators to forecast influent variability, anticipate process upsets, and optimize aeration or chemical dosing, thereby reducing risks and improving treatment outcomes. Similarly, real-time monitoring technologies (e.g., UV/Vis spectral analysis for COD and BOD) facilitate rapid process adjustments, ensuring regulatory compliance and protecting environmental quality. Automated reporting systems further streamline compliance obligations, reduce administrative burdens, and support long-term strategic planning through instant access to historical operational data.

Equally important is the cultivation of a strong safety culture that integrates workforce training, leadership commitment, and participatory risk management. Such engagement enhances employee morale, reduces human error, and fosters organizational resilience. Together, these measures ensure that WWTPs can operate reliably, achieve efficiency gains, and deliver sustainable outcomes without compromising safety. Together, these measures ensure that WWTPs can operate reliably, achieve efficiency gains, and deliver sustainable outcomes without compromising safety.

5 Governance, OSH Integration, and Stakeholder Confidence

Effective implementation of circular economy principles in WWTPs extends beyond process-level interventions to include organizational governance, OSH integration, and stakeholder engagement. Corporate policies, transparent management practices, and compliance with ESG standards are critical enablers for ensuring that operational safety and resource recovery efforts are consistently maintained and recognized.

This section explores how governance structures and OSH frameworks intersect to enhance accountability, regulatory compliance, and public trust. Section 5.1 addresses corporate governance and reputation, highlighting the role of leadership, policy enforcement, and organizational culture. Section 5.2 focuses on the integrative perspective of OSH in WWTP operations, detailing mechanisms for embedding safety into daily processes and decision-making. Finally, Section 5.3 examines how proactive safety management and transparent reporting build stakeholder confidence and strengthen the nexus between circular economy objectives and social, environmental, and corporate responsibilities. Together, these discussions illustrate that achieving sustainable circularity in WWTPs requires coordinated organizational strategies alongside technical process management.

5.1 Corporate Governance and Reputation

Reliability in operations does more than sustain circular processes, it directly strengthens corporate governance and organizational reputation. When WWTPs integrate OSH practices into day-to-day management, they not only safeguard workers but also demonstrate accountability to regulators, investors, and communities. This transparent approach to safety and sustainability fosters trust, supports ESG compliance, and reduces both financial and reputational risks. Embedding OSH into governance structures also provides organizations with a stronger "social license to operate," particularly when implementing innovative circular economy solutions such as co-digestion, nutrient recovery, or advanced biogas utilization.

Case studies from European WWTPs illustrate this link: in Poland, the Tychy–Urbanowice plant achieved energy self-sufficiency through co-digestion while maintaining strict OSH protocols, ensuring worker protection alongside renewable energy gains [31]. Similarly, in Germany, enhanced thermal hydrolysis and phosphorus recovery systems not only improved resource recovery efficiency but also demonstrated compliance with safety and environmental standards, securing stakeholder confidence [32]. At the industry level, companies like Veolia and Xylem increasingly disclose ESG performance linked to safety integration, reflecting a broader recognition that corporate resilience in the water sector depends on transparent, safe, and sustainable practices [33]. In this way, governance and reputation become strategic enablers of circular economy initiatives, ensuring that WWTPs maintain credibility while advancing both environmental and social objectives.

5.2 Integrative Perspective of OSH into WWTP Operations

The preceding sections highlight how OSH considerations intersect with circular processes in WWTPs, first at the operational level through safe resource recovery (Sect. 5.1), then in enabling reliable and efficient plant performance (Sect. 5.2), and finally in strengthening corporate governance and stakeholder trust (Sect. 5.3). Taken together, these dimensions demonstrate that OSH is not a peripheral requirement but a unifying framework for advancing sustainable, circular wastewater management.

5.2.1 Strengthening Corporate Governance through OSH Integration

Corporate governance in WWTPs is significantly strengthened when OSH practices are embedded into organizational culture and strategic decision-making. Beyond mere compliance, this means that safety becomes central to leadership priorities, enabling ethical, resilient, and data-informed operations. For instance, research conducted in a Chinese sewage treatment plant utilized a WSR-based risk evaluation model (integrating equipment, safety systems, and operator training), leading to targeted safety improvements and enhanced governance through structured risk management [34]. Similarly, a Slovak WWTP study demonstrated how comprehensive risk assessment, including technological, human, and environmental threats, fortifies operational resilience and supports governance frameworks attuned to both safety and community expectations

[35]. Addressing these gaps through targeted capacity upgrades and safety improvements not only mitigated risks but also strengthened governance by institutionalizing evidence-based and accountable decision-making.

5.2.2 Ensuring ESG Compliance through Proactive Safety Measures

OSH integration plays a pivotal role in meeting ESG criteria, particularly across the social and environmental dimensions. A study of four Italian WWTP operators highlighted that proactive management of biological hazards, such as airborne pathogens, through risk assessment, health surveillance, and site-specific safety practices is essential for safeguarding worker health while ensuring regulatory compliance [36]. Complementing this, structured safety management tools—including audits and hazard identification frameworks, provide systematic means to document compliance and enhance transparency in ESG performance reporting [28].

5.2.3 Building Stakeholder Confidence through Transparent Safety Management

Transparency in OSH not only meets regulatory requirements but also builds stakeholder engagement and public trust. Workers in WWTPs face risks such as confined-space entry, slips, hazardous chemicals, and airborne pathogens, making proactive monitoring and safety communication essential. Periodic occupational health monitoring programs, supported by awareness campaigns and proper use of protective devices, have been highlighted as critical safeguards [37]. Similarly, first-aid preparedness, through well-equipped stations, trained personnel, and regular emergency drills, remains a core employer responsibility in ensuring timely medical response [38].

Practical guidance is consolidated in technical manuals. Spellman and Welsh [39] emphasize structured safety training, workplace audits, hazard labeling in compliance with Occupational Safety and Health Administration (OSHA) requirements, and first-aid training. OSHA guidelines further strengthen this approach by mandating systematic hazard inspections, covering stations, equipment, storage, and maintenance, supported by tailored checklists for specific tasks such as electrical work, fire protection, and ergonomics [40]. These procedures reinforce transparency, demonstrating measurable commitment to safety. Finally, personal protective equipment (PPE) remains the most visible and non-negotiable marker of stakeholder trust. Risks in WWTPs can be substantially reduced by equipping workers with respirators, masks, goggles, gloves, boots, and protective clothing [41]. The consistent use and transparent reporting of such measures signal to regulators, communities, and investors that safety is prioritized, thereby strengthening the social license to operate and consolidating stakeholder confidence in WWTP operations.

5.3 Nexus of Circular Economy and Safety in WWTP Operations

Safety in WWTPs underpins the successful implementation of CE strategies across the water–energy–food–climate nexus. Resource recovery and reuse—from nutrient reclamation to biogas production—carries inherent biological, chemical, and physical risks that must be managed through robust OSH practices. Safe wastewater reuse supports agricultural productivity and public health, while energy recovery via anaerobic digestion requires strict controls to prevent explosions or environmental releases. Integrating OSH into these processes ensures that circular economy benefits are realized without compromising worker safety or ecological integrity. Table 1 summarizes the relationships between key resources, CE opportunities, safety concerns, and operational challenges.

Table 1 Circular economy opportunities, safety concerns, and implementation challenges in WWTP resource recovery

Resource/process	Circular economy opportunities	Safety concerns	Challenges/barriers	Reference
Nutrient recovery	Nutrient recovery from wastewater, biosolids (P as struvite, N & K recovery); use of microalgae for fertilizer and feed; reduces fertilizer demand	Exposure to chemicals (acids, bases, flocculants); biohazard risks from pathogens; odour and allergen issues	public acceptance of biosolids and water separation; high costs (operational/chemical); limited uptake of aqua-species technologies	[8, 42]
Water reuse	Freshwater conservation; improved effluent quality; reuse in agriculture, industry, urban sectors; direct/indirect potable reuse	Risk of pathogen or chemical exposure; potential contamination if safety protocols fail	High infrastructure and operational costs; Social resistance to direct potable reuse	[10, 13]
Energy recovery	Biogas (~ 6.5 kWh/m^3); enhanced AD via pretreatment; co-digestion with organics; heat recovery from effluent/sludge	Explosion and fire hazards due to methane; burns and heat stress from thermal processes; mechanical and electrical hazards	Many WWTPs lack energy optimization; integration requires specialized safety monitoring	[12, 43]

(*continued*)

Table 1 (*continued*)

Resource/process	Circular economy opportunities	Safety concerns	Challenges/barriers	Reference
Other resource recovery	Sewage sludge ash in construction; recovery of Cu/Ag/Au; bioplastics (PHA) from biomass; electricity via biogas	Chemical hazards from ash/bioplastics; electrical risks in BFCs; handling hazards in metal recovery	Emerging technologies at pilot scale; economic feasibility and scaling remain limited	[7, 25]

Embedding OSH into these recovery streams enhances systemic resilience. Effective hazard assessment, PPE provision, and transparent safety reporting build stakeholder trust, essential for public acceptance of water reuse, biosolids, and energy applications. Safety thus functions as both a prerequisite and guarantor of circularity, enabling resource loops to operate efficiently while preventing unintended socio-environmental consequences. In this way, a WWTP's transition toward circular economy alignment is inseparable from its commitment to rigorous safety management and organizational governance.

6 Integrative Discussion and Future Perspectives

The preceding sections have demonstrated that achieving CE objectives in WWTPs requires an integrated approach spanning both process-level operations and organizational governance. Resource circularity, operational efficiency, and occupational safety are deeply interdependent: fluctuations in influent quality, sludge handling, or energy recovery processes can directly affect worker safety, process reliability, and environmental performance [44]. At the same time, robust governance structures, transparent OSH integration, and proactive stakeholder engagement are essential to ensure that technical measures translate into sustainable outcomes. From an operational perspective, safe resource recovery and reuse, including nutrient reclamation, biogas production, and water recycling, relies on continuous hazard assessment, monitoring, and adherence to safety protocols. These measures not only protect workers and downstream users but also maintain high effluent quality and ecological integrity. Operational efficiency further enhances circularity by optimizing aeration, sludge recirculation, and energy recovery, ensuring that process performance is resilient to variability in influent loads and other dynamic factors.

At the organizational level, governance mechanisms, ESG compliance, and transparent safety reporting play a critical role in embedding CE principles within the institutional framework of WWTPs. The CE–safety nexus underscores that circularity cannot be achieved solely through technological or process interventions; it requires alignment between operational practices, risk management, and stakeholder trust. Effective OSH integration serves as both a prerequisite and guarantor of circularity, preventing the transfer of hazards to communities, ecosystems, or downstream users. Despite these opportunities, several challenges remain. Variability in influent composition, the pilot-scale nature of emerging recovery technologies, and social resistance to direct potable reuse or biosolid applications can limit the realization of CE objectives. High capital and operational costs, gaps in technical expertise, and regulatory uncertainty further constrain adoption. Addressing these barriers necessitates coordinated efforts in process optimization, policy alignment, and workforce training. To advance safe circularity in WWTPs, the following recommendations are proposed:

(i) Implement continuous training, hazard assessment, and safety monitoring integrated with resource recovery processes.
(ii) Align governance and operational frameworks with ESG objectives to strengthen public trust and ensure regulatory compliance.
(iii) Promote research and pilot studies on emerging resource recovery technologies, embedding safety considerations from the outset.
(iv) Evaluate social, behavioural, and policy factors influencing stakeholder acceptance of reused water, biosolids, and energy outputs.

Overall, the sustainable transition of WWTPs toward circular economy alignment requires a holistic, nexus-based approach that integrates technical operations, safety management, and organizational governance. By embedding OSH into process-level strategies and institutional frameworks, WWTPs can ensure that resource loops function efficiently, risks are mitigated, and environmental and public health outcomes are safeguarded. This integrative approach provides a roadmap for resilient, sustainable wastewater management that reconciles economic, social, and ecological objectives.

7 Conclusion

This study highlights that achieving circular economy objectives in WWTPs requires the integration of technical operations, occupational safety, and organizational governance. Resource circularity—including water reuse, nutrient recovery, and energy generation—cannot be realized safely or sustainably without robust OSH measures embedded in both process-level practices and institutional frameworks. Operational efficiency, combined with vigilant monitoring and hazard prevention, ensures that recovered resources are safe, reliable, and environmentally sound. At the organizational level, governance structures, ESG compliance, and transparent safety reporting are critical enablers for translating circular economy principles into practice. The CE–safety nexus underscores that circularity and resilience are mutually dependent: effective safety management safeguards workers, communities, and ecosystems while facilitating the efficient functioning of resource loops. Overall, the findings demonstrate that WWTPs must adopt a holistic, nexus-based approach to align operational, organizational, and societal objectives. By embedding safety into circularity strategies, operators can achieve sustainable, resilient, and socially accepted wastewater management systems. Future research and policy efforts should focus on optimizing emerging technologies, strengthening governance frameworks, and fostering stakeholder trust to fully realize the benefits of safe, circular WWTP operations.

Acknowledgements. The corresponding author wishes to thank the C.V. Raman Global University (CGU), for their support in carrying out this phenomenal work.

References

1. Biswas, A., et al.: Water scarcity: a global hindrance to sustainable development and agricultural production—a critical review of the impacts and adaptation strategies. Cambridge Prism: Water. **3**, e4 (2025)
2. Ghosh, P., Chakravarty, T.: Sustainable development goals: challenges, opportunities, and the way forward. Sustain. Sci. policy, Pract. India Challenges Oppor. **235–241** (2024)
3. Geissdoerfer, M., Savaget, P., Bocken, N.M.P., Hultink, E.J.: The circular economy—a new sustainability paradigm? J. Clean. Prod. **143**, 757–768 (2017)
4. Water, U.N.: Waste Water: the Untapped Resources. Facts and Figure; the United Nations World Water Report 2017. UNESCO Publishing, Paris, France (2017)
5. Sato, T., Qadir, M., Yamamoto, S., Endo, T., Zahoor, A.: Global, regional, and country level need for data on wastewater generation, treatment, and use. Agric. Water Manag. **130**, 1–13 (2013)
6. Newell, R.G., Qian, Y., Raimi, D.: Global Energy Outlook 2015. National Bureau of Economic Research (2016)
7. Guven, H., Ersahin, M.E., Dereli, R.K., Ozgun, H., Isik, I., Ozturk, I.: Energy recovery potential of anaerobic digestion of excess sludge from high-rate activated sludge systems co-treating municipal wastewater and food waste. Energy. **172**, 1027–1036 (2019). https://doi.org/10.1016/j.energy.2019.01.150
8. Bunce, J.T., Ndam, E., Ofiteru, I.D., Moore, A., Graham, D.W.: A review of phosphorus removal technologies and their applicability to small-scale domestic wastewater treatment systems. Front. Environ. Sci. **6**, 8 (2018)
9. Mohan, S.V., Hemalatha, M., Chakraborty, D., Chatterjee, S., Ranadheer, P., Kona, R.: Algal biorefinery models with self-sustainable closed loop approach: trends and prospective for blue-bioeconomy. Bioresour. Technol. **295**, 122128 (2020)
10. Das, A., Mishra, S., Ayejoto, D.A.: Harnessing biotechnology for sustainable wastewater treatment and circular economy. Earth Syst. Environ. **1–14** (2025)
11. Lima, D., Appleby, G., Li, L.: A scoping review of options for increasing biogas production from sewage sludge: challenges and opportunities for enhancing energy self-sufficiency in wastewater treatment plants. Energies. **16**(5), 2369 (2023)
12. Mishra, S.: Anaerobic digestion as a sustainable approach to convert organic solid waste to bio energy. Shodhaganga (2024). [Online]. Available: http://hdl.handle.net/10603/595274
13. Maura, J., Atreya, S., Arshi, A.: The treatment of wastewater, recycling and reuse-past, present, and in the future. Int. J. Sci. Res. **12**(11), 210–222 (2023)
14. Maher, T., Antar, C., Alshrari, A., Ali, H.: Advancing environmental sustainability and consumption security through wastewater reuse in arid regions. Eur. J. Sustain. Dev. **14**(2), 797 (2025)
15. Meher, S., Saha, A., Philip, L.: Biotechnological innovation in valorization of sludge from wastewater for environmental sustainability. In: Smart Waste and Wastewater Management by Biotechnological Approaches, pp. 157–177. Springer, Heidelberg (2025)

16. Yesil, H., Tugtas, A.E., Çalli, B.: Beyond landfills: transforming biodegradable waste into climate solutions and valuable resources. Rev. Environ. Sci. Bio/Technology. **1–25** (2025)
17. Haitsma Mulier, M.C.G., van de Ven, F.H.M., Kirshen, P.: Circularity in the urban Water-energy-nutrients-food nexus. Energy Nexus. **7**, 100081 (2022). https://doi.org/10.1016/j.nexus.2022.100081
18. Silva-Afonso, A., Pimentel-Rodrigues, C.: Water–energy–nutrients nexus of urban environments. Water. **16**(6) (2024). https://doi.org/10.3390/w16060904
19. Mugoni, E., Shumbanhete, B., Nyagadza, B.: The nexus between circular economy (CE) and sustainable supply chain management (SSCM): a systematic literature review and mapping of future research agenda. Circ. Econ. Sustain. **1–38** (2025)
20. Zahran, S.: Investigating the nexus between green supply chain practices and sustainable waste management in advancing circular economy. Sustainability. **16**(9) (2024). https://doi.org/10.3390/su16093566
21. Hoosain, M.S., Paul, B.S., Doorsamy, W., Ramakrishna, S.: The influence of circular economy and 4IR technologies on the climate–Water–energy–food nexus and the SDGs. Water (Switzerland). **15**(4), 1–18 (2023). https://doi.org/10.3390/w15040787
22. Dhar, A.R.: Building climate-resilient food systems through the water–energy–food–environment nexus. Environments. **12**(5), 167 (2025)
23. Ulu, F., Kobya, M.: Ammonia removal from wastewater by air stripping and recovery struvite and calcium sulphate precipitations from anesthetic gases manufacturing wastewater. J. Water Process Eng. **38**, 101641 (2020). https://doi.org/10.1016/j.jwpe.2020.101641
24. Elia, V.J., et al.: Hazardous chemical exposure at a municipal wastewater treatment plant. Environ. Res. **32**(3), 360–371 (1983). https://doi.org/10.1016/0013-9351(83)90118-4
25. Prithiviraj, N., Praveen Kumar, K.: Occupational health and safety in wastewater treatment: preventing chemical hazards. Int. Res. J. Adv. Eng. Hub. **3**(03), 1021–1028 (2025). https://doi.org/10.47392/irjaeh.2025.0145
26. Marijolaine: Risks and Safety Measures for Anaerobic Digestion: how Can you Make your Plant Safer?, pp. 1–17. BiogasWorld (2018). [Online]. Available: https://www.biogasworld.com/news/safety-precautions-anaerobic-digestion-systems/
27. Hegazy, H., Saady, N.M.C., Zendehboudi, S.: Safety in biogas plants: an analysis based on international standards and best practices. Process. Saf. Environ. Prot. **200**, 107390 (2025). https://doi.org/10.1016/j.psep.2025.107390
28. Dannoun, Y., Nouban, F.: Occupational health hazards and risk assessments in wastewater treatment plant. Int. J. Adv. Eng. Sci. Appl. **2**(2), 21–25 (2021). https://doi.org/10.47346/ijaesa.v2i2.83
29. Benson, C., Obasi, I.C., Akinwande, D.V., Ile, C.: The impact of interventions on health, safety and environment in the process industry. Heliyon. **10**(1), e23604 (2024). https://doi.org/10.1016/j.heliyon.2023.e23604
30. Das, A.: Multivariate statistical approach for the assessment of water quality of Mahanadi basin, Odisha. Mater. Today: Proc. **65**, A1–A1 (2022)
31. Zubrowska-Sudol, M., Gieleciak, Z.: Circular economy implementation in the water sector: wastewater treatment plant tychy–urbanowice case study. In: Smol, M., Prasad, M.N.V., Stefanakis, A.I. (eds.) Water in Circular Economy, pp. 235–240. Springer International Publishing, Cham (2023). https://doi.org/10.1007/978-3-031-18165-8_18
32. Kleyböcker, A., Kraus, F., Meyer, S., Heinze, J., Gromadecki, F., Remy, C.: Towards carbon neutrality and circular economy: an innovative combination of enhanced biogas production and nutrient recovery from sludge dewatering liquor at a municipal wastewater treatment plant in Germany. Water Sci. Technol. **90**(3), 680–695 (2024). https://doi.org/10.2166/wst.2024.247
33. Grand View Research: Water and Wastewater Treatment Industry ESG Thematic Report. **2022** (2022)

34. Liu, X., Wang, Z., Pan, D., Zhang, Y., Zhang, J., Chen, Y.: Evaluation of operational safety risk in wastewater treatment plants based on WSR and matter–element extension theory. Water. **16**(20) (2024). https://doi.org/10.3390/w16202925
35. Tušer, I., Oulehlová, A.: Risk assessment and sustainability of wastewater treatment plant operation. Sustainability. **13**(9), 5120 (2021)
36. Porru, S., Carta, A., Bellina, B., Arici, C.: Occupational health practices for wastewater treatment workers. BMJ J. **210** (2018)
37. Tiwari, R.: Occupational health hazards in sewage and sanitary workers. Indian J. Occup. Environ. Med. **12**(3), 112–115 (2008). https://doi.org/10.4103/0019-5278.44691
38. Ahmed Elsayed, S., Elahmady Sarhan, A., Abd-Elrazek Mahmoud, A.: Occupational health hazards among workers in sewage treatment plants. J. Nurs. Sci. Benha Univ. **5**(1), 719–735 (2024)
39. Spellman, F.R., Welsh, K.: Safe Work Practices for Wastewater Treatment Plants. CRC Press (2018)
40. Spellman, F.R.: Occupational Safety and Health Simplified for the Industrial Workplace. Bernan Press (2015)
41. Sulojeva, J., Percovs, A., Maļukova, J., Urbane, V.: Occupational safety management aspects on municipal waste water treatment plant. Saf. Technog. Environ. **1**, 62–68 (2011)
42. Uddin, M.S., et al.: Coupling nexus and circular economy to decouple carbon emissions from economic growth. Sustainability. **15**(3) (2023). https://doi.org/10.3390/su15031748
43. Martins, F.F., Castro, H., Smitková, M., Felgueiras, C., Caetano, N.: Energy and circular economy: nexus beyond concepts. Sustainability. **16**(5), 1728 (2024)
44. Das, A.: An optimization based framework for water quality assessment and pollution source apportionment employing GIS and machine learning techniques for smart surface water governance. Discover Environ. **3**(1), 1–35 (2025)

Effects of Elevated Humidity and Temperatures on Tropospheric Ozone Content in Iraq

Fahmi A. Mohammed and Zaidoon T. Abdulrazzaq(✉)

Scientific Research Commission, Baghdad, Iraq
zaidoon.t.abdulrazzaq@src.edu.iq

Abstract. This work, Iraq, is used as a case study to evaluate how temperature and relative humidity affect the amount of ozone in the troposphere at mid-latitudes. Studies were conducted over the three cities for eleven year's period (2010–2020) using approximate monthly mean troposphere ozone column data from the OMI and MLS instruments on board the Aura satellite, along with relative humidity and temperature data from the Food and Agriculture Organization of the United Nations (FAO). The maximum and minimum temperatures correlated positively with ozone in every station. The results showed that relative humidity and ozone have negative correlations in Erbil, Baghdad, and Basra, respectively. It was discovered that the association between temperature, relative humidity, and ozone was stronger the higher the altitude above sea level— the relationship was strongest from Basra to Erbil. Additionally, the governorate of Baghdad experiences stronger relationships between temperature (maximum and minimum) and ozone, relative humidity, and ozone during the Spring and Autumn seasons. For the three governorates of Erbil, Baghdad, and Basra, as well as for the period (2010–2020), the seasonal changes in the troposphere ozone column were also analyzed in this study. The changes were consistent for all four seasons for all three governorates. There is no excessive air pollution in the study region that poses a serious hazard to human health.

Keywords: Relative humidity · Troposphere ozone · Maximum temperature · Minimum temperature

1 Introduction

A significant secondary pollutant, ground-level ozone is created when sunlight, volatile organic compounds (VOCs), and nitrogen oxides (NO_x) interact photo chemically [1–3]. Because of its extremely potent oxidative properties, troposphere ozone has a negative impact on ecosystems, human health (affecting lung function and causing premature death), and both. Due to industrial and urban expansion, high levels of environmental ozone have been detected in numerous cities [4, 5]. Variation in the ozone concentration at a specific site is influenced by things like the presence of sources that emit ozone precursors, the topography, and the region's prevailing weather patterns [6, 7]. Humidity affects not only the chemistry but also the meteorological processes influencing atmospheric pollutants. Atmospheric water vapor induces the formation of clouds and haze

Y. Zhu (Ed.): ISESIE 2025, LNCE 000, pp. 255–264, 2026.
https://doi.org/10.1007/978-981-92-2389-3_20

that reduces solar radiation and slows photochemical reactions [8]. The role of humidity on ambient ozone is complex. Water vapor mediates chemical reactions related to ozone photochemistry by modulating the formation of radicals, altering the efficiency of photochemical ozone production and changing the atmospheric lifetime of ozone precursors. Furthermore, atmospheric humidity influences meteorological and boundary layer processes in different ways altering the ambient concentrations of pollutants [9].

The MIROC Atmospheric Chemistry-Climate Model studied temperature and surface ozone relationships in a future 2 Ã— CO_2 climate. Simulation of biogenic emissions and photolysis rates suggests the control of ozone production chemical regimes, varying with megacity proximity. Ozone mixing ratios increased across the temperature range measured, and a linear relationship between daily maximum ozone and temperature remains after excluding local meteorology effects. At a given temperature, higher ozone levels associate with low anticyclone city, as stagnant air allows pollutant accumulation, enhancing ozone production. Positive effects of temperature on biogenic emissions and photochemistry, combined with recirculation, produce the linear relationship.

Ozone levels often peak in rural areas owing to nitrate radical- and NO-assisted degradation in urban centers, where NO concentrations are elevated. Ozone forms photochemically on warm, sunny days; as daily maximum temperature increases, the ozone concentration also rises. Both production and dispersion are temperature-dependent, with a stronger correlation observed during heat waves [10]. Sunlight-driven processes generate ozone, secondary particles (e.g., sulfate, nitrate), and primary pollutants. Although human activities concentrate pollutants in urban areas, PM10 levels increase during heat waves in Birmingham, particularly in the city center. Elevated temperatures and slow wind speeds during persistent high-pressure conditions enhance ozone, NO2, and secondary aerosol concentrations. These meteorological and temperature conditions favor the accumulation of secondary pollutants. Ozone has a brief lifetime in the atmosphere (a few hours), but in much polluted cities where its precursors are present in large quantities, it can live for several weeks [11]. The generation, dispersion, transportation, and dilution of air pollutants are all significantly influenced by meteorology. Surface ozone concentrations and its precursors are significantly impacted by changes in local climatic conditions, including wind speed, direction, temperature, and relative humidity [12]. The temporal fluctuation in surface-level ozone (O_3) observed in Gummidipoondinear Chennai, Tamilnadu, was presented by Mohan and Saranya [13]. Belan and Savkin examined the connection between changes in the ozone concentration in the near-surface air layer and the absolute humidity using data collected at the Troposphere Ozone Research (TOR) station of the Institute of Atmospheric Optics, Siberian Branch, Russian Academy of Sciences, between 1994 and 2017 [14]. Allu et al. tracked the normal temporal variation of O_3 and its precursors, such as NOx and CO, at the Tata Institute of Fundamental Research-National Balloon Facility (TIFR-NBF; 17.47°N, 78.58°E) in Hyderabad in 2016. They also recorded meteorological parameters, such as temperature, relative humidity, solar radiation, and wind speed. Correlation study was done for 2016 using metrological data and the hourly averaged concentration of trace gases [15].

In Lanzhou, Western China, from 2013 to 2017, Yuxia Ma and colleagues evaluated the intricate nonlinear interactions between O_3 concentration and the factors affecting O_3 concentration using a generalized additive model (GAM) [16]. Ozone concentration changes over time (by hours, months, and seasons) and the relationship between ozone concentration and climatic variables (temperature, air pressure, relative humidity, wind speed, and wind direction) are examined by Predrag ILIC et al. The data was collected at a station in the Republic of Srpska (Banja Luka), Bosnia and Herzegovina, between January and December of 2016 [17]. According to Liu Zihao and colleagues' analysis of the temporal and spatial variations in Hefei's air pollution and its correlations with meteorological variables, air pollutants have an oblique relationship with humidity and wind speed, a positive correlation with air pressure, and an inverse relationship with temperatureand sunshine [18].

The study is conducted in the mid-latitudes, using Iraq as a case study. Three locations in Iraq were chosen from among others to achieve a balanced distribution: Basra, Baghdad, and Erbil. This study will establish the association between seasonal fluctuations in ozone concentration in the troposphere and temperature and humidity in Iraq.

2 Materials and Methods

The approximate monthly mean troposphere ozone data (from ground to approximately 16 Km, pressure 100 hpa) for the stations were considered; Basra, Baghdad, and Erbil, which are embedded in different zones in Iraq, were obtained from AURA OMI/MLS, this is obtained from a NASA public data: https://acd-ext.gsfc.nasa.gov/Data_services/cloud_slice/new_data.html. Relative humidity and temperature data were obtained from the Food and Agriculture Organization of the United Nations (FAO). The data span is for 11 years (2010–2020). Fig. 1 shows seasonal variation of troposphere ozone column (histogram) during 2010–2020 for Erbil, Baghdad and Basra.

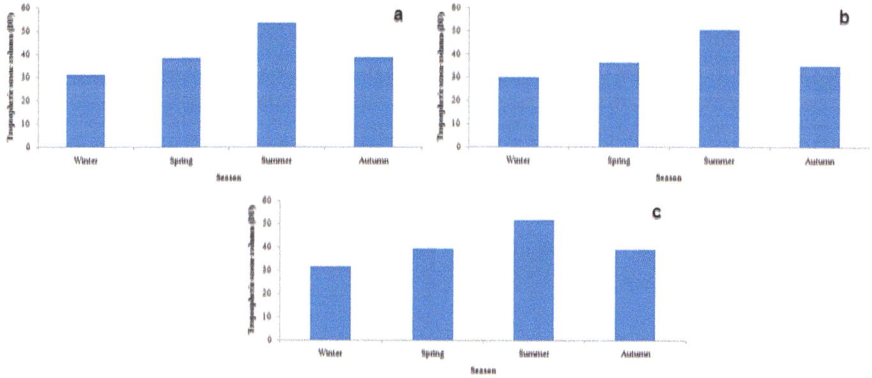

Fig. 1. Seasonal variation of troposphere ozone column during 2010–2020 for (a) Erbil, (b) Baghdad and (c) Basra.

3 Results and Discussion

The seasonal characteristics of O_3 pollution in Erbil for the 2010–2020 period are Summer> Spring > Autumn > Winter, with the greatest pollution concentrations (50,682 DU) occurring in Summer, as a result of photochemical reaction conditions (Fig. 1a).

Summer >Fall> Spring > Winter in Baghdad. Peak pollution concentrations are highest in the summer (53,527 DU), as shown in Fig. 1b. In Basra, the sequence is Summer> Spring > Autumn > Winter. The maximum pollution concentrations (51,836 DU) occur throughout the summer (Fig. 1c). Karem et al. investigated the air pollution in the West Qurna-2 oil field in southern Iraq. They discovered that the following is the arrangement of the seasonal O_3 concentration (in PPb units): Spring >Autumn> Summer. The O_3 gaseous pollutant has greater average concentrations in the winter than in the summer. The variations in weather, particularly in wind direction and speed, air temperature, and humidity, may be the main cause of this [19].

Due in large part to the local NO_2 photochemistry and the dry high-pressure circumstances, the summer season showed the highest O_3 concentrations. The maximum troposphere ozone concentrations typically happen during the summer and at midday because photochemical reactions are most intense on sunny, warm days. Ozone concentrations are lower in the Winter due to beneficial weather conditions and the lower sun angle that typically keeps ozone concentrations down than in Spring and Summer. Ozone concentrations are almost equal in the spring and autumn seasons. This is because the temperatures are almost equal or close in these two seasons. Also, the period of these two seasons is short, so that changes in ozone concentrations cannot be clearly observed, unlike what happens during the summer and winter seasons. Fig. 2 shows the linear relationship between monthly troposphere ozone column and max. temperature during 2010–2020 for Erbil, Baghdad and Basra, respectively. Figure 3 shows the linear relationship between monthly troposphere ozone column and min. temperature during 2010–2020 for Erbil, Baghdad and Basra, respectively. Figure 4 shows the linear relationship between monthly troposphere ozone column and relative humidity during 2010–2020 for Erbil, Baghdad and Basra, respectively.

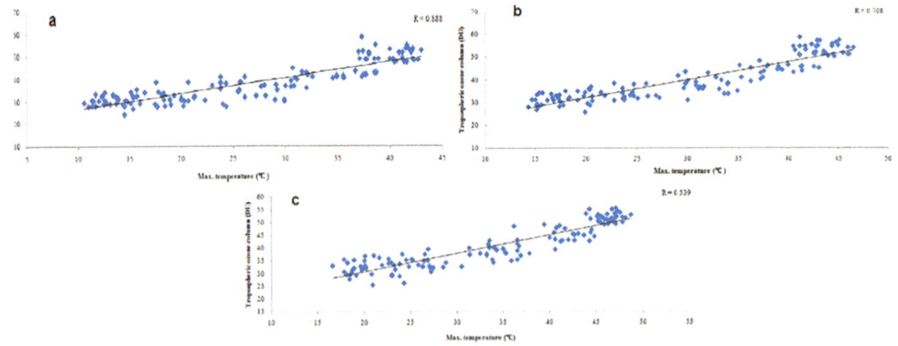

Fig. 2. Scatter plots of monthly maximum temperature and monthly troposphere ozone concentration for (a) Erbil, (b) Baghdad and (c) Basra for the period of 2010–2020.

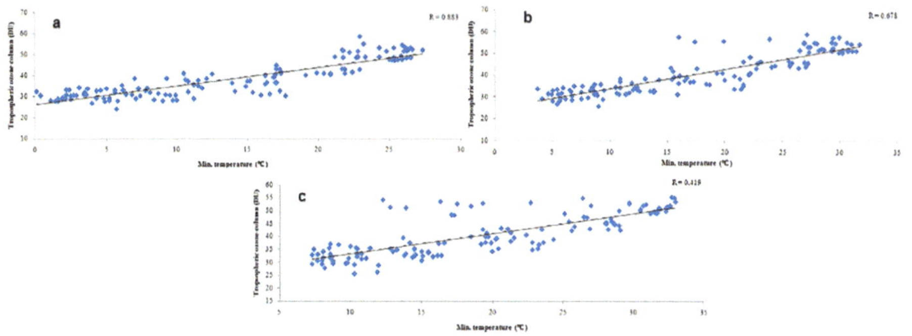

Fig. 3. Scatter plots of monthly minimum temperature and monthly troposphere ozone concentration during 2010–2020 for (a) Erbil, (b) Baghdad and (c) Basra.

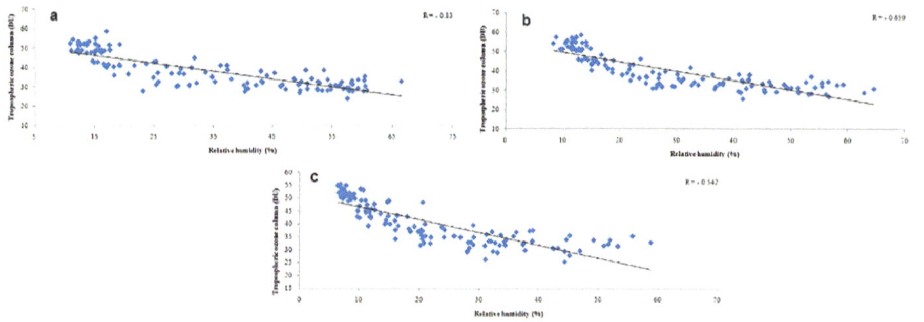

Fig. 4. Scatter plots of monthly relative humidity and monthly troposphere ozone concentration during 2010–2020 for (a) Erbil, (b) Baghdad and (c) Basra.

The data display sporadic points due to the erratic variation in wind speed and direction throughout the year. Ozone concentration levels and the ambient air temperature (minimum and maximum) were shown to be linearly correlated. The results fit a linear model to describe the association between monthly troposphere ozone concentration and monthly maximum and minimum temperatures for the years 2010–2020, according to the scatter plot (Figs. 2 and 3). The correlation coefficient between minimum temperature and troposphere ozone for Baghdad, Basra, and Erbil is 0.678, 0.419, and 0.883, respectively, while the correlation coefficient between maximum temperature and troposphere ozone for Baghdad, Basra, and Erbil is 0.708, 0.539 and 0.888, respectively. Because it directly affects chemical kinetic rates and the mechanistic pathway for ozone generation, high temperatures were thought to be an ozone enhancement factor because of their positive correlation with ozone. However, this is not always the case because the rate of increase also depends on the concentration levels of VOCs and NOx. Since temperature affects the speeds of chemical reactions required for the creation of O_3, temperature and O_3 exhibited a positive link. Ozone concentration increases with temperature because high temperatures lower the concentration of one of the ozone precursors, peroxyacyl nitrate (PAN), which in turn causes high ozone concentration. Furthermore, rising temperatures are frequently accompanied by rising radiation, falling water vapor, and naturally occurring isoprene emissions, all of which raise the concentration of ozone. O_3 and relative humidity had a negative relationship, indicating that wet deposition and/or the nighttime contributions of NOx and black carbon are the main causes of the decrease in O_3 concentration in the atmosphere. The low ozone levels caused by less intense photochemical production and, potentially, by ozone deposition on water droplets are generally linked to the rainy, wet weather with high relative humidity. Increases in cloud cover, wind speed, and other meteorological factors that are detrimental to ozone production and pollution buildup frequently accompany increases in relative humidity, which lowers the concentration of ozone. The rising of hotter air at noon and the mixing with dryer air masses caused changes in the boundary layer height, as demonstrated by a positive correlation of temperature and a negative correlation of relative humidity. Figure 5 shows linear variation of troposphere ozone column with max. temperature during 2010–2020 for Winter, Spring, Summer and Autumn, respectively, for Baghdad. Where Fig. 6a–d shows linear variation of troposphere ozone column with relative humidity during 2010–2020 for Winter, Spring, Summer and Autumn, respectively, for Baghdad.

Meteorological influences on ozone concentrations were relatively weak in Summer when ozone concentrations were the highest. It is noted from Fig. 5a, c that the maximum temperatures do not affect the troposphere ozone column, meaning that the relationship between them is weak ($R = 0.094$, $R = 0.077$) in the Winter and Summer, respectively. As for the Spring and Autumn, the relationship between the maximum temperatures and the troposphere ozone column is strong ($R = 0.779$, $R = 0.834$), respectively; the ozone increases with the increase in temperature. In the sense that temperatures have a strong effect on the ozone column in the Spring and Autumn, Fig. 5b, d. The effect of relative humidity is also strong on the ozone column, where the relationship between them is inverse ($R = -0.641$, $R = -0.707$) in the Spring and Autumn, respectively, Fig. 6b, d. This is due to several factors:

1. Temperature: Tropospheric ozone formation is a complex process involving the interaction of sunlight, nitrogen oxides (NOx), volatile organic compounds (VOCs), and oxygen. Warmer temperatures in spring and autumn enhance the photochemical reactions responsible for ozone formation. Higher temperatures increase the rate of ozone formation, leading to elevated ozone levels.
2. Atmospheric Stability: In spring and autumn, the atmosphere often exhibits greater temperature variations, resulting in more frequent atmospheric inversions. These inversions trap pollutants, including ozone precursors, near the surface, preventing their vertical dispersion and leading to higher ozone concentrations.
3. Meteorological Conditions: Spring and autumn are characterized by weather patterns that favor the accumulation and transport of pollutants. Stagnant weather conditions, such as light winds and low atmospheric mixing, can lead to the accumulation of pollutants, including ozone, in specific regions.
4. Biogenic Emissions: During spring, there is an increase in biogenic emissions from vegetation as plants resume growth and release volatile organic compounds. These compounds, in the presence of sunlight and nitrogen oxides, can contribute to ozone formation.

Figure 7 shows a histogram of yearly variation of troposphere ozone column during (2010–2020) for Baghdad, Basra and Erbil.

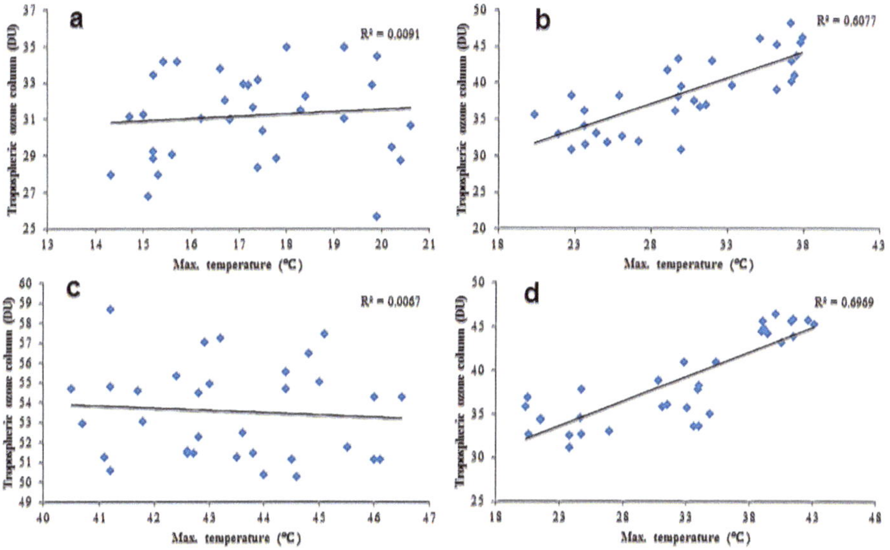

Fig. 5. Linear relationship between troposphere ozone column and max. temperature during 2010–2020 for (a) Winter (b) Spring (c) Summer and (d) Autumn for Baghdad.

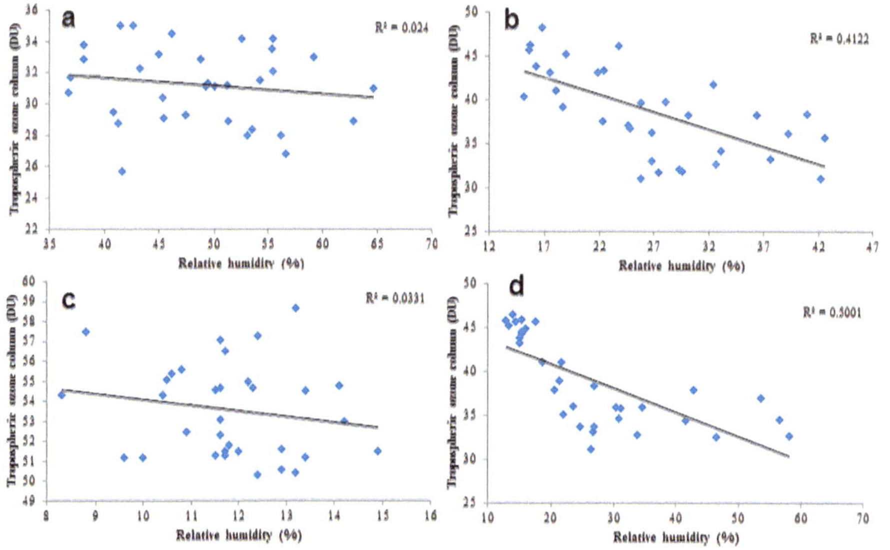

Fig. 6. Linear relationship between troposphere ozone column and relative humidity during 2010–2020 for (a) Winter (b) Spring (c) Summer and (d) Autumn for Baghdad.

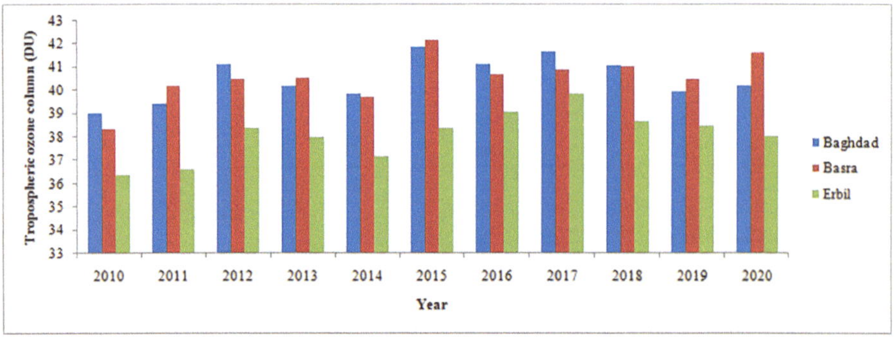

Fig. 7. Yearly variation of troposphere ozone column during (2010–2020) for Baghdad, Basra and Erbil.

It is noted from the above figure that the troposphere ozone column for the cities of Baghdad and Basra is higher (41.85 DU and 42.18 DU respectively) in 2015 compared to the rest of the years. There are many reasons for this, but the most prominent is the massive overcrowding in Baghdad's capital. Its failure to accommodate the large numbers of displaced people from the various governorates of Iraq as a result of the war against terrorism, as a result of the destruction of their areas and the delay in the process of rebuilding their cities and villages and returning them to them, in addition to those who abandoned their farms and pastures due to waves of the severe drought that struck the governorates of Iraq in general over the past years, and its impact was evident in the central and southern governorates in particular. Government statistics indicate that the population of Baghdad has reached nearly 10 million people. So this considerable increase in population numbers leads to high pressure on the infrastructure of Iraqi cities, as it is not equipped to accommodate such huge numbers. In Erbil, the troposphere ozone column was higher (39.88 DU) in 2017 compared to the rest of the years, as illustrated in Fig. above. Figure 8a–d shows the variation of the troposphere ozone column during Winter (2010–2020), Spring (2010–2020), Summer (2010–2020), and Autumn (2010–2020) for Baghdad, Basra, and Erbil.

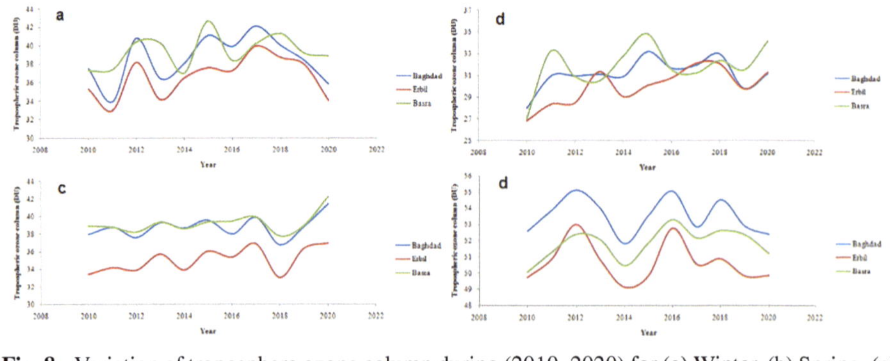

Fig. 8. Variation of troposphere ozone column during (2010–2020) for (a) Winter, (b) Spring, (c) Summer and (d) Autumn for Baghdad, Basra and Erbil.

The change of the troposphere ozone column for the three governorates between 2010 and 2020 is regular and similar, Fig. 8b–d. The troposphere ozone column values for the city of Baghdad are located between (28–33.2 DU), (34–42.1 DU), (51.8–55.1 DU) and (36.8–41.4 DU) for the years (2010–2015), (2011–2017), (2014–2016) and (2018–2020) for Winter, Spring, Summer, and Autumn, respectively, and for the city of Basra between (27.1–34.8 DU), (37–42.6 DU), (50.1–53.3 DU) and (37.7–42.2 DU) for the years (2010–2015), (2014–2015), (2010–2016) and (2018–2020) for Winter, Spring, Summer, and Autumn respectively, while for the city of Erbil, it falls between (26.9–32.1 DU), (33–39.9 DU), (49.1–53 DU) and (33–36.9 DU) for the years (2010–2017, 2018), (2011–2017), (2014–2012) and (2018–2020) for Winter, Spring, Summer, and Autumn respectively. Figure 8a–d. It is noted from the above figure that the troposphere ozone change curve for Baghdad is higher than that for Basra, and for Basra, is higher than that for Erbil (this is evident in the Summer). Also, it is noted from the exact figure that the troposphere ozone change curve for Baghdad and Basra is higher than that for Erbil. The highest concentration (58.7 DU = 0.022 PPm = 22 μgrm/m^3) was recorded in June2016 for Baghdad and Erbil, while (55.4 DU = 0.0205 PPm = 20.5 μgrm/m^3) in July/2018 for Basra. The maximum values of ozone pollution for the three governorates and all years of the study are below the permissible limit of the EPA (0.07 PPm).

4 Conclusions

This study examined the regional and temporal fluctuations in ozone concentrations in Baghdad, Basra, and Erbil between 2010 and 2020. Regression analysis was also used to examine the connection between ozone concentrations and climatic factors, including the monthly average temperature and relative humidity. Since ozone levels are mostly dependent on solar radiation and high temperatures, they are higher in the summer than in the winter. Temperature and relative humidity, two climatic indicators, were also successfully monitored and connected to the ozone concentrations. Ozone and relative humidity were found to be negatively correlated, meaning that as relative humidity dropped, ozone rose. On the other hand, ozone and temperature were found to be positively correlated, meaning that the concentration of ozone rose as the temperature rose (for three cities). The higher the altitude above sea level, the stronger the relationship between temperature and ozone, relative humidity, and ozone (the further we go from Basra to Baghdad up to Erbil, the stronger the connection becomes).At the seasonal scale, meteorological influences on ozone concentrations were stronger in Spring and Autumn than in Summer and Winter in Baghdad. Ozone pollution levels are higher in Baghdad and Basra than in Erbil for all years.

References

1. Silva-Quiroz, R., Rivera, A., Ordoñez, P., Gay-Garcia, C., Frank, A.: Atmospheric blockages as trigger of environmental contingencies in Mexico City. Heliyon. **5**(7) (2019)
2. Monks, P.S. et al.: Tropospheric ozone and its precursors from the urban to the global scale from air quality to short-lived climate forcer. Atmos. Chem. Phys. **15**(15), 8889–8973 (2015)

3. Paraschiv, S., Constantin, D.E., Paraschiv, S.L., Voiculescu, M.: OMI and ground-based in-situ tropospheric nitrogen dioxide observations over several important European cities during 2005–2014. Int. J. Environ. Res. Public Health. **14**(11), 1415 (2017)
4. Yu, S. et al.: Spatiotemporal characterization and regional contributions of O3 and NO2: an investigation of two years of monitoring data in Henan, China. J. Environ. Sci. **90**, 29–40 (2020)
5. Paraschiv, S., Paraschiv, L.S.: Analysis of traffic and industrial source contributions to ambient air pollution with nitrogen dioxide in two urban areas in Romania. Energy Procedia. **157**, 1553–1560 (2019)
6. Alahmadi, S., Al-Ahmadi, K., Almeshari, M.: Spatial variation in the association between NO2 concentrations and shipping emissions in the Red Sea. Sci. Total Environ. **676**, 131–143 (2019)
7. Wang, J. et al.: Influence of meteorological conditions on explosive increase in O3 concentration in troposphere. Sci. Total Environ. **652**, 1228–1241 (2019)
8. Et, A., Masoudi, M.: Status and prediction of ozone as an air pollutant in Ahvaz City, Iran. Casp. J. Environ. Sci. **12**(2), 215–224 (2014)
9. Ihedike, C., Mooney, J.D., Fulton, J., Ling, J.: Evaluation of real-time monitored ozone concentration from Abuja, Nigeria. BMC Public Health. **23**(1), 496 (2023)
10. Kalisa, E., Fadlallah, S., Amani, M., Nahayo, L., Habiyaremye, G.: Temperature and air pollution relationship during heatwaves in Birmingham, UK. Sustain. Cities Soc. **43**, 111–120 (2018)
11. Celik, M., Kadi, I.: The relation between meteorological factors and pollutants concentrations in Karabuk City. Gazi Univ. J. Sci. **20**(4), 87–95 (2007)
12. Elminir, H.K.: Dependence of urban air pollutants on meteorology. Sci. Total Environ. **350**, 225–235 (2005)
13. Mohan, S., Saranya, P.: Assessment of tropospheric ozone at an industrial site of Chennai megacity. J. Air Waste Manage. Assoc. **69**(9), 1079–1095 (2019)
14. Belan, B.D., Savkin, D.E.: The role of air humidity in variations in near-surface ozone concentration. Atmos. Ocean. Opt. **32**(5), 586–589 (2019)
15. Allu, S.K., Srinivasan, S., Maddala, R.K., Reddy, A., Anupoju, G.R.: Seasonal ground level ozone prediction using multiple linear regression (MLR) model. Model. Earth Syst. Environ. **6**(4), 1981–1989 (2020)
16. Ma, Y., Ma, B., Jiao, H., Zhang, Y., Xin, J., Yu, Z.: An analysis of the effects of weather and air pollution on tropospheric ozone using a generalized additive model in Western China: Lanzhou, Gansu. Atmos. Environ. **224**, 117342 (2020)
17. Ilić, P., Popović, Z., Markić, D.N.: Assessment of meteorological effects and ozone variation in urban area. Ecol. Chem. Eng. **27**(3), 373–385 (2020)
18. Zihao, L., Jianwu, H., Deya, K.: Analysis on the temporal and spatial changes of air pollution in Hefei and its correlation with meteorological factors. Environ. Sci. Manag. **44**(2), 43–48 (2019)
19. Karem, D.S., Kadhim, H.A., Al-Saad, H.T.: Study the air pollution in west Qurna-2 oil field southern Iraq. J. Pharm., Chem. Biol. Sci. **4**(3), 416–430 (2016)

Data Governance and AI in Smart Cities: Reducing Infrastructure's Lifecycle Carbon Footprint with ESG Implementations

Yu Zhou[1,2,3](✉)

[1] Peking University HSBC Business School, Shenzhen, GD 518000, China
yu.zhou@pku.org.cn
[2] The University of Melbourne, Parkville, VIC 3052, Australia
[3] Veritas & Consilio Enterprise, Melbourne, VIC 3000, Australia

Abstract. The recent advancements in generative artificial intelligence (Gen AI) and the subsequent escalation of associated data governance requirements have had a profound influence on both technological, environmental and economic landscapes in smart cities. This study examines the unique nexus between data governance and AI, particularly in simulating and optimising infrastructure with AI and data governance for Pareto Optimum. With a case study of a complex energy infrastructure in Australia, the study analysed the potential of data governance and Gen AI in energy infrastructure. The combined capabilities of AI and data governance to reduce the lifecycle greenhouse gas emissions and lifecycle costs are further discussed. Subsequently, computing infrastructure, a pivotal component in the training of Gen AI, related standards, regulations, and ESG (environmental, social and governance) implementation are also discussed, together with potential challenges. The study ultimately seeks to provide valuable insights and guidance to policymakers, industry leaders, and researchers who are striving to forge a more sustainable and equitable future.

Keywords: Lifecycle Greenhouse Gas Emissions · Artificial Intelligence · Data Governance · Infrastructure Optimisation · ESG

1 Introduction

The demand for energy worldwide continues to rise [1], driven by economic growth, population expansion, and technological advancements [2]. Simultaneously, there is an increasing awareness of the environmental impact of traditional energy sources [3], leading to a significant transformation in the energy landscape. Traditional fossil fuel-based energy sources are being supplemented and, in some cases, replaced by renewable energy alternatives [4], such as solar, wind, and geothermal energy resources. Currently, various countries including Australia and China are currently undergoing push for a significant energy transition, towards cleaner, sustainable and renewable energy sources [1–6]. This transition to cleaner energy is vital to combat climate change, save energy consumption, improve energy efficiency and reduce the ecological footprint of energy

Y. Zhu (Ed.): ISESIE 2025, LNCE 000, pp. 265–277, 2026.
https://doi.org/10.1007/978-981-92-2389-3_21

production and improve the livability in cities [5–9]. However, this transition is not without its challenges. This shift is driving a growing demand for energy infrastructure globally, highlighting the necessity for more efficient and effective management of these complex systems. Sub-optimal investment by failing to consider whole of system environmental impact could cost global economy billions of dollars.

Renewable energy sources are intermittent [5–10], and the integration of these sources into existing energy systems requires detailed simulation, modelling of and optimisation due to its complexity, involving processing large dataset [11–13], which are time-consuming and resource-intensive. This study aims to address these challenges, with focuses on optimising energy systems [5, 12–15] through the integration of Artificial Intelligence [16–18], with the ultimate goal of reducing the lifecycle emissions and costs [8]. Generative Artificial Intelligence (Gen AI) is a new type of AI that generates new and original content by learning from realistic large-scale datasets [19]. Gen AI can improve the optimisation process of renewable energy systems by replacing comprehensive simulation and modelling [18].

The current understanding of energy infrastructure heavily depends on detailed simulation tools, such as Finite Element Modelling simulations, which is both time-consuming and resource-intensive. Table 1 summaries the function of different methods currently available and the estimated computing time. The lack of detailed simulation efficient methods further hampers the accessibility of the knowledge and applications of energy infrastructure.

Table 1. Methods and limitations

	Detailed simulation	Simplified simulation	LCA	LCCA	Computing time
AI (ANN)	Y	Y	Y (TBD)	Y (TBD)	secs
COMSOL	Y	Y	N	N	days
TRNSYS	N	Y	Y	Y	hours
HOMER pro	N	Y	Y	Y	hours
GLD	N	Y	N	N	minutes

In recent years, extensive research has been conducted on developing AI tools for enhancing simulation capabilities. As presented in Table 2, these AI tools are predominantly concentrated within the academia. They are primarily employed to enhance simulation efficiency in specific technological aspects within case studies. Hence, there are currently no publicly available applications that can utilise AI for multi-objective optimisation of the energy infrastructure on the lifecycle scale, which requires cross-disciplinary expertise. The research gap is identified as the absence of an industry—viable tool for analysis and optimisation of energy infrastructure. To address this, the proposed methods that integrate ANN with lifecycle assessment models, will be discussed in the next section.

Table 2. Summary of related research

Reference	Methodology	Applications
Topic: Data Governance and Simulation based on AI		
Guo et al. [14]	Automated machine learning (AutoML) Analyzer with genetic algorithm-based topology optimisation algorithms	Increase power conversion efficiencies in thin-film organic solar cells (OSCs)
Cao et al. [27]	Integrates the core strengths of Long Short-Term Memory (LSTM) and Transformer, introducing hybrid architecture	Real-time multi-task prediction in engineering systems
Afsari et al. [28]	Using legged robots with the adaptability and mobility to capture on-site image	Monitor construction progress
Esen et al. [29]	Artificial neural network (ANN) and adaptive neuro-fuzzy inference system (ANFIS)	Forecasting performance of a solar-GSHP system
Xiang et al. [30]	Mobile projective AR (MPAR) framework	Project virtual construction information
Sirmacek and Vinuesa [31]	Combine satellite remote sensing imaging and artificial intelligence	Measuring climate adaptation of cities
Zhou et al. [16]	Verifying a TRNSYS model for a typical horizontal GHE and Artificial Neural Network (ANN) to create a model for GCHP design	Heat Pumps with Horizontal Ground Heat Exchangers
Topic: Life Cycle Assessment (LCA) in the realms of infrastructure		
Santero et al. [32]	Profiling the status of existing pavement LCA problems and improvement recommendations	Pavements infrastructure
Liljenström et al. [33]	Comparing LCA to determine the length, maintenance frequency, and effects of the analysis on infrastructure performance	Road and rail infrastructure

(*continued*)

Table 2. (*continued*)

Reference	Methodology	Applications
Saxe and Kasraian [34]	Assessing the life cycle impacts to reflect the long lifetimes, durability, persistence, and feedback loops of transport infrastructure	Transport infrastructure
Lucas et al. [35]	Analysing the life cycle of transport infrastructure by focusing on electricity and on hydrogen supply through centralised steam methane reforming and on-site electrolysis	Electric and hydrogen transportation systems
Wohlschlager et al. [36]	Using a two-step methodological approach, including evaluating research projects	Smart energy system
Zhou et al. [8]	Proposing configurations of hybrid geothermal–solar–gas heating systems to minimise lifecycle costs and GHG emissions by finding the Pareto front	LCA of hybrid geothermal–solar (PV)–gas heating systems
Ma and Zhou [22]	Integrating LCCA, Analogous Estimating Method, and the Emission Factor Method to quantify environmental impact	Computing infrastructure

2 Methods

The Pareto Optimum [20], a concept borrowed from economics, represents a state where no further improvements can be made in one aspect without making something else worse off. In the context of energy systems, the Pareto Optimum signifies the ideal balance between energy efficiency, cost-effectiveness, and environmental impact [8, 21]. Finding these optimal configurations is a complex task due to the multidimensional nature of the problem. Life cycle assessment (LCA) is a broadly used tool and framework [22, 23] in environment science and engineering area, to assess service or product environmental and potential impacts through whole life cycle or between the setting life boundaries [23, 24]. Life cycle cost analysis (LCCA) could estimate and quantify economic performance of a project or service [25], which would cover all processes of project including establishment, implementation, operation, maintenance, and post-retirement disposal, considering future cost [23–25]. LCCA could provide clearly and quantitative economic comparison for decision-making. The specific formulations for the LCCA and LCA objectives are outlined below:

Lifecycle Cost in Net Present Value [8, 23–26]:

$$LCC = C + \sum_{n=1}^{t} C_r \left(\frac{1}{1+d}\right)^n \quad (1)$$

where, C is the initial cost, Cr is the recurring cost, d is the real discount rate, and t is the project lifetime in years [8, 23–26].

Lifecycle Greenhouse Gas (GHG) Emissions [8, 23–26]:

$$\mathrm{LGE} = \mathrm{ME} + \sum_{\mathrm{n}=1}^{\mathrm{t}} \mathrm{OE_n} \quad (2)$$

where, LGE is the lifecycle GHG emission ,ME is the manufacture GHG emission, $\mathrm{OE_n}$ is operational GHG emission in year n t is analysis period.

Stage 1: Establish a data pool to analyse instrumented data and modelling data of current energy infrastructure installed on various projects in Australia;

The first aim of this research endeavor is to curate a comprehensive data pool that encapsulates both instrumented operational data and theoretical modeling outcomes of Ground Source Heat Pump (GSHP) integrated with Solar Energy systems currently deployed across diverse energy infrastructure projects in Australia. This data pool will serve as the cornerstone for our investigation, providing insights into the performance characteristics, operational efficiencies, and environmental interactions of these hybrid renewable energy systems. By consolidating data from a wide array of installations, we aim to capture the variability inherent in real-world applications, enabling a more accurate analysis of energy infrastructure' performance across different climatic and geographical contexts. In this phase of the research, this study uses established software tools—specifically, GLD (Ground Loop Design), TRNSYS (Transient System Simulation Tool), and COMSOL Multiphysics—to simulate and evaluate diverse energy system configurations. These software packages, though rooted in traditional methodologies, exhibit varying levels of accuracy and simulation efficiency, making them valuable in comprehensively assessing and optimising energy infrastructures.

Stage 2: Develop a predictive Deep Learning model based on expertise knowledge of AI and infrastructure practices, with cross-validation and training using the data pool;

Building upon the established data pool, the second aim is to leverage the expertise in AI and energy infrastructures to develop a predictive Deep Learning model. This model will be designed to capture the complex interactions and non-linear relationships within the hybrid energy infrastructure, utilizing advanced machine learning techniques. Examples of the related models developed are shown as below (Fig. 1):

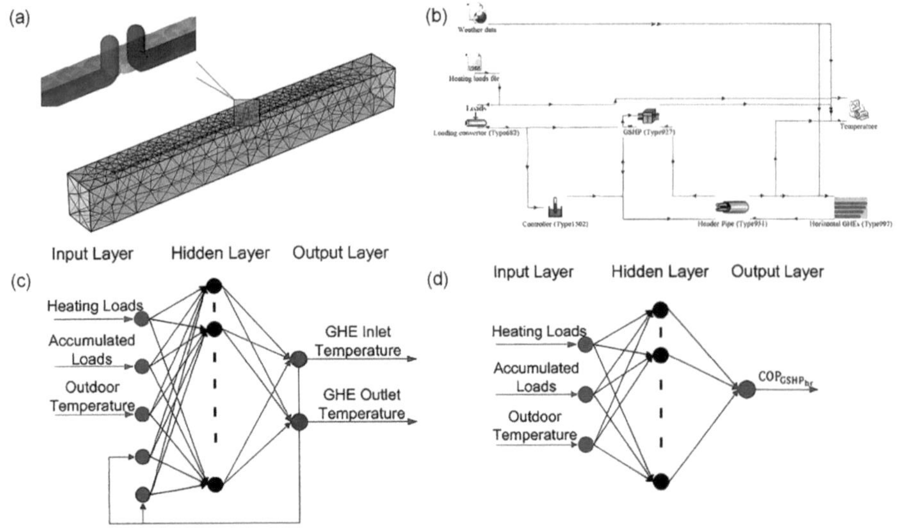

Fig. 1. Examples of models developed: (a) FEM [11]; (b) TRNSYS [16]; (c) AI (NARX) [16]; (d) AI (Feed forward) [16]

The model's development will involve cross-validation and training procedures, ensuring its robustness and accuracy in predicting system behavior under varying conditions. The Deep Learning model will integrate recent advancement in AI [16, 27] to solve transitionally unsolvable problem of these hybrid systems with even limited data inputs (loading, climate, site (soil) conditions and configuration factors as examples), providing a powerful tool for predicting performance, identifying areas for optimisation, and informing future design iterations. In the second phase of the research, the focus shifts towards leveraging the vast amount of data generated from the simulations and instrumentation to develop a predictive and optimisation framework through the integration of machine learning techniques. This phase aims to enhance the understanding of energy system performance, identify patterns, and ultimately, facilitate the design of more efficient and sustainable energy systems.

Stage 3: Conducting multi-objective optimisation with cost and GHG emissions based on the Deep Learning model, depicting the economic and environmental impact of energy infrastructure;

The ultimate goal of this research is to conduct a multi-objective optimisation analysis based on the predictive ANN model developed in the previous aim. This analysis will focus on minimising both the economic costs and GHG emissions associated with the operation of hybrid energy infrastructures. By framing the optimisation problem in this manner, we aim to holistically assess the economic, environmental, and potential social impacts of deploying these systems. The multi-objective approach recognizes the inherent trade-offs between cost savings and emissions reductions, enabling us to explore various scenarios and identify optimal configurations that balance these competing objectives. The outcomes of this analysis will provide valuable insights for policymakers, industry stakeholders, and the research community, guiding the development and implementation of hybrid energy infrastructures which integrates geothermal and solar energy that are both economically viable and environmentally sustainable.

Model Integration and Optimisation: Incorporate the thoroughly trained machine learning model into the research framework, enabling dynamic, real-time predictions and optimisations driven by data streams. Design an optimisation algorithm that capitalizes on the predictive capabilities of the machine learning model to identify optimal and sustainable energy system configurations. To address both economic and environmental aspects, adopt multi-objective functions within the optimisation process. The primary objective will prioritize minimising the lifecycle cost, emphasizing the financial aspect. The secondary objective will target minimising the lifecycle GHG emissions, highlighting the environmental concerns. This approach will not only answer the question of how to optimise a hybrid energy infrastructure based on economic and environmental factors but also deepen our understanding of such complexity in the energy and infrastructure's dynamics.

Optimisation parameters and Validation: A multitude of potential system configurations will be analyzed on a scenario-by-scenario basis, considering product prices, installation labor costs, operation and maintenance expenses, product performance, electricity production, delivered diesel prices, load, load profiles, local solar radiation, and so forth. These configurations will then be ranked according to their financial and environmental performance, which includes metrics such as payback period, net present value, internal rate of return, and levelized cost of energy. The optimisation framework will be validated using a subset of test data from comparable sustainable infrastructure projects, and the deep learning model will be refined accordingly. The optimised solutions will be compared with existing systems to showcase their potential for enhanced performance and sustainability. The next section will demonstrate preliminary results based on case studies in Australia.

3 Preliminary Results

With industry cases of hybrid energy infrastructure, detailed in [16–18], the preliminary study used a hybrid simulation model combining ANNs, TRNSYS, and COMSOL Multiphysics. These models solve 3D ground and GHE models via implicit finite difference for accurate performance assessment of complex GSHP applications and can be used for building energy simulations. ANNs were trained with data from TRNSYS simulations of hybrid energy infrastructure in the Australian rural industry. Among three Australian locations, NSW case was the training case for ANNs, and the others were for verification [16]. The results indicated that ANNs can accurately predict performance which can be further used for optimisation, with significant reduction of the computing time (up to 10,000 times faster).

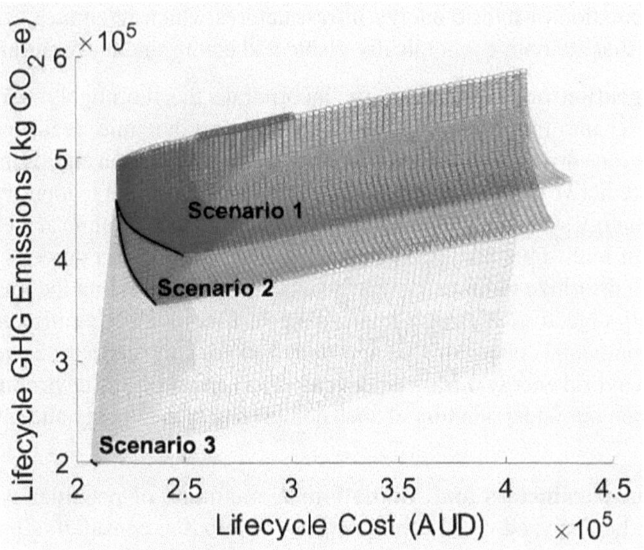

Fig. 2. Examples of optimisation results [8]

Backed by case studies via Pareto Optimum analysis of Australian locations (references [4, 7, 8, 11, 16–18]; Fig. 2), a well-optimised energy infrastructure can reduce the 40-year lifecycle costs and emissions by up to 55% and 50%, if other renewable energy resources (geothermal, solar and gas) are integrated and managed properly. Industry, including construction, contributes to 27.4% of the Australian Gross Domestic Product, equivalent to AU$658.6 billion [37]. If only 3% of Australian industry can be supported by hybrid energy infrastructures, it will bring up to AU$10 billion economic benefit to Australia, as an indirect outcome from this project, bringing profound economic advantages for consumers, businesses, and governments alike.

4 Discussions

AI, particularly Gen AI, revolutionizes simulation and modeling processes by learning from large-scale datasets to generate detailed, context-aware simulations. The section further discussed the Environmental, Social and Governance (ESG) benefits, emissions of computing infrastructure for AI training as well as associated regulations and challenges.

4.1 ESG Benefits and Sustainable Finance

The environmental impact of this study is equally remarkable. By promoting the rational complementation and substitution of traditional energy sources with renewable ones, it can significantly reduce the environmental footprint (up to 50% in 40-year lifecycle, approx. 30,000 kg CO_2-e per project as case study in Fig. 2) of energy consumption. This, in turn, contributes to mitigating climate change and preserving natural resources for future generations. Additionally, the study aligns with cultural values of environmental stewardship, fostering a cultural identity deeply rooted in the responsible use of land and resources.

Beyond economic and environmental benefits, this study also holds immense social significance. More affordable and reliable energy will undoubtedly enhance the overall quality of life for citizens, particularly in remote and vulnerable regions where energy security is paramount. Furthermore, the study anticipates several significant outcomes, including the identification of Pareto Optimum and a comprehensive understanding of the quantifiable benefits of hybrid energy infrastructure, leading to policy recommendations that will guide which technologies will replace the retiring fossil fuel infrastructure, towards a more sustainable energy future.

The aforementioned benefits also align with the study's emphasis on ESG criteria within sustainable finance. It is also important to note that finance can play a significant role in fostering technological development and innovation, thereby promoting industrial transformation [38]. By incorporating ESG factors, investors can prioritize projects that deliver both financial returns and measurable environmental benefits. This dual focus on profitability and sustainability positions ESG-aligned finance as a catalyst for scaling renewable energy solutions.

4.2 Emissions of Computing Infrastructure for AI Training

Computing infrastructure in enabling the training of Gen AI, a cornerstone of modern technological advancement. Computing infrastructure, encompassing servers, storage, GPUs, and distributed systems, forms the backbone of AI development. However, this progress is accompanied by significant energy consumption and carbon emissions, with AI training processes demanding massive computational power. For instance, training large language models can emit CO_2equivalent to the lifetime emissions of multiple cars, highlighting the urgency of aligning computing infrastructure with sustainability goals.

Similar quantitative approach should also be adapted to evaluate the lifecycle cost and environmental impact of computing infrastructure. As an example, a typical data center in Shenzhen could generating over 30 million kg CO2 per year [22]. These emission is highly related to the regional energy structure, where electricity-related carbon emissions account for approximate 90% in Shenzhen, which may be slightly less as approx. 80% in the United States [22]. Future studies should also take such impact in consideration with analysis the benefits of AI.

4.3 Regulations and Challenges

Standardisation and regulations are vital for enabling AI in energy systems, providing interoperability and transparency. The international standards, such as *ISO (International Standard Organisation) 2022:37170 Smart community infrastructures: Data framework for infrastructure governance based on digital technology in smart cities* [12], provides a framework for data governance in smart cities, and to provide secure and efficient management of urban data. The study acknowledges that there are several international and local standards that must be adhered to, but also points out, as discussed above, the regulatory gaps that fail to address long-term uncertainties associated with environmental issues.

Regulations like GDPR (General Data Protection Regulation) in Europe, CCPA (The California Consumer Privacy Act) in the US and China's PIPL (Personal Information Protection Law) impose strict compliance with data privacy requirements, with severe penalties for non-compliance [39]. However, while regulations have been established ensuring adequate protection for individual privacy rights, a major challenge remains striking the correct balance between using this data for training AI systems and ensuring individual privacy rights. At the same time, the challenge also lies in the availability of quality data—fragmented and poor-quality datasets will compromise the predicted capabilities of the AI.

4.4 Applicability and Future Directions

This study proposes a comprehensive framework at how data governance and AI can be used to optimise smart cities' energy infrastructure while aligning with environmental, social, and governance principles. While it is backed by case studies in Australia as presented in Sect. 3, future research could be further conducted on a broader scale, including applying it in different countries and regions [40].

Furthermore, future research can be conducted on integrating this framework with diverse types of infrastructure, including traditional urban infrastructure and cutting-edge technological infrastructures. Moreover, future studies can also evaluate whether it is capable of being fine-tuned to operate effectively even in regions where data availability is scarce, testing whether its benefits are limited to data-rich areas. Additionally, it is also encouraged to the expansion of incorporating a broader range of considerations, such as addressing issues of social justice and enhancing community resilience. This comprehensive future research will ensure that the framework not only meets current needs but also remains relevant and adaptable to future challenges, thereby fostering a more inclusive and robust system.

5 Conclusions

In conclusion, this study demonstrates that data governance and AI can transform infrastructure in smart cities. The study offers a centralized data pool that collects and evaluates operational output and theoretical modeling results to demonstrate the importance of data governance. The data governance represents the principles of data quality, accuracy, and security that provide the foundation for training robust AI models. In this study, AI models such as ANNs were trained to predict energy system performance under varying conditions, achieving multi-objective optimization that minimized lifecycle costs and GHG emissions. The case study on hybrid heating systems in Australia exemplifies this potential, demonstrating how integrating ANNs with tools like TRNSYS and COMSOL Multiphysics reduces lifecycle costs by up to 55% and emissions by 50%. Such outcomes validate AI's capacity to balance technical and environmental objectives, a critical advantage in renewable energy transitions.

In addition, computing infrastructure, an essential component in Gen AI training, applicable standards, regulations and ESG implementation are discussed. The study concludes to provide timely insights, findings and guidance to policymakers, industry leaders, and researchers who are aiming to build a more sustainable and equitable future. While this study proposes a comprehensive framework, with case studies in Australia, future research could be conducted on a broader scale, including different countries and regions, as well as various types of infrastructures.

References

1. Ahmad, T., Zhang, D.: A critical review of comparative global historical energy consumption and future demand: the story told so far. Energy Rep. **6**, 1973–1991 (2020)
2. Van Ruijven, B.J., De Cian, E., Sue Wing, I.: Amplification of future energy demand growth due to climate change. Nat. Commun. **10**(1), 2762 (2019)
3. Qazi, A., et al.: Towards sustainable energy: a systematic review of renewable energy sources, technologies, and public opinions. IEEE access. **7**, 63837–63851 (2019)
4. Shrestha, A., et al.: Evolution of energy mix in emerging countries: modern renewable energy, traditional renewable energy, and non-renewable energy. Renew. Energy. **199**, 419–432 (2022)
5. Zhou, Y., et al.: Hybrid geothermal-gas and geothermal-solar-gas heating systems for poultry sheds. Proc. Zero Energy Mass Cust. Home, 545–554 (2018)
6. Olówósejéjé, S., Leahy, P., Morrison, A.P.: Optimising photovoltaic-centric hybrid power systems for energy autonomy. Energy Rep. **7**, 1943–1953 (2021)
7. Zhou, Y., et al.: Optimisation of a hybrid geothermal-solar-gas system: a case study for a typical poultry shed in New South Wales, Australia. In: 4th Building Simulation and Optimisation Conference, Cambridge, UK (2018)
8. Zhou, Y., et al.: Achieving pareto optimum for hybrid geothermal–Solar (PV)–Gas heating systems: minimising lifecycle cost and greenhouse gas emissions. Sustainability **16**(15), 6595 (2024)
9. Montoya, M.A., et al.: Renewable and non-renewable in the energy-emissions-climate nexus: Brazilian contributions to climate change via international trade. J. Clean. Prod. **312**, 15 (2021)
10. Bloomfield, H.C., et al.: Quantifying the sensitivity of European power systems to energy scenarios and climate change projections. Renew. Energy. **164**, 1062–1075 (2021)
11. Zhou, Y., Bidarmaghz, A., Makasis, N., Narsilio, G.: Ground-source heat pump systems: the effects of variable trench separations and pipe configurations in horizontal ground heat exchangers. Energies. **14**, 3919 (2021)
12. International Organization for Standardization.: ISO 37170:2022 Smart community infrastructures: Data framework for infrastructure governance based on digital technology in smart cities (2022)
13. Zhang, Y., Qi, H., Zhou, Y., Zhang, Z., Wang, X.: Exploring the impact of a district sharing strategy on application capacity and carbon emissions for heating and cooling with energy infrastructure. Appl. Sci. **10**(16), 5543 (2020)
14. Guo, R.Q., et al.: In silico design of freeform solar cell structures from high-throughput artificial intelligence-generated configurations. Sol. RRL, 8 (2023)
15. Ogunmodede, O., et al.: Optimising design and dispatch of a renewable energy system. Appl. Energy. **287**, 17 (2021)
16. Zhou, Y., et al.: Artificial neural networks for predicting the performance of heat pumps with horizontal ground heat exchangers. Front. Energy Res., 12 (2024)

17. Zhou, Y., Narsilio, G., Makasis, N., Aye, L.: A machine learning approach for the performance prediction of GCHPs with horizontal ground heat exchanger. XVI Panamerican Conference on Soil Mechanics and Geotechnical Engineering, 17–20 November 2019, Cancún, Mexico (2019)
18. Zhou, Y., et al.: An AIpowered simulation and optimisation framework for energy infrastructure investment to achieve Pareto optimum solutions. Sustain. Emerg. Energy Technol.: Chall. Oppor. Perspect. **2025**, eISBN: 978-3-031-92417-0, published by springer nature (2025)
19. Shi, Y., Shang, M.Y., Qi, Z.Q.: Intelligent layout generation based on deep generative models: a comprehensive survey. Inf. Fusion. **100**, 24 (2023)
20. Petchrompo, S., et al.: A review of Pareto pruning methods for multi-objective optimisation. Comput. Ind. Eng. **167**, 108022 (2022)
21. Tezer, T., Yaman, R.: A Pareto optimum approach and a power management strategy for a stand-alone wind turbine-PV-hydrogen with reformer hybrid renewable energy system. Environ. Prog. Sustain. Energ. **41**(4), 17 (2022)
22. Ma, K., Zhou, Y.*.: A comprehensive quantitative lifecycle cost and environmental impact analysis model for computing infrastructure. MethodsX. **13**, 103009 (2024)
23. Huang, M.Y., et al.: LCA and LCCA based multi-objective optimisation of pavement maintenance. J. Clean. Prod. **283**, 10 (2021)
24. Ott, D., et al.: LCA as decision support tool in the food and feed sector: evidence from R&D case studies. Environ. Syst. Decis. **43**(1), 129–141 (2023)
25. Altaf, M., et al.: Evaluating the awareness and implementation level of LCCA in the construction industry of Malaysia. Ain Shams Eng. J. **13**(5), 9 (2022)
26. Khan, U., et al.: Life cycle cost analysis (LCCA) of Stirling-cycle-based heat pumps vs. conventional boilers. Clean. Environ. Syst. (2023)
27. Cao, K., Zhang, T., Huang, J.: Advanced hybrid LSTM-transformer architecture for real-time multi-task prediction in engineering systems. Sci. Rep. **14**, 4890 (2024)
28. Afsari, K., et al.: Identification of indicators for effectiveness evaluation of four-legged robots in automated construction progress monitoring. p. 610–620 (2022)
29. Esen, H., Esen, M., Ozsolak, O.: Modelling and experimental performance analysis of solar-assisted ground source heat pump system. J. Exp. Theor. Artif. Intell. **29**(1), 1–17 (2017). https://doi.org/10.1080/0952813X.2015.1056242
30. Xiang, S., Wang, R., Feng, C.: Mobile projective augmented reality for collaborative robots in construction. Autom. Constr. **127**, 103704 (2021)
31. Sirmacek, B., Vinuesa, R.: Remote sensing and AI for building climate adaptation applications. Results Eng. **15**, 100524 (2022)
32. Santero, N.J., Masanet, E., Horvath, A.: Life-cycle assessment of pavements. Part I: critical review. Resour. Conserv. Recycl. **55**(9), 801–809 (2011)
33. Liljenström, C., Björklund, A., Toller, S.: Including maintenance in life cycle assessment of road and rail infrastructure—a literature review. Int. J. Life Cycle Assess. **27**(2), 316–341 (2022)
34. Saxe, S., Kasraian, D.: Rethinking environmental LCA life stages for transport infrastructure to facilitate holistic assessment. J. Ind. Ecol. **24**(5), 1031–1046 (2020)
35. Lucas, A., Neto, R.C., Silva, C.A.: Energy supply infrastructure LCA model for electric and hydrogen transportation systems. Energy. **56**, 70–80 (2013)
36. Wohlschlager, D., Fröhling, M., et al.: Overcoming challenges in life cycle assessment of smart energy systems—a map of solution approaches. J. Clean. Prod. **423**, 138584 (2023)
37. World Bank.: Industry (including construction), value added (% of GDP). Retrieved Oct 16, 2024 from https://data.worldbank.org/indicator/NV.IND.TOTL.ZS
38. Han, X., et al.: Finance for science and technology: from innovation to entrepreneurship and industrial transformation. Nov res. Sci. **16**(2), NRS.000884 (2024)

39. Belengar, F.M., Zhou, Y.: Navigating data privacy challenges in the lifecycle of large AI models for banks: a comparative analysis of GDPR, CCPA/CPRA, and PIPL Regulations. In: 37th Australasian Finance and Banking Conference, Sydney Australia (2024)
40. Zhou, Y.: Direct Geothermal Energy: Reducing Poultry Industries' Carbon Footprint. PhD Thesis,. Dept. of Infrastructure Engineering, the University of Melbourne (2019)

Data Anomaly Identification and Noise Reduction Application of Eco-Environment IOT Monitoring System Based on Wavelet Transform Theory

Yue Deng[1,2], Zheng He[1,2(✉)], Rui Xiao[3], Zihao Guo[4], Xun Zhang[5], Xiyue Yang[6], and Mingyuan Gao[6]

[1] Technology Innovation Center for Land Spatial Ecological Protection and Restoration in Great Lakes Basin, Ministry of Natural Resources, Nanchang 330025, China
jxlyhz@126.com
[2] Jiangxi Institute of Land Space Survey and Planning, Nanchang 330025, China
[3] School of Remote Sensing and Information Engineering, Wuhan University, Wuhan 430079, China
[4] College of Jiyang, Zhejiang A&F University, Shaoxing 311800, China
[5] PowerChina Huadong Engineering Corporation Limited, Hangzhou 311122, China
[6] Jiangsu Province Hydrology and Water Resources Investigation Bureau, Nanjing 210029, China

Abstract. The proliferation of Internet of Things (IoT) sensors in ecological monitoring presents critical challenges related to data quality. Noise and anomalies in raw sensor data can compromise the accuracy of subsequent analyses and early-warning systems. This paper proposes and validates a two-stage methodology to address this issue. First, a Z-Score analysis is employed for statistical anomaly detection. Subsequently, a multi-scale Discrete Wavelet Transform (DWT), coupled with a soft thresholding procedure, is used for robust signal denoising before reconstruction via the Inverse DWT. The efficacy of this approach was validated using an in-situ pH monitoring dataset from Shijiu Lake. The results demonstrated a marked improvement in data quality. Compared to the original data, the denoised signal showed significant reductions in key metrics when benchmarked against ground-truth data: Standard Deviation (SD) decreased by 13.69%, Root Mean Square Error (RMSE) by 13.62%, and Mean Absolute Error (MAE) by 15.75%. These findings confirm that the proposed methodology provides an effective and practical tool for enhancing the integrity and reliability of real-time environmental monitoring data, establishing a stronger foundation for scientific analysis and decision-making.

Keywords: Ecological Environment Monitoring, Wavelet Transformation, Data Anomaly Identification, Data Noise Reduction

Y. Zhu (Ed.): ISESIE 2025, LNCE 000, pp. 278–289, 2026.
https://doi.org/10.1007/978-981-92-2389-3_22

1 Introduction

With the rapid growth of the global economy, ecological-environment problems have become increasingly acute, prompting China to set higher standards for environmental monitoring [1]. In 2015, the General Office of the State Council issued the Plan for Building an Ecological-Environment Monitoring Network, introducing the concept of a space-ground-integrated monitoring network. In 2018, the CPC Central Committee and the State Council released the Opinions on Comprehensively Strengthening Ecological-Environment Protection and Resolutely Winning the Battle against Pollution, which called for establishing "an independent, authoritative and efficient ecological-environment monitoring system" and constructing a space–ground integrated network.

Subsequently, the Ministry of Ecology and Environment published the Outline for Ecological-Environment Monitoring (2020–2035) in 2020, stressing that China's monitoring approach must shift from traditional field sampling to an integrated, automated, intelligent and interoperable space–ground system. The 14th Five-Year Plan for Ecological-Environment Monitoring (2022) further mandated the coordinated deployment of monitoring networks across land, sea, sky, surface and subsurface domains. Most recently, the General Office of the State Council, in its 2024 Opinions on Deepening Reform of Environmental Monitoring and Improving Data Quality, emphasized the deployment of big data, artificial intelligence, satellite remote sensing and other advanced technologies to raise the scientific caliber of environmental monitoring.

Against this policy backdrop, building an ecological-environment monitoring framework that integrates multi-source IoT data to deliver multi-scale, three-dimensional assessments of environmental quality has become an indispensable means of implementing the national mandate to develop a space–ground integrated monitoring system [2, 3].

Nevertheless, building and operationalizing an integrated ecological-environment monitoring system encounters several technical challenges when multiple data sources are combined. First, data obtained from disparate monitoring platforms—satellite remote sensing, unmanned aerial vehicles (UAVs) and in-situ sensors—differ markedly in spatial scale, temporal resolution and spectral characteristics, leading to heterogeneity in data quality, error structures and file formats [4, 5]. Second, the pronounced spatiotemporal dynamics of environmental variables introduce inconsistencies in sampling intervals and transmission latency, which, in turn, generate registration errors in time-series datasets [6, 7]. Third, insufficient quality control and inadequate handling of missing values lower the reliability of ecological monitoring and decision-support outputs [8, 9].

Consequently, it is imperative to integrate a suitable data quality monitoring methodology within the construction and operationalization of IoT-based ecological monitoring systems. Such an approach is critical not only for addressing practical challenges like data anomalies and missing values—thereby reducing noise, mitigating bias, and ensuring the consistency and reliability of multi-source data for enhanced overall accuracy [10]—but also for detecting subtle variations within datasets, which this capability enables the capture of minor ecological shifts and abrupt events, providing a scientific basis for real-time early-warning systems and risk assessment [11].

Therefore, to address the current limitations in the acquisition and processing of multi-source ecological data, this study puts forward a methodology for anomaly detection and processing rooted in wavelet transform theory. This approach is specifically designed for data transmitted from a multi-scale network of in-situ sensors—monitoring parameters such as water quality, noise, and atmospheric conditions—where the prevalence of missing, noisy, or erroneous data presents a significant challenge. The methodology employs the Z-Score method for initial data screening, followed by wavelet-based transforms for robust data denoising. In doing so, this research offers a feasible technical pathway for the high-efficiency quality control of ecological big data and provides direct technical support for the national mandate to strengthen the application of advanced technologies in environmental monitoring.

2 Data Anomaly Detection

In IoT-based ecological monitoring networks, the susceptibility of individual sensors to environmental interference can introduce significant data noise, making rigorous data cleaning and standardization indispensable. The process of data cleaning focuses on discrete errors, including the removal of spurious data, imputation of missing values, correction of inaccuracies, and elimination of duplicate records. Data standardization then addresses systemic discrepancies by mitigating biases that stem from source heterogeneity, thereby improving overall data accuracy and consistency. Ultimately, the systematic filtering of anomalous data from each source dataset is a critical prerequisite for ensuring that the integrated data provides a high-fidelity representation of the true environmental state.

The Z-Score method is a standard statistical technique used to identify outliers in large datasets [12, 13]. Its fundamental principle is predicated on the properties of a normal distribution, wherein the majority of data points cluster symmetrically around the mean. This distribution is characterized by the empirical rule, which dictates that approximately 68% of data lie within one standard deviation of the mean, 95% lie within two standard deviations, and 99.7% lie within three standard deviations. Consequently, any data point situated beyond this three-standard-deviation range can be identified as a potential outlier. The Z-Score is calculated with the following formula:

$$Z = \frac{x - \mu}{\sigma}$$

In this formula, x represents an individual data point, μ is the mean of the historical dataset, and σ is the standard deviation of that same dataset. The calculation of the Z-Score thus quantifies how many standard deviations a data point is from the mean. If the absolute value of a calculated Z-Score exceeds a predefined threshold (e.g., $|Z| > 3$), the corresponding data point is classified as an outlier. Such outliers are subsequently flagged for appropriate handling, which typically involves either removal from the dataset (filtering) or replacement via data imputation.

In the context of ecological and environmental monitoring, the Z-Score method is primarily applied for anomaly detection, enabling the statistical identification of data points that may be spurious due to collection errors or environmental interference. The key advantages of this technique are its simplicity and ease of implementation, which provide a foundational layer of quality control for subsequent data processing stages. However, a significant limitation of the Z-Score method is its reliance on the assumption of a normal data distribution. For datasets that deviate markedly from normality, its detection accuracy can be compromised, often necessitating the use of complementary techniques to ensure robust performance.

3 Processing of Data Anomalies

Denoising via wavelet transform is a signal processing technique rooted in wavelet theory, designed to extract an underlying signal from data corrupted by noise [14]. The fundamental methodology comprises three sequential stages: decomposition, thresholding, and reconstruction. First, the original signal is decomposed across multiple scales using a wavelet transform. In this multi-resolution representation, the low-frequency components (approximations) capture the signal's primary trend, while the high-frequency components (details) predominantly contain the noise. Second, in the wavelet domain, noise typically manifests as coefficients with small amplitudes. A thresholding function is applied to these coefficients to selectively filter out those presumed to represent noise. Finally, the inverse wavelet transform is applied to the modified set of coefficients to reconstruct the original signal with the noise component significantly suppressed [15]. This technique is widely employed in big data applications for both data denoising and signal enhancement. It is particularly well-suited for mitigating noise in multi-source monitoring data that originates from sensor inaccuracies or external interference, thereby enhancing overall data quality and analytical precision [16–18].

To further mitigate the impact of data noise, the methodology in this study employs the Daubechies 4 (db4) wavelet to perform a 3-level decomposition of the signal. The fundamental equation governing this wavelet decomposition is as follows:

$$c_{j,k} = \int_{-\infty}^{+\infty} f(t)\psi_{j,k}(t)dt$$

$$\psi_{j,k}(t) = 2^{\frac{j}{2}}\psi\left(2^{j}t - k\right)$$

where $\psi_{j,\,k}(t)$ represents the wavelet basis function. This decomposition yields the third-level low-frequency approximation coefficients A_3 as well as the high-frequency detail coefficients for each respective level (D_1, D_2, D_3).

Subsequently, a soft thresholding procedure is applied to the high-frequency detail coefficients (D_1, D_2, D_3). This technique is designed to filter out noise—which typically manifests as low-amplitude coeffients—while preserving the key information that reflects the signal's true dynamics. The core principle of soft thresholding involves a shrinkage function: coefficients with an absolute value less than a specific threshold (λ) are set to zero, whereas coefficients exceeding the threshold are shrunk towards zero by the threshold value. A standard formula for calculating this threshold is:

$$\lambda = \sigma\sqrt{2\ln n}$$

where n is the length of the signal, and σ represents the standard deviation of the noise. The value of σ is robustly estimateusing the Median Absolute Deviation (MAD) of the first-level detail coefficients, as follows:

$$\sigma = \frac{MAD(|D|)}{0.6745}$$

For each detail coefficient ω, the soft thresholding transformation is given by:

$$\omega' = \text{sign}(w) \cdot \max(|\ \omega\ | - \lambda, 0)$$

In this equation, λ is the predeterminedhreshold, and $sign(w)$ is the signum function, which returns the sign of the detail coefficient ω. The function operates as follows: if the absolute value of a coefficient is less than or equal to the threshold ($|\omega| \leq \lambda$), the transformed coefficient, $\omega^{'}$, is set to zero. Conversely, if the absolute value exceeds the threshold ($|\omega| > \lambda$), the coefficient is shrunk by subtracting the threshold value, while its original sign is preserved.

This approach smoothly attenuates the smaller, noise-related coefficients while minimizing distortion to the large-magnitude coefficients that represent the signal's essential features. As a result, noise is effectively removed, and the principal structure of the signal is preserved.

The final stage in the denoising procedure is signal reconstruction. This step involves synthesizing the modified detail coefficients, which have undergone soft thresholding, with the original, unmodified approximation coefficients. The Inverse Discrete Wavelet Transform (IDWT) is applied to perform this synthesis, which yields a new, smoothed data signal with the noise component significantly attenuated [19]. Mathematically, this reconstruction process is represented as:

$$f_{recon} = \sum_{k} A_L(k)\varnothing_{L,k}(t) + \sum_{j=1}^{L}\sum_{k} D_j^{'}(k)\psi_{j,k}(t)$$

In this representation, $A_L(k)$ denotes the low-frequency approximation coefficients and $D_j^{'}(k)$ represents the thresholded detail coefficients at decomposition level j. The terms $\varnothing_{L,\,k}(t)$ d $\psi_{j,\,k}(t)$ are the scaling function and the wavelet basis function, respectively.

The significance of this complete process lies in its dual effect: it preserves the principal trend of the signal, which is embedded in the approximation coefficients, while simultaneously eliminating the high-frequency noise that was filtered from the detail coefficients. This procedure yields a data foundation of significantly higher quality, suitable for all subsequent processing and analysis.

4 Performance Evaluation

To validate the efficacy of the proposed wavelet denoising technique, this section evaluates its performance in enhancing data quality, smoothing signal fluctuations, and preserving principal signal features within ecological monitoring data. The evaluation relies on a quantitative comparison of statistical metrics derived from the data before and after the denoising procedure. For this purpose, the primary indicators selected are the Root Mean Square Error (RMSE) and the Standard Deviation (SD).

The RMSE quantifies the average magnitude of the discrepancy between two datasets. In this context, it is used to compare the denoised signal against the original signal. The formula is as follows:

$$\text{RMSE} = \sqrt{\frac{1}{n}\sum_{i=1}^{n}(x_i - y_i)^2}$$

In the equation above, x_i represents the i-th observation from the ground-truth signal, while y_i is the corresponding value from the denoised signal, and n is the total number of data points.

In the context of ecological monitoring, raw data is typically a superposition of this true signal and unwanted noise. The objective of denoising is to eliminate the noise component while preserving the signal's principal trends. The RMSE value provides a direct measure of how effectively this is achieved; a smaller value signifies a lower average discrepancy between the denoised signal and the true signal. Therefore, a significant reduction in the RMSE after processing is a key indicator of successful noise suppression.

5 Case Study

5.1 Data Anomaly Detection

To illustrate the proposed data processing methodology, this study presents a case study using a continuous series of in-situ monitoring data. The selected study area is Shijiu Lake, a large regulating lake situated on the border of Jiangsu and Anhui provinces. Given its critical role in regional water security and flood control, the lake is equipped with a well-established automatic water quality monitoring system.

For this analysis, we selected a segment of 60 consecutive raw pH readings collected beginning on January 1, 2024. While this dataset exhibits a generally stable trend, it contains subtle anomalies resulting from factors such as environmental interference, fluctuations in instrument precision, and data transmission errors. These anomalies are not readily apparent through manual inspection but are significant enough to compromise the accuracy of subsequent data fusion and early-warning models, thus necessitating a formal detection procedure.

Applying the Z-Score methodology, the mean and standard deviation of this data segment were calculated to be 8.13 and 0.26, respectively. A decision threshold of $|Z| = 2.0$ was selected for this demonstration, which stipulates that approximately 95% of the data points are expected to fall within the normal range. Data points exceeding this threshold are consequently identified as anomalies, as illustrated in Fig. 1.

As illustrated in Fig. 1, the anomaly detection results highlight two distinct types of observations. First, data points #12, #16, #38, and #52 exhibit Z-scores that approach the decision threshold of $|Z| = 2.0$. While these points are classified as normal according to the selected criterion, their proximity to the boundary suggests they are borderline cases that warrant closer inspection. In contrast, points #1, #40, and #47—with pH values of 7.6, 7.6, and 8.7, respectively—are identified as clear anomalies. Although

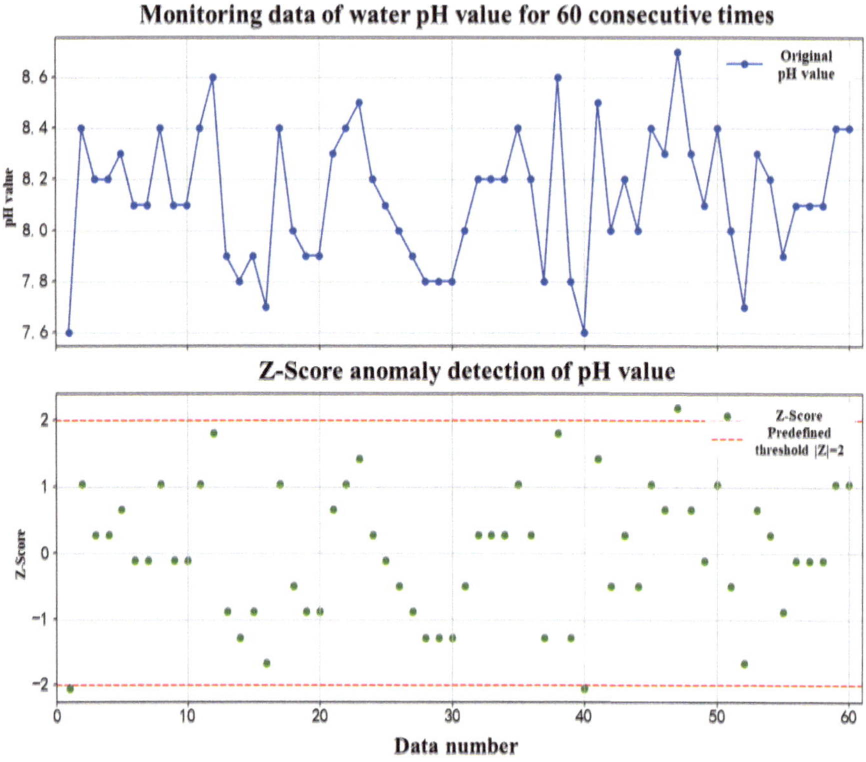

Fig. 1. Z-Score method for data anomaly identification.

their absolute pH values are not extreme, their Z-scores indicate a significant deviation from the central tendency of the dataset. The identification of such anomalies prompts investigation into potential root causes, such as sensor malfunction, data transmission errors, or abrupt environmental events. Following this diagnosis, appropriate corrective actions can be implemented, including data imputation, outright removal (filtering), or targeted investigation to determine the source of the anomaly. This process is critical for enhancing the overall integrity of the dataset and ensuring the reliability of subsequent early-warning systems and predictive analyses.

5.2 Data Denoising

This section addresses the issue of minor deviations within the generally stable dataset, which are characteristic of instrument noise or ambient environmental interference. Following the initial anomaly identification (as described in Sect. 5.1) and once it has been determined that these deviations are not caused by significant environmental events, a denoising procedure is applied. The first step involves decomposing the original time-series signal to separate its principal trend from high-frequency noise components

For the 60-point pH dataset from Shijiu Lake, we performed a 3-level Discrete Wavelet Transform (DWT) using the Daubechies 4 (db4) wavelet. This decomposition yielded the low-frequency approximation coefficients(A_3), representing the data's primary trend, alongside the high-frequency detail coefficients for each respective level (D_1, D_2, D_3), which contain the noise and finer signal details. The results of this multi-resolution decomposition are visualized in Fig. 2.

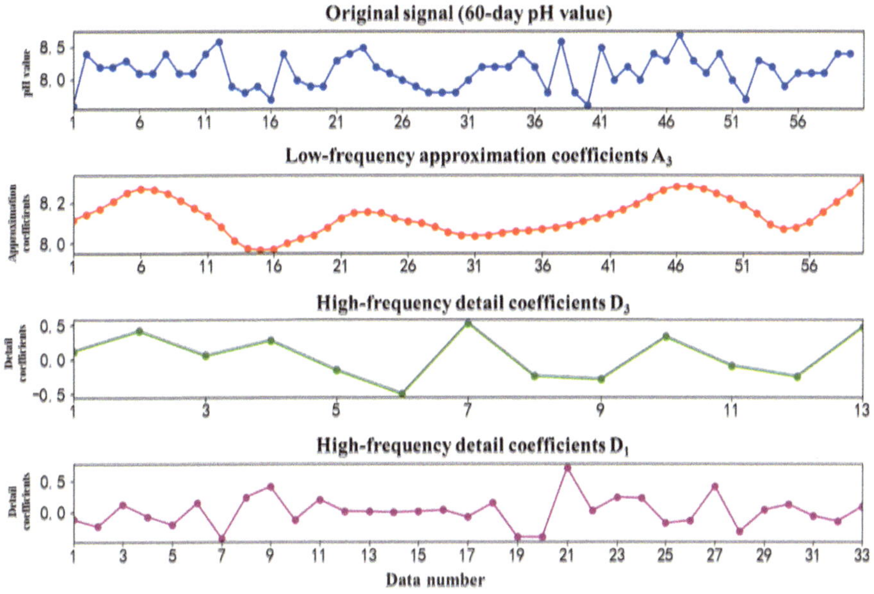

Fig. 2. Multi-scale decomposition of wavelet transform.

The four subplots in Fig. 2 visually illustrate the multi-scale decomposition of the 60-point pII datasct via the wavelet transform. From top to bottom, each subplot represents a different component of the signal:

Original Signal: The first subplot displays the raw time-series of the 60 pH values. While the data appears generally stable, it contains subtle fluctuations and noise components that are the primary targets of the denoising procedure.

Approximation Coefficients (A_3): The second subplot shows the reconstructed 3rd-level approximation signal. This represents the low-frequency component of the data, capturing the primary trend or overall pattern. Isolating this component facilitates the analysis of long-term trends by filtering out a significant portion of the high-frequency noise.

Detail Coefficients (D_3): The third subplot presents the 3rd-level detail coefficients. These capture finer fluctuations and local variations within the mid-to-low frequency bands. Compared to the approximation signal, this component is more sensitive to localized, non-transient disturbances.

Detail Coefficients (D_1): The bottom subplot displays the 1st-level detail coefficients. This level reflects the highest-frequency components in the signal, corresponding to rapid, short-term variations and instantaneous noise. Analyzing these high-frequency details is crucial for identifying transient anomalies or abrupt events.

Following the multi-scale decomposition, the soft thresholding procedure is applied to the detail coefficients derived from the raw monitoring data. As previously described, this process eliminates noise by applying a "soft" shrinkage function to these coefficients while aiming to preserve the signal's authentic features. Figure 3 presents a comparative visualization of the detail coefficients before and after this soft thresholding process is applied.

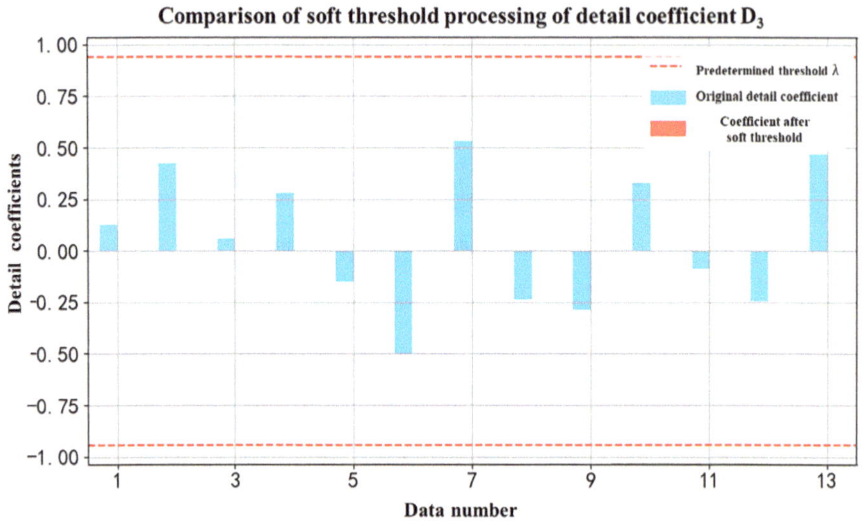

Fig. 3. Comparison of soft threshold processing of detail coefficient.

Figure 3 presents a comparison of the 3rd-level detail coefficients D_3 before and after the soft thresholding procedure. The horizontal axis represents the coefficient index corresponding to each data point, while the vertical axis indicates the coefficient amplitude.

The blue bars represent the original D_3 coefficients, which embody the local fluctuations and noise components inherent in the signal. In contrast, the red bars display the same coefficients after the thresholding function has been applied. In this particular instance, the relatively low noise level of the input data resulted in the calculation of a high threshold value. Consequently, the majority of the original coefficients fell below this threshold and were effectively shrunk to zero during the process, an outcome that is visually confirmed by the red bars in the figure.

The final step in the procedure is signal reconstruction via the Inverse Discrete Wavelet Transform (IDWT). In this stage, the original low-frequency approximation coefficients (A_3) are synthesized with the newly thresholded detail coefficients (D_1, D_2, D_3). This process yields the final, denoised signal, which preserves the principal trend of the original data while significantly reducing noise interference. The resulting improvement in the signal's smoothness and reliability is visualized in Fig. 4.

Figure 4 presents a direct comparison between the original signal and the reconstructed signal following the wavelet denoising procedure. The horizontal axis indicates

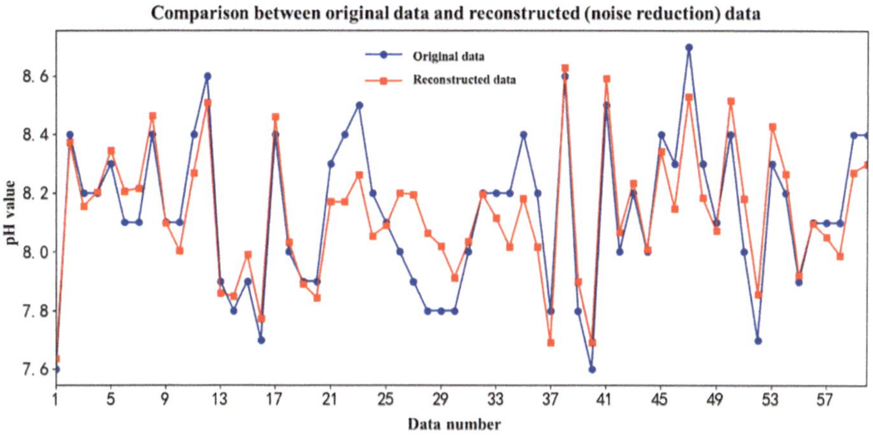

Fig. 4. Comparison between original data and reconstructed (noise reduction) data.

the data point index, while the vertical axis represents the water body's pH value. The blue line plot corresponds to the original, raw signal, exhibiting both the data's natural fluctuations and superimposed high-frequency noise. In contrast, the red line plot represents the final denoised signal. It is visually evident that the reconstructed signal is significantly smoother; minor oscillations and jitters present in the original data, likely attributable to instrument noise or external interference, have been effectively suppressed. Notably, even at points where the original data exhibits subtle anomalous fluctuations, the denoised signal successfully preserves the underlying trend and core information of the dataset. This process thus yields a more reliable data foundation for subsequent applications and advanced ecological analysis.

5.3 Quantitative Comparison and Evaluation

This section presents a quantitative evaluation of the wavelet denoising method's performance, using the 60-point pH dataset from the Shijiu Lake case study. For this comparative analysis, the raw data from the in-situ IoT sensor is defined as the "Original Signal". The result of our wavelet processing is termed the "Denoised Signal". To serve as a benchmark for accuracy, concurrent data from manual monitoring is used as the "Ground Truth". The statistical comparison of these datasets is presented in Table 1 below:

Table 1. Comparison of statistical indexes between original data and reconstructed (noise reduction) data.

Statistical index	Standard deviation	Root mean square error	Mean absolute error
Original data	0.2594	0.2578	0.2108
Reconstructed data	0.2239	0.2227	0.1776

A comparison of the Standard Deviation before and after processing reveals a decrease from 0.2594 for the Original Signal to 0.2239 for the Denoised Signal, a reduction of 13.69%. This decrease in variance indicates that the denoising procedure successfully removed erratic noise components, resulting in a significantly smoother and more stable signal. Furthermore, error metrics calculated against the Ground Truth data validate the method's efficacy in improving accuracy. The Root Mean Square Error between the sensor data and the ground truth was reduced from 0.2578 to 0.2227, a 13.62% improvement. Similarly, the Mean Absolute Error decreased from 0.2108 to 0.1776, a 15.75% reduction. These significant reductions in both variance and error metrics conclusively demonstrate that the proposed wavelet denoising procedure is highly effective. It successfully minimizes the deviation between the sensor data and the true environmental state, thereby substantially enhancing the overall quality and reliability of the monitoring data.

6 Conclusion

This research addressed the critical challenge of noise interference and data anomalies within ecological Internet of Things (IoT) monitoring systems by proposing and validating a robust methodology rooted in wavelet transform theory. The approach first employs a Z-Score analysis to statistically identify subtle deviations, followed by a multi-scale wavelet decomposition. A soft thresholding procedure is then applied to the high-frequency detail coefficients to effectively separate the underlying signal from noise before reconstruction via the Inverse Discrete Wavelet Transform (IDWT). This process yields a smoother data series with enhanced stability, priming it for more accurate multi-source fusion and predictive analysis.

The efficacy of this methodology was demonstrated through a case study using in-situ pH monitoring data from Shijiu Lake, where the procedure successfully suppressed noise while preserving the principal signal trend. Quantitative evaluation confirmed these findings, with significant reductions in both Standard Deviation (SD) and Root Mean Aabsolute Error (MAE) post-processing. These results validate the proposed method as an effective and practical tool for enhancing the quality and reliability of environmental monitoring data.

In summary, this research provides a dual contribution: it enhances the theoretical framework for processing multi-source environmental data and establishes a practical technical pathway for data management and denoising within ecological IoT networks.

Future work can build upon this foundation in several key directions. These include the further optimization of wavelet parameters, such as the choice of the mother wavelet and decomposition level, and the integration of more advanced algorithms for dynamic weight adjustment and data fusion. Extending this methodology to a broader spectrum of ecological monitoring applications represents a valuable next step, which will provide more robust data support for effective environmental management and evidence-based scientific decision-making.

Acknowledgement. This work was supported by the Open Research Fund of the Technology Innovation Center for Land Space Ecological Protection and Restoration in Great Lake Basins, Ministry of Natural Resources (Grant No. JXCXZX2024001).

References

1. Huang, B., Wei, Z., Jiang, Y., et al.: Study on the progress and prospect of environmental and health work carried out by environmental departments in China. Environ. Sustain. Dev. **44**(5), 5–8 (2019)
2. Wang, J.: Research on architecture and key Technologies of air-space-ground integrated big data platform along the Yellow River Basin. Yellow River. **43**(4), 6 (2021)
3. Liu, B., Song, L., Zhao, Y., et al.: Research and application of space-air-ground integrated environmental monitoring system. Ecol. Environ. Monit. Three Gorges. **8**(2), 17–25 (2023)
4. Wang, Q.: Progress of environmental remote sensing monitoring technology in China and some related frontier issues. National Remote Sensing Bulletin. **25**(1), 25–36 (2021)
5. Zhang, L., He, J., Yang, Q., et al.: Data-driven multi-source remote sensing data fusion:progress and challenges. Acta Geodaetica et Cartographica Sinica. **51**(7), 13–17 (2022)
6. Ren, C., Ju, H., Zhang, H., et al.: Multi-source data for Forest land type precise classification. Sci. Silvae Sin. **52**(6), 54–65 (2016)
7. Li, P., Li, Z., Li, T., et al.: Wide-swath InSAR geodesy and its applications to large-scale deformation monitoring. Geomatics and Information Science of Wuhan University. **42**(9), 1195–1202 (2017)
8. Sun, T., Pan, J.: Problems and countermeasures in water environment testing. Nat. Sci. Res. **1**(3), 50–52 (2024)
9. Xu, W., Wu, F.: IOT-based smart sensors for environmental monitoring applications. J. Proj. Manag. **5**(11), 215–217 (2024)
10. Gao, X., Zhao, Z., Yin, H., et al.: Development and application of space target situation awareness and multi-source data fusion. Vac. Cryog. **29**(6), 543–554 (2023)
11. Xu, Q., Dong, X., Li, W.: Integrated space-air-ground early detection, monitoring and warning system for potential catastrophic Geohazards. Geomatics and Information Science of Wuhan University. **44**(7), 957–966 (2019)
12. Wang, Z., Hong, J., Liu, P., et al.: Voltage fault diagnosis and prognosis of battery systems based on entropy and Z-score for electric vehicles. Appl. Energy. **196**, 289–302 (2017)
13. Sun, Z., Wang, Z., Chen, Y., et al.: Modified relative entropy-based lithium-ion battery pack online short-circuit detection for electric vehicle. IEEE Trans. Transp. Electrification. **8**(2), 1710–1723 (2021)
14. Zhang, Z., Ding, F., Luo, Y., et al.: Application of wavelet threshold denoising in processing of divertor spectral signal. Chin. J. Quantum Electron. **39**(3), 307 (2022)
15. Yuan, Z.F., Liu, X.G., Xia, M.Y.: Deep learning network for magnetic anomaly target signal denoising. In 2024 IEEE 17th International Conference on Signal Processing (ICSP), pp. 69–73. IEEE (2024)
16. Liu, W., Tan, L., Xu, Q., et al.: Water quality anomaly detection approach based on a neural network prediction model. Desalin. Water Treat. **121**, 351–356 (2018)
17. Liu, H., Zhang, X., Dong, H., et al.: Magnetic anomaly detection based on energy-concentrated discrete cosine wavelet transform. IEEE Trans. Instrum. Meas. **72**, 1–10 (2023)
18. Barzegar, R., Adamowski, J., Moghaddam, A.A.: Application of wavelet-artificial intelligence hybrid models for water quality prediction: a case study in Aji-Chay River, Iran. Stoch. Environ. Res. Risk Assess. **30**(7), 1797–1819 (2016)
19. Zhou, J., Zhang, B., Fan, L., et al.: Aeromagnetic anomaly detection under low SNR conditions using multiscale wavelet energy accumulation. In: 2020 IEEE 20th International Conference on Communication Technology (ICCT), pp. 1641–1644. IEEE (2020)

Assessment of the Reliability of Thin-Walled Steel Profiles in High-Temperature Conditions

Vladimir Dmitrienko(✉), Stanislav Maslennikov, Natalya Merenkova, and Olga Pashkova

Institute of Service and Business (branch) of DSTU in Shakhty, 346500 Shakhty, Russia
vadmitrienko@rambler.ru

Abstract. In low-rise construction, the technology of building objects from light steel profiles is very popular. To assess the safety of objects, the temperature distribution in the fence was studied by modeling during prolonged heating from a fire source. Laboratory studies of the deformability of thin-walled profile samples at temperatures of 20 and 290 °C were conducted using a special method. A 13.4% decrease in the deformation modulus of a thin-walled steel pipe at a temperature of 290 °C was observed compared to loading at a normal temperature (20 °C). For modeling by the finite element method, a two-storey cottage with a frame made of profiled steel pipes and a fence made of foam or gas concrete blocks was adopted. In the software package "Lira", studies of the stress-strain state of the building frame at temperatures of 20 and 290 °C were carried out. As a result of the calculations, diagrams of longitudinal and transverse forces, bending moments, and deformations of the frame elements were obtained, the values of which are significantly lower than the permissible values. It is shown that the maximum displacements in the vertical elements of the frame structure increase by 1.4 times at a temperature of 290 °C, but the calculated value of longitudinal flexibility does not exceed the maximum permissible value. Therefore, the use of thin-walled steel profiles in the frames of low-rise buildings is acceptable, taking into account the possible increase in temperature.

Keywords: Building frames · thin-walled profiles · modulus of elasticity · modulus of deformation · and stresses

1 Relevance of the Research

The products of the construction industry should retain their functions for tens of years. Hence the most important requirements for construction objects: reliability; safety and durability during the whole operation period.

A distinctive feature of fast-erected buildings is construction with the use of metal frame [1]. By the end of the twentieth century in Russia in the construction of buildings and structures began to be widely used light steel profiles (LSTS), which in addition to a very good ratio "price / quality" have the following advantages: less weight, ease of installation and dismantling, the ability to install large blocks, transportability, non-combustibility, high load-bearing capacity, durability and ease of disposal [2, 3]. In

Y. Zhu (Ed.): ISESIE 2025, LNCE 000, pp. 290–301, 2026.
https://doi.org/10.1007/978-981-92-2389-3_23

combination with lightweight fencing and roofing structures, metal frame buildings reduce the load on the foundation and reduce the cost of construction [4, 5].

Profiles made of galvanized steel, for a long period of time retain an attractive appearance without additional maintenance, easy enough to install [6–8]. Many years of experience in the operation of buildings made of LSTS proves their efficiency [9]. However, buildings with a frame made of LSTS are more susceptible to progressive destruction and accidents of such buildings occur more often than those made of conventional profiles [10, 11].

Since the requirements to the reliability of residential buildings are quite high, the design of frames with the use of LSTK requires taking into account all the negative factors that contribute to changes in the stress-strain state (STS) of the frame structures. Very important in this case is the change in the characteristics of metal in conditions of high temperatures and possible increase in load during operation [12].

When assessing the mechanical properties of steel, static tensile tests are carried out according to GOST 1497–84 at a temperature of $20 \pm 5°$. For this purpose cylindrical specimens of solid cross-section with a diameter of 5–10 mm are used, specimens with a diameter of 10 mm are considered to be the main type. The working length of cylindrical specimens should be not less than two diameters plus the length necessary to fix the specimen in the clamps of the testing machine.

Analysis of the work of elements of building frames indicates that most of the structures are subject to compression and bending, so given the use of hollow sections as load-bearing elements in most cases, it is advisable to test them in compression. In this case it is possible to take into account the influence of longitudinal stiffness on the deformation characteristics of the profile.

In addition, it is well known that the mechanical characteristics of steel change with temperature, so thin-walled profiles are more susceptible to loss of stability due to lower thermal inertia. This can lead to reduced load-bearing capacity and consequently insufficient fire resistance of buildings.

Thus, the study of the performance of thin steel sections in conditions of elevated temperatures is important in assessing the reliability of building objects, that is, the task of improving the design of building structures is still relevant.

2 Methods and Materials

Analytical and numerical methods are used in the design of construction objects [13]. Building objects consist of many elements, with complex links and interaction, that is, they are very complex i n structure and differing in mechanical properties and laws of deformation systems, so in the design it is necessary to take into account the characteristics and behavior of different materials, to calculate and predict their stress-strain state (STS) to ensure the stability and reliability of buildings [14].

To ensure the required stability of buildings, even self-supporting structures use many elements (connections), which complicate the configuration and lead to uneven distribution of mechanical and temperature fields. This leads to the fact that the calculation by analytical formulas, only conditionally describes the heat transfer resistance of the enclosing structure and the stress-strain state of the building elements. Therefore, it is most appropriate to use modeling [15. 16].

Modern methods of building design are based on the wide application of program complexes. Currently, for the design of building objects, special programs such as LIRA-SADS, LIRA-SOFT and a number of others have been created, which are capable of modeling individual structures and entire buildings with different characteristics. The basis of these programs is the finite element method, the essence of which is to divide structures into a set of separate small- sized elements (rectangles, triangles) connected to each other in nodes. This provides continuity of displacements of finite elements, which may have different mechanical properties [14]. The elastic moduli of the materials used are used as input data in these programs. In the case of using steel structures for the frame of the objects, it is necessary to introduce the elastic modulus of steel. Therefore, an important design step to ensure the reliability of the object is the determination of the mechanical characteristics of steel.

As a rule, the modulus of elasticity *E is* determined based on the results of experimental studies. The computational method of determination consists in determining, using a strain gauge or strain meter, the change in the length of the specimen during its stepwise loading. For this purpose, the specimen is placed on the testing machine and loaded in equal steps up to the value of stress equal to 70–80% of the limit of proportionality. The magnitude of the loading step should be 5–10% of the maximum value.

The number of specimens to evaluate the average value of mechanical characteristics should be at least five, but if the difference in the determined characteristics does not exceed 5%, you can limit yourself to three specimens.

The finite element method can be used to model not only mechanical fields, but also thermal fields. In this case it is advisable to use the ELCUT program, with the help of which it is possible to study the temperature distribution in enclosing structures.

To determine the deformation characteristics of thin-walled profiles at a temperature of 20oC the methodology is adopted, which provides for: manufacturing of profile pipe samples; labeling them with marks; loading of samples is carried out on a hydraulic press E 160 (Fig. 1); with fixation of deformations at each stage of loading by an optical cathetometer KM-8 with a division value of 0.001 mm. The magnitude of the load is controlled by the electronic display of the press (Fig. 2).

In order to determine the deformation characteristics of thin-walled profiles at high temperatures, a special method was developed, which includes: manufacturing of profile pipe samples; winding on their outer surface of heaters made of constantan wire with a diameter of 0.3 mm, on top of insulation (Fig. 3).

The sample was placed inside the insulation block. The second heater was placed inside the pipe. Temperature control was carried out by a thermocouple included in a multimeter. Voltage was applied to the heaters 82 B. After heating up to the required temperature, the inner heater was removed and the sample in the heat-insulating block was placed on the press plates (Fig. 4). Temperature stability was maintained by the external heater by adjusting the voltage.

Fig. 1. Testing of a profile pipe specimen on E160 press.

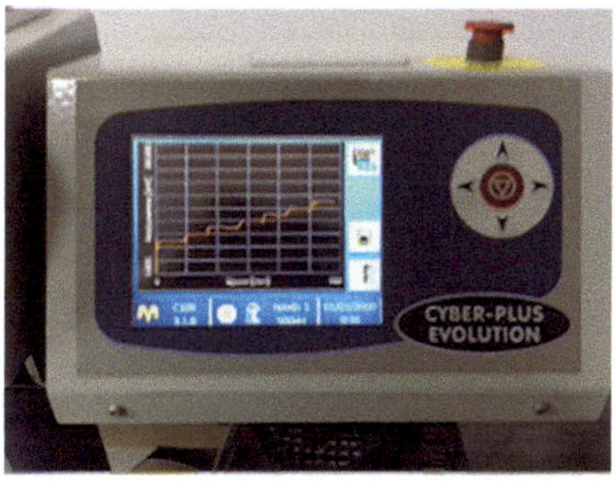

Fig. 2. Stepwise growth of load on the sample.

3 Research Results and Discussion

At the first stage of experimental research loading of each sample was carried out on the press at a temperature of 20 °C in steps of 5 kN up to a maximum value of 50 kN with fixation of deformations at each step of loading. Since it was not possible to fix the load on the press with high accuracy at each stage, there were deviations of ±1 kN.

Fixation of deformations was carried out by cathetometer KM-8 on the marks glued on the specimens, and the rate of deformations by the indicator of hour type ICH-50.

After testing three specimens, the results of one of them are presented in Table 1 and in Fig. 5, a graph of strain-stress dependence is shown.

After checking the maximum deviations, the graph of dependence (Fig. 6) of relative deformations of three specimens on stresses for the areas with linear dependences was plotted.

Maximum deviation from the average value of the strain modulus at the of one of the loading stages amounted to 5.06%, and in general did not exceed 5.0%.

According to the obtained results, the relative deformation of specimens and their strain moduli were calculated. The average value of the deformation modulus of the profile specimens at normal temperature for the linear section was 106987.1 MPa.

Fig. 3. Profile pipe sample with a heater

Fig. 4. Testing of specimens at high temperature.

The obtained result is almost 2 times less than the modulus of elasticity of steel of the corresponding grade, which indicates the influence of thin walls of the profile on the

Table 1. Test results for sample II.

Load on sample, kN	Stresses, MPa	Deformation of the specimen, mm	Relative deformation of the specimen	Strain modulus, MPa
22,53	12,729	0.0000	0.0000	
70,59	39,881	0.0380	0.0005	56084
11,795	66,638	0.0610	0.0008	91239
16,516	93,311	0.0850	0.0011	87109
21,253	120,073	0.1080	0.0014	91149
25,902	146,339	0.1260	0.0016	114245
30,518	172,418	0.1410	0.0018	136064
35,157	198,627	0.1580	0.0020	120605
40,018	226,090	0.1860	0.0024	76685
44,805	253,136	0.2250	0.0029	54171
49,711	280,853	0.2840	0.0036	36652

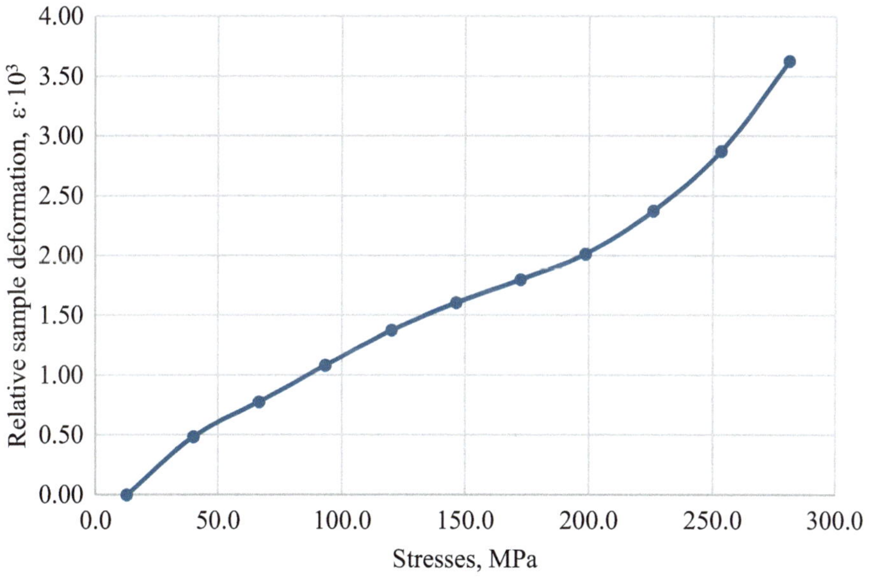

Fig. 5. Graph of dependence of relative deformations of specimen II on stresses.

magnitude of deformations of samples in contrast to the tests of steel of a continuous section.

The study of deformation characteristics of profile specimens at temperature 290 °C was carried out according to the methodology given earlier. The test temperature was

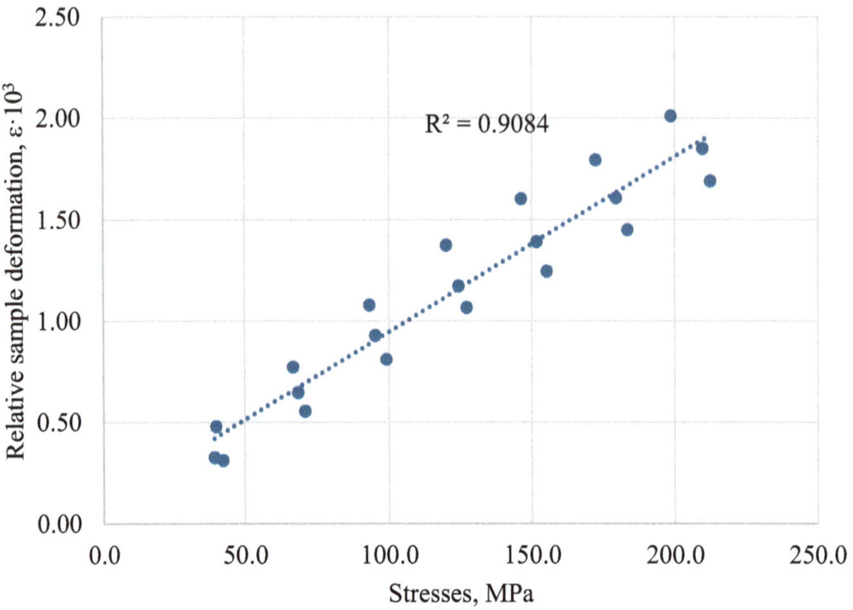

Fig. 6. Graph of linear sections of the dependence of the relative strains of three specimens from stresses tested at temperature 20 °C.

justified by modeling the building envelope in the ELCUT program. Calculation of several variants of change of approximate internal temperature at fire and external air. Possible temperature of steel column behind the envelope of foam concrete blocks was in the range of 230÷340 °C. For testing of profile specimens the temperature of 290 °C was accepted.

The results of tests of one of the samples are presented in Table 2, on the basis of which the graph of dependence of relative deformations on stresses (Fig. 7) and the general graph of linear sections for three samples (Fig. 8), by which the modulus of deformation of profiles at temperature 290 °C was determined.

To justify the use of lightweight steel frames in the design of buildings, the research was carried out by modeling in the program complex "Lira-SADR steel frame of a two-storey cottage. The size in plan of the cottage is accepted 10 × 10 m with the height of the middle columns 6,54 m. According to the condition of ultimate flexibility for the columns the profile steel pipe with cross-section 80 × 80 × 3 mm according to GOST 27772.88 is used.

Loads acting on the frame structures are represented by: own weight; normative (household); snow and wind loads. For the erection of wall enclosures 200 mm thick aerated concrete blocks are used, the slab is monolithic concrete, the roof covering is profiled steel sheets with the device: vapor and heat insulation; water insulation layer; leveling screed; waterproofing; membrane.

The reliability of the building frame at 290 °C was determined by determining the VAT of steel elements in the Lira program complex. For this purpose, the obtained values

Table 2. Test results of sample V.

Load on sample, kN	Stresses, MPa	Deformation of the specimen, mm	Relative deformation of the specimen	Strain modulus, MPa
32,16	18,169	0.0000	0.0000	
79,82	45,096	0.0360	0.0005	58697,64
12,786	72,237	0.0640	0.0008	76008,11
17,587	99,362	0.0900	0.0011	81747,48
22,383	126,458	0.1170	0.0015	78584,69
27,182	153,571	0.1400	0.0018	92250,44
31,918	180,328	0.1620	0.0021	95122,92
36,572	206,621	0.1840	0.0023	93423,47
41,258	233,096	0.2110	0.0027	76598,30
45,915	259,407	0.2470	0.0032	57047,27
50,631	286,051	0.2850	0.0036	54677,76

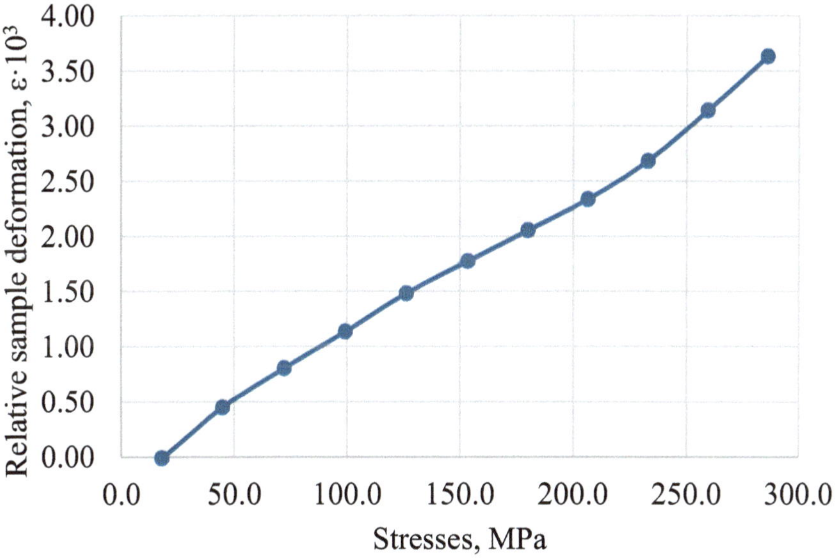

Fig. 7. Graph of dependence of relative strains of specimen V on stresses.

of deformation modulus at 20 and 290 °C were set in the compiled calculation scheme of the building frame. As a result of the calculations, the epuples of longitudinal (Fig. 9) and transverse forces, bending moments (Fig. 10), deformations of the frame elements (Fig. 11) were obtained.

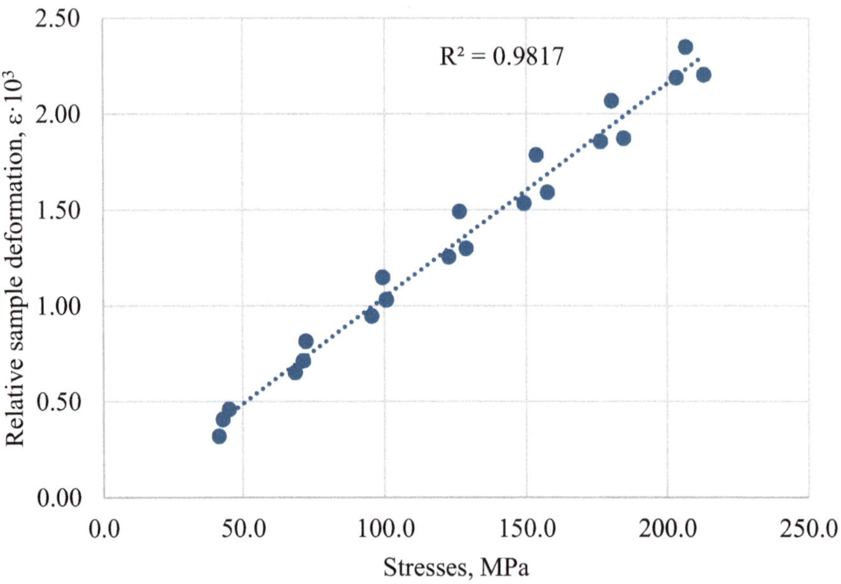

Fig. 8. Graph of linear plots of dependence of relative strains of three samples on stresses tested at 290 °C.

To assess the VAT of the frame structures, we select the most loaded elements and compare them with the limit normative values.

Based on the performed calculations, Table 3 of maximum values of frame parameters for two variants of deformation moduli at temperature 20 and 290 °C compiled.

Table 3. Design parameters of frame elements.

Test temperature,°C	Maximum force in the column N_{pr}, kN	Maximum bending moment M_y, kN· m	Maximum stresses σ, MPa	Maximum design displacements F_z, mm
20	29,43	6,20	112,58	31,47
290	33,40	6,77	116,19	43,55

4 Conclusions

By finite element modeling, the temperature reduction at the surface of a steel column protected by a foam concrete enclosure has been established. The average surface temperature of the thin-walled tube column can be 290 °C when fire occurs.

A 13.4% reduction in strain modulus of steel pipe specimens at 290 °C has been experimentally determined.

By finite element modeling, a significant increase in the deformation of columns and beam deflection by 1.4 times with increasing temperature was found, but this is much

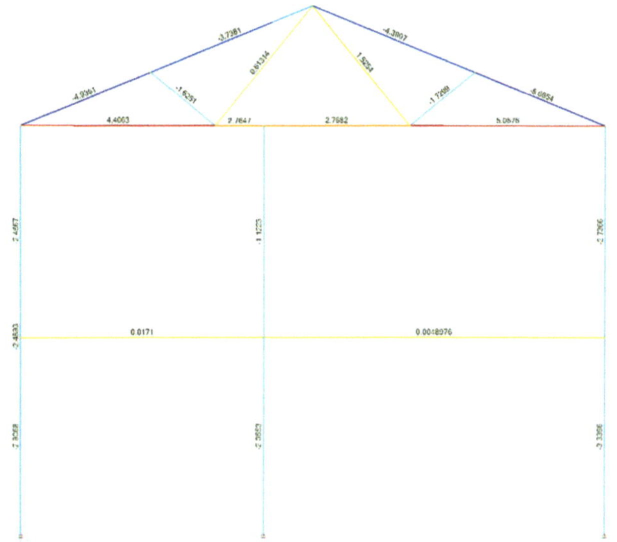

Fig. 9. Diagrams of longitudinal forces in frame elements.

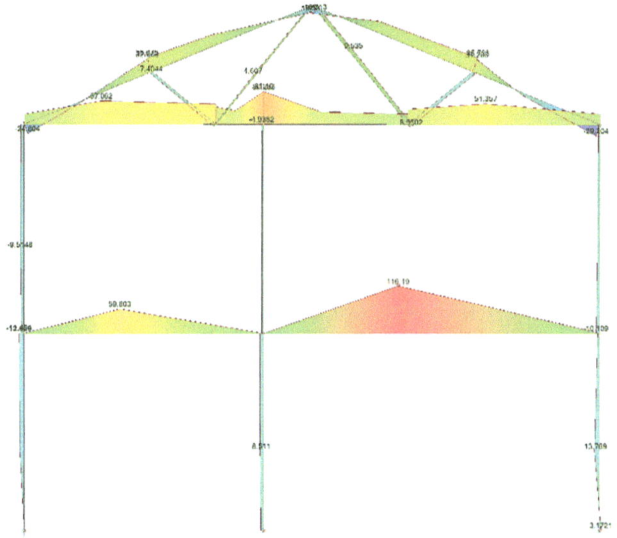

Fig. 10. Stress diagrams in frame elements.

less than the allowable displacement, which confirms the possibility of using thin-walled sections for load-bearing structures of buildings.

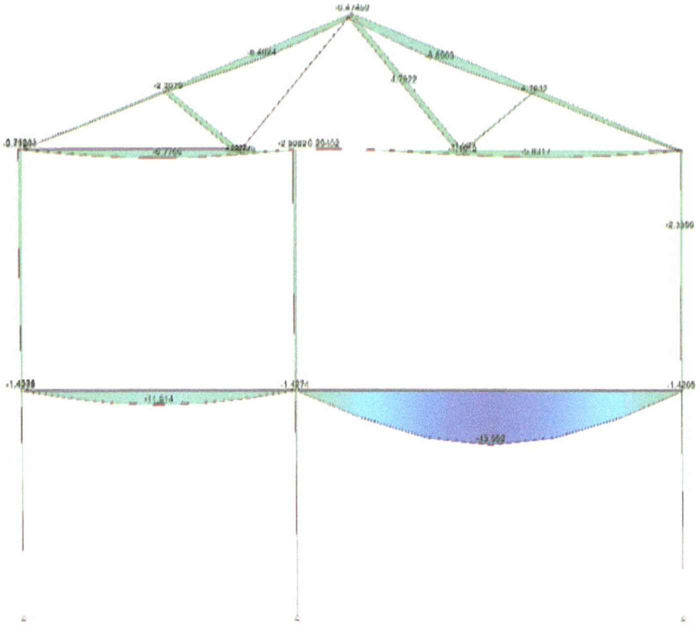

Fig. 11. Maximum design displacement diagrams in frame elements.

References

1. Semenov, V.S., Grekhovodov, A.M., Kondrashov, A.V.: Rapidly constructed low-rise buildings made of light steel thin-walled structures. Bull. Kyrg.-Russ. Slav. University. **8**(9), 154–158 (2008)
2. Nikolaeva, E.L., Kazeykin, V.S., Baronin, S.A., Chernykh, A.G.: Problems and Trends in the Development of Low-Rise Housing Construction in Russia. INFRA-M, Moscow (2012)
3. Asaul, A.N., Kazakov, Y.N., Bykov, V.J.L., Prince, I.P., Erofeev, P.Y.: Theory and practice of using rapidly constructed buildings. Humanistics, St. Petersburg. (2004)
4. Badin, G.M., Sychev, S.A., Makaridze, G.D.: Technologies of Construction and Reconstruction of Energy-Efficient Buildings. BHV, Saint Petersburg (2017)
5. Head P.R. Construction materials and technology: a look at the future. Proceedings of the ICE—Civil Engineering 144(3), 113–118 (2001).
6. Knaack, U., Chung-Klatte, S., Hasselbach, R.: Prefabricated Systems: Principles of Construction. De Gruyter (2012)
7. Sychev, S.A.: Promising high-tech construction Systems for Rapidly Constructible Transformable Multi-Storey Buildings. Hous. Constr. **4**, 36–40 (2018)
8. Sychev, S.A.: Technological principles of accelerated house construction, prospects for automated and robotic assembly of buildings. Ind. Civ. Eng. **3**, 66–70 (2016)
9. Nikonova, E.V., Vechtomov, P.O., Ladnykh, I.A.: Technical and economic indicators of enclosing structures for low-rise construction. Hous. Constr. **7**, 47–50 (2018)
10. Larionov, V.V.: Two aspects of the mechanical safety of buildings and structures. Ind. Civ. Eng. **6**, 11–13 (2011)
11. Mezentseva, E.A., Lushnikov, S.D.: Rapidly constructed buildings from light steel structures. Vestn. MGSU. Spec. Issue. **1**, 62–64 (2009)

12. Bagerman, A., Troitsky, A., Leonova, I.: Young's modulus of iron and nickel in steels and alloys. Trans. Krylov State Res. Cent. **2**(396), 67–72 (2021)
13. Valuyskikh, V.P., Zaikin, V.G.: Methodological foundations of using computational complexes in the calculation and design of structures. In: Proceedings of the International Scientific and Practical Conference "Results of Construction Science", pp. 124–131, Vladimir, Vladimir State University (2010)
14. Arleninov, P.D., Krylov, S.B.: The role of the load application scheme in ensuring the load-bearing capacity of building structures. Hous. Constr. **4**, 30–33 (2018)
15. Zaikin, V.G., Valuiskikh, V.P., Miroshnikov, N.N.: Effectiveness increase of application programme complex calculation of building constructions in mass projecting on the systematic approach basis. In: Telichenko, V. (Ed.). et al. Abstract Volume "Abstract of the 14th International Conference on Computing in Civil and Building Engineering", pp. 448–449. Moscow (2012)
16. Zaikin, V.G.: On the ambiguous assessment of calculations of building structures. Build. Kazakhstan. **16**(17), 4–6 (2006)

Study of Solar Panel Performance During Cooling

Mihail Lemeshko, Vladimir Dmitrienko(✉), and Liana Babenko

Institute of Service and Business (branch) of DSTU in Shakhty, 346500 Shakhty, Russia
vadmitrienko@rambler.ru

Abstract. The article is devoted to the issue of increasing the efficiency of solar panels, in particular, methods for improving the energy performance of their operation by cooling them. In order to minimize the negative impact on the environment of the technologies and materials used in the production of solar panels, it is necessary to increase their energy efficiency. The results of experimental studies of cooling a solar panel with fans from the back of the solar panel under various operating conditions are presented. The data on measuring the performance of a solar panel when moistening its back side with water from nozzles are presented. A criterion for the rational use of various methods of cooling the solar panel has been developed, taking into account the costs of the cooling process. The studies were carried out on specialized equipment in the form of research stands. A description of these stands and the methods of their use in experimental studies is given. The operating principle of the cooling system is described, the accents of the features of the decrease in the productivity of solar panels with an increase in temperature above the nominal value are shown. The expediency of cooling the panel with different air flows from fans, depending on the thermal load, is substantiated. A comparative assessment of air and evaporative cooling was carried out.

Keywords: Solar panel · Efficiency improvement · Solar panel Cooling · Cooling methods

1 Introduction

Solar energy has been developing and improving especially actively in the last decade. The use of solar energy in a number of advanced, economically developed countries currently reaches 17–20% of the total level of energy generation by various methods [1, 2]. Obviously, interest in alternative energy sources is associated with environmental problems caused by the use of hydrocarbon fuel as an energy source. Researchers in many countries around the world are intensively studying solar energy converters into electricity, and the first place among such converters are solar panels, batteries, power plants based on photovoltaic cells. Much attention is paid to solar panels, which can be used as components of solar batteries or used in powerful solar power plants. Solar panels, relatively low power are increasingly used by local consumers, for household conditions.

Y. Zhu (Ed.): ISESIE 2025, LNCE 000, pp. 302–313, 2026.
https://doi.org/10.1007/978-981-92-2389-3_24

The study of the features of solar panels revealed the following circumstances. Research on improving solar panels is constantly being carried out in the world, in particular, methods of cooling the surfaces of the solar panel are being studied. In this regard, the study of various methods of cooling solar panels at relatively high ambient temperatures is an urgent area of research.

For powerful solar power plants, the use of water or other technologies that improve the performance of solar panels is not always rational. However, for home use of this type of electricity sources, various methods of cooling solar panels can be useful, including original and non-typical upgrades.

It is well known that the share of alternative energy sources is increasing every year. Traditional (non-renewable) energy sources based on the combustion of hydrocarbons are gradually being replaced by renewable energy sources, including those based on the use of wind and solar power. The advantages of this replacement are known and obvious. Electricity generation technologies based on the use of photoelectric converters in the form of solar panels are constantly being improved.

At present, the issue of decreasing the efficiency of photoelectric converters with increasing temperature of these converters themselves is poorly studied [3]. Apparently, the photoelectric process of converting light energy into electricity is an exothermic process.

Further, it has been experimentally shown that in the cold period of the year, the temperature of the back surface of the solar panel is higher than the ambient air temperature and higher than the temperature of the front surface of the solar panel. The issue of cooling solar panels, which affects the energy performance of their work in the hot period of time, has been relatively little studied.

Obviously environmentalists, and the population of the countries of our planet are interested in a faster replacement of traditional energy sources with alternative sources, are interested in increasing the efficiency of solar energy, in improving solar panels.

2 Introduction Justification of the Research

Various literary sources provide data on the power of solar energy that does not always coincide. It is believed that the power of solar energy at the upper boundary of the atmosphere is 1350 W/m^2. Some scientists call this value the "solar constant". Other sources note that the power of solar energy does not coincide at the equator and at the poles. After the absorption of solar energy by the atmosphere, its dispersion and reflection at the earth's surface, the power of solar energy is 1000–1200 W/m^2 [4].

This value depends on the position of the sun and other factors. As the sun approaches the horizon, the radiation power decreases. So, at a 45% tilt of the sun's rays to the horizon, the power drops to 450–500 watts. The value of 1000 watts is a relatively large value of solar energy. However, modern solar panels convert only 12–25% of this energy into electricity. Silicon solar panels have the highest efficiency of photoelectric conversion, providing from 17 to 22% of this conversion. During operation, solar panels often overheat, especially in the summer. Depending on the ambient temperature, depending on the power of the consumer or the load on the solar panel, the temperature of the solar panel can exceed the ambient temperature. It is known that the advertising energy characteristics of solar panels correspond to their operating conditions at 25 °C. At a temperature of 25 °C and below, the power of the solar panel is maximum and corresponds to the passport values. As the ambient temperature increases, the operating power of the solar panel decreases and differs from the advertising value [5, 6].

To calculate the power of a solar panel when its surface temperature rises above 25 °C, it is proposed to use the following expression (1):

$$P_{op} = P_{25} \cdot \left[1 - \left(\beta \cdot \left(t_{sp} - 25\right)\right)/100\right], \tag{1}$$

where P_{op} is the operational power, P_{25} is the rated power of the solar panel, at its temperature of 25 °C, β is the temperature coefficient, t_{sp} is the temperature of the solar panel.

The temperature coefficient of a crystalline solar panel, according to the source [4], is 0.45 (β = 0,45). As an example, let us estimate the operating power of a crystalline solar panel with a power of 100 W, at a solar panel surface temperature of 75 °C (1):

$$\begin{aligned} P_{op} = \mathrm{P}_{25} \cdot \left[1 - \left(\beta \cdot \left(t_{sp} - 25\right)\right)/100\right] = 100\Big(1 - (0,45 \cdot 50)/100 \\ = 0,775 \cdot P_{25} = 77,5 \text{ W} \end{aligned} \tag{2}$$

For a 100 W solar panel, for the given calculation conditions, its actual power will be only 77.5 W. The power drop is 22.5%.

The given example roughly characterizes the actual power of a solar panel when it is heated to 750C. The surface temperature of the solar panel itself depends on the operating conditions and consumption load. Heat inflow and heat removal of the solar panel depend on many factors and therefore, under operating conditions, the actual value of the actual power of the solar panel will be different. In summer, the surface of the solar panel can heat up to high temperatures.

In Russia, in the region of 55° north latitude, the solar panel heats up to an average of 70 °C. In the equatorial regions of the Earth – up to 80–90 0C. However, in the absence of wind and high solar activity, the surface temperature of the solar panel (in the absence of a cooling system) can reach 100 °C [7].

It is advisable to maintain the operational power of a solar panel at high air temperatures or minimize its decrease by cooling. As is known, passive and active methods of cooling solar panels are used in the summer to maintain the optimal operating temperature of panels in warm regions of the country. It is obvious that the energy costs for cooling a solar panel should not exceed the value of the power drop from its heating without cooling. When planning our research, we chose methods for cooling the solar panel mainly on its rear (back) surface. One of the research priorities is to study heat removal from the surface of the rear (back) surface of the solar panel by blowing an air flow over its surface. For any method of cooling a solar panel, the usefulness of cooling can only be in the case when the cooling costs are less than the increase in the power of the solar panel caused by this cooling.

A criterion for assessing a method or device for cooling a solar panel is proposed, at which the costs of cooling will be equal to the efficiency of cooling. Such a criterion is the value of the surface temperature of the solar panel, at which a specific cooling method will not be unprofitable. The costs of cooling will be equal to the additional power from cooling the solar panel. Fig. 1 interprets the above reasoning and shows the dependence of the solar panel power on its surface temperature, with and without cooling.

Fig. 1 interprets the above reasoning and shows the dependence of the solar panel power on its surface temperature, with and without cooling.

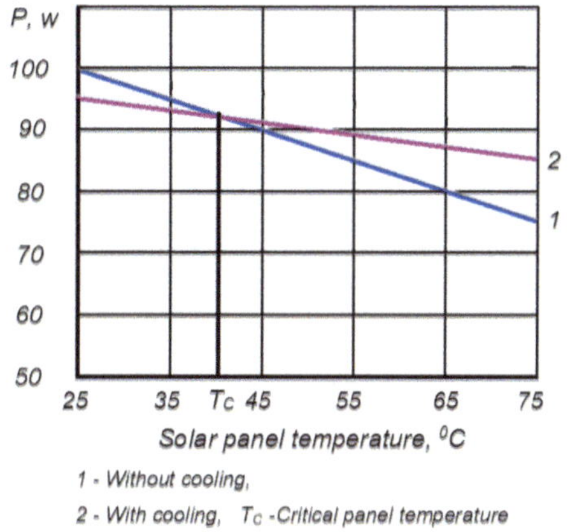

Fig. 1. Dependence of solar panel power on its temperature.

Fig. 1 shows the chane in the power of a solar panel without cooling (line 1) and the change in power with cooling (line 2). The intersection point of these lines corresponds to the situation when the cost of cooling is equal to the benefit of cooling. This situation corresponds to the critical value of the solar panel temperature (Tc).

The given quantitative example shows the possible heating of the solar panel surface to 75 °C, this value exceeds 25 °C and corresponds to an excess of the rated value of the solar panel operating temperature by 50 °C. Conventionally, using the example, we can predict an increase in the efficiency of a solar panel with cooling, when the temperature of the solar panel surface increases.

When the solar panel temperature exceeds the critical value, it is advisable to use active cooling. Obviously, the developed system and cooling method must be effective.

Below is a brief overview of the methods and solutions related to cooling solar panels.

Cooling the front surface of the solar panel:—spraying from fine dispersion nozzles,—watering with a free-flowing liquid, watering with liquid under pressure, immersion in a cooling liquid. Cooling the back (rear) surface of the solar panel:—making flat heat sink fins (plate radiator),—air cooling, blowing with fans,—humidification with fine dispersion nozzles, evaporative cooling (humidification and blowing with fans),—using heat exchangers with an external cooling circuit, using the refrigeration cycle of compression refrigerators, hybrid designs of a solar collector combined with a solar panel. There are other options for solving problems.

Fig. 2 shows an example of cooling the front surface of solar panels with water from a pipe, with its free flow. Designs for placing miniature nozzles on the front side of the solar panel are known.

Fig. 2. Cooling of solar panels from water nozzles.

A hybrid solar collector is known [9], in which the heat removed from the solar panel is used for domestic needs. At the same time, the authors embodied the idea in a real design solution. By installing a cooling system on the roof of a multi-story building. A method for cooling solar panels is known, implemented by several technical solutions, which use a cooling system with a coolant and with channels for removing heat from the solar panel from its back side. The method is implemented in a multi-purpose solar battery [10]. In such a design of the solar panel cooling system, thermal resistance is formed between the heat-dissipating tubes on the back surface of the solar panel, reducing the efficiency of heat removal.

In the development of various methods for cooling solar panels from water nozzles, the details of these methods vary, including solutions that improve the reliability of the systems. For example, an automatic cleaning and cooling system for a solar photovoltaic power generation system [11]. This system uses a high-pressure water pump and a high-pressure automatic spray device. When the temperature detected by the temperature sensor reaches a predetermined threshold, the water pump from the water source supplies water to the spray devices.

The above-described methods and designs for cooling solar panels are used by us in the study and development of effective methods for cooling a solar panel.

For the liquid method of cooling a solar panel, a method for monitoring the operation of sprayers has been developed [12]. Such a device is used to check the performance of the sprayer of the cooling liquid used to cool and clean the solar module.

A solar panel with a cooling device in the form of nozzles is also known [12]. The device contains an inclined solar panel, a cooling device located on the bottom side of the solar cell panel. The disadvantage of such a design and cooling method is that when spraying the coolant from the bottom up, increased power is required, as well as the complexity of setting the optimal angle of inclination of the spray nozzle, which complicates the automation of the process of cooling the solar module and turning the plane of the solar panel perpendicular to the sun's rays.

Also known are a method and device for cooling photovoltaic modules [13]. The device contains a controller, a current sensor, a solar radiation sensor, a wind speed sensor, and a frequency-controlled pump. One of the temperature sensors is tightly fixed to the back of the photovoltaic module, and the other is located in the shadow of the photovoltaic module. The controller processes the sensor data. In this case, water evaporated by natural convection is used as a cooling agent, which does not effectively remove heat from the surface of the solar panel and limits the applicability of this and similar solar panel designs with devices for cooling the solar panel, which is a disadvantage. Patent analysis made it possible to clarify the research objectives and take into account the research results of other researchers.

3 Methodology and Organization of Research

Research was carried out in natural, field conditions and in the laboratory, under stable experimental conditions. Two specialized research stands were developed to study solar panel cooling options. One stand was used to conduct research in the field, in different weather conditions and under different natural conditions. The other stand was used in stationary, deterministic measurement conditions.

The stand for field research is made in the form of a portable case, in the upper part of which a solar panel is fixed. The design provides for changing and fixing the angle of inclination of the solar panel to the horizon. The stand includes load elements in the form of a battery and incandescent lamps of different power. Measuring equipment allows obtaining data for constructing volt-ampere characteristics. Fig. 3 shows a schematic diagram of this stand. In the course of research, the stand is constantly being modernized.

The field test bench is designed to measure the parameters of the solar panel under study and its cooling method in natural conditions, with variable illumination, random wind speed with gusts, with natural and variable humidity, shading, with changes in solar radiation intensity and ambient air temperature. The stationary test bench provided a stable value of all external factors and more reliable values of the obtained dependencies and parameters. In the conditions of the stationary test bench, stable values of the experimental conditions were maintained and the studied parameters were changed with relatively accurate values.

The test bench for research in stationary conditions includes a heat-insulated room, a device for vertical mounting of the solar panel with the ability to change and fix the angle of inclination of the solar panel to the horizon and to the direction of the light source, a system for regulating and maintaining stable measurement conditions, a system for measuring and recording the results. The temperature in the heat-insulated room was set using 2 kW thermal fans and a split system.

The main experiments were performed using a polycrystalline solar panel "DELTA SV 100-12P" with a peak power of 100 W and a voltage of 12 V, which is a source of constant voltage.

We have carried out studies of the effect on the thermal energy performance of the solar panel in various conditions, at different ambient temperatures. Much attention was paid to cooling the solar panel with highly efficient fans, by forced convective heat removal. Evaporative cooling was studied, when the back surface of the solar panel,

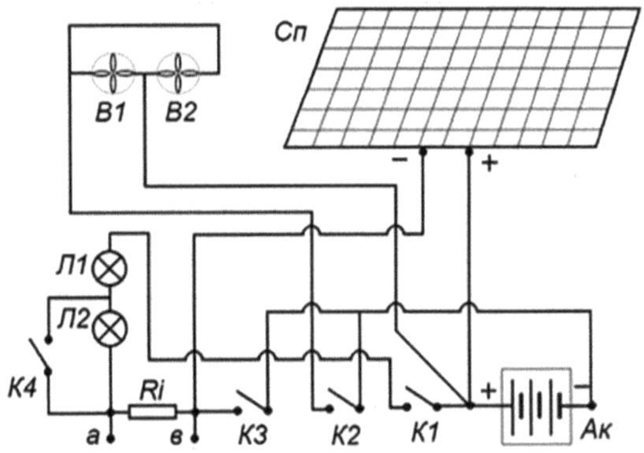

Fig. 3. Research stand diagram.

moistened from nozzles, was blown by an air flow from fans. In all experiments, cooling was performed by the back (rear) surface of the solar panel.

To form a controlled air flow, a shielding plate with side plates adjacent to the frame of the solar panel is attached to the back side of the solar panel at some distance from the plane of the solar panel. Thus, an air tunnel is formed through which air is blown by fans.

Fig. 4 shows a simplified diagram of the device for studying air cooling of a solar panel. The figure shows: 1–a solar panel with an aluminum frame, 2–a screening plate made of flat heat-insulating plastic. The sides of the solar panel are covered with side plates 3. Between elements 1, 2, 3–an air tunnel 4 is formed, through which air is blown by fans. Fans 5 are located near the end of the solar panel between elements 1 and 2. In the studies, fans from personal computers (coolers) with adjustable performance and a power of 1.4, 1.8 W were used.

Using fans according to the scheme shown in Fig. 4 provides regulation of fan performance and solves the problem of minimizing cooling costs. Heat removal from the back surface of the solar panel must be rational.

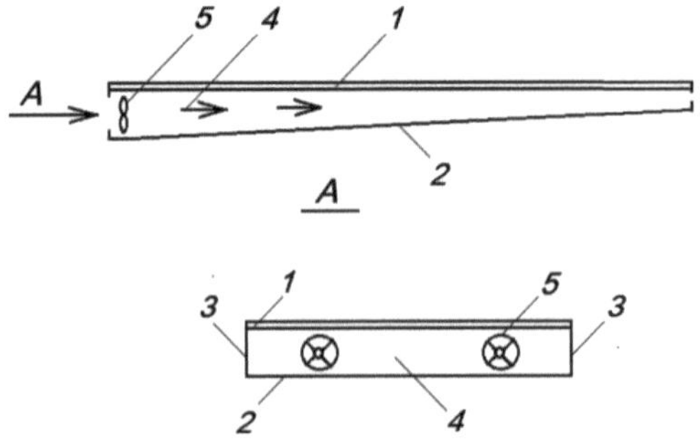

Fig. 4 Using fans to cool the solar panel: 1–solar pane l, 2–shielding plate, 3–side plates, 4–air tunnel, 5–fan.

The given scheme provides the ability to study heat removal from the back surface of the solar panel at different air flow powers at different thermal loads, depending on the operating conditions of the solar panel.

The studies were carried out under different lighting, different ambient air temperatures, and different light power.

4 Results

The studies were carried out in different periods of the year, when the length of the day changes, and the intensity of solar radiation on the surface of the solar panel changes, associated with the position of the sun relative to the horizon, estimated by the AM indicator "air mass".

Fig. 5 shows the averaged information on the results of surface illumination measurements at the research location in different periods of time.

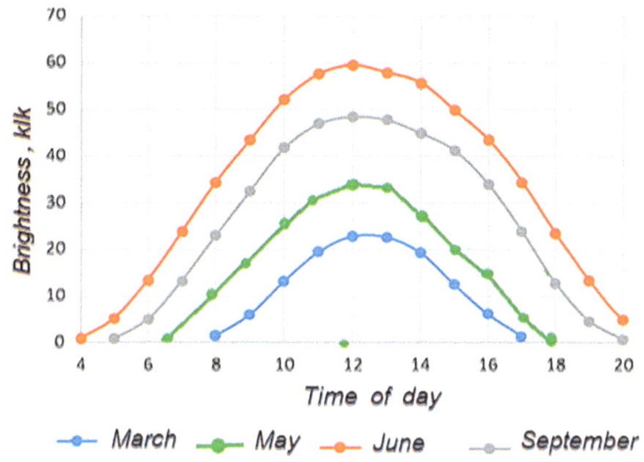

Fig. 5. Results of observations of illumination at different periods of the day and year.

The location of the research has the following geolocation data: 47°43′13″ N, 40°13′03″ E. The data from observations conducted from 2024 to 2025 were used to analyze the research.

Below are some results of the study of the solar panel in various modes. The results of measurements of the solar panel parameters depending on the degree of illumination of its surface, at an ambient air temperature of 30°C are given in Table 1. The measurement was performed at noon in field conditions with the solar panel operating without a load.

As can be seen from the measurement results, the voltage of the solar panel in the "no-load" mode does not change with changes in illumination. One of the important tasks of the study is to clarify the question: "Which surface of the solar panel is more rational to cool. The front (frontal) or the back (rear)?" This question has been preliminarily resolved using a thermal imager, and a number of instrumental measurements have been performed. During the study period, the temperature of the back surface was almost always higher than the temperature of the front surface of the solar panel. Therefore, it is necessary to cool mainly the back surface of the solar panel.

Table 1. Effect of illumination level on solar panel temperature.

Light level	Front wall temperature (°C)	Back wall temperature (°C)	Voltage
In the sun (~ 100клк)	14	16	21B
Penumbra (~61 клк)	5	8	21B
Shadow (~53 клк)	3	5	21B

Table 2 shows the results of temperature measurements of the front and rear surfaces of the solar panel at different ambient temperatures.

Table 2. Temperatures of the front and back surfaces of the solar panel.

Air flow temperature (°C)	Front wall temperature (°C)	Temperature of the rear wall (°C)	Voltage
9	26	28	21B
23	37	43	21B

As can be seen from the data in Table 2, the ambient air temperature does not affect the generation of electricity (in the no-load mode), but it does affect the surface temperatures of the solar panel; the temperature of the back surface of the solar panel is higher than the temperature of the front surface. In the field, studies were conducted on the effect of ambient air temperature on the surface temperatures of the solar panel.

Fig. 6 shows the values of the back surface temperature of the solar panel when the ambient air temperature changes, with a load on the solar panel of 50% of the power. We have previously indicated that the performance of a solar panel is significantly affected by external operating conditions. Wind gusts have a large effect on temperature indicators. Fig. 6 shows the average values of the solar panel

temperature and the maximum and minimum temperatures for some ambient air temperature conditions.

As measurements have shown, at an ambient air temperature of 40 0C, the temperature of the back surface of the solar panel varied from 630C to 810C. For all experimental conditions used, the temperature of the back surface was higher than the temperature of the front side of the solar panel.

Studies of solar panel cooling were performed using humidification of the back surface of the solar panel from finely dispersed nozzles, under conditions of cooling the back surface of the solar panel by natural convection. In this case, the average wind speed was 5.4 m/s. Table 3 shows the measurement results.

As can be seen from the results of measurements under the same conditions, when the solar panel is humidified, the power of the solar panel increases by 27.8%. Studies were conducted on the operation of the solar panel with and without fan cooling. Experiments were conducted to study evaporative cooling, and after humidifying the surface of the solar panel, it was blown with air. Table 4 presents some of the results of these studies.

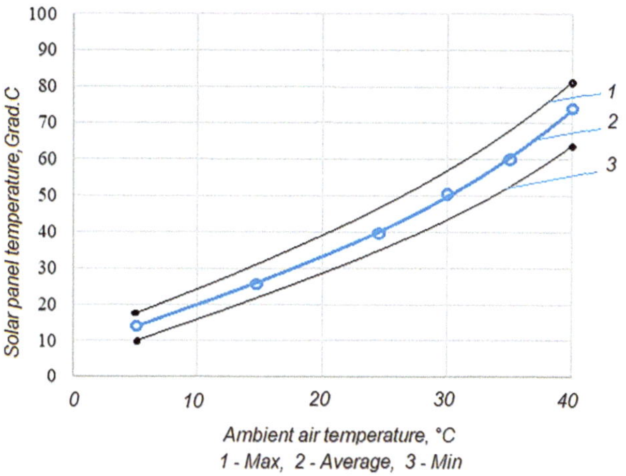

Fig. 6. Solar panel temperature dependence on ambient air temperature.

Table 3. Cooling of the solar panel when its back surface is moistened.

Test conditions	Ambient air temperature °C	Back surface temperature °C	Current, (A)	Voltage (V)
Without cooling	30	38	1,6	12
Humidification and natural convection	30	32	1,9	14

The table shows the results of evaporative cooling of the surface of the solar panel with fans without taking into account the costs of humidification. Research in this area is ongoing. An important point is the cost of humidification. Various options for economical humidification of the surface of the solar panel will be tested. However, given the high efficiency of the cooling method with blowing on the humidified surface, it can be assumed that this method will be the most promising method for cooling solar panels.

Table 4. Solar panel performance with and without cooling.

Test conditions	Ambient air temperature °C	Back surface temperature °C	Current (A)	Voltage (V)
Without cooling	35	48	1,8	14
Fan cooling	35	38	2,4	16
Humidification and blowing with fans	35	27	3,5	19

As can be seen from the measurement results, for the same conditions, cooling only with fans (air cooling) provides an increase in the solar panel power from 25.2 W to 38.4 W, and the use of humidification and fan blowing (evaporative cooling) from 25.2 W to 66.5 W. Summarizing the series of experiments with different conditions, the investigated solar panel cooling modes can be assessed using the diagram shown in Fig. 7.

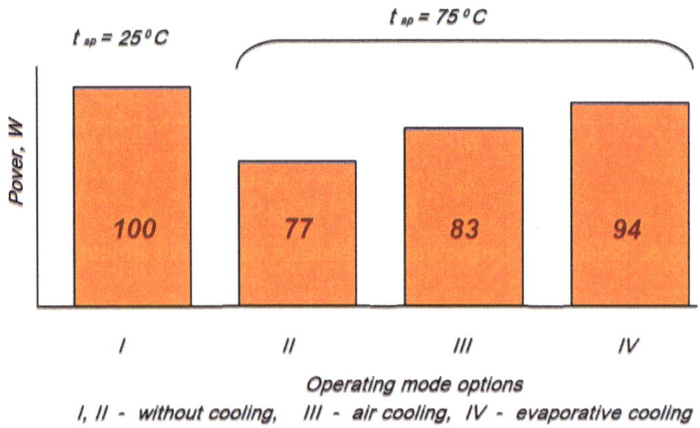

Fig. 7. Solar panel performance indicators.

The operation of a solar panel without cooling is characterized by the following operating mode indicators: I (at an ambient air temperature of 25 °C) and II (at an ambient air temperature of 75 °C). In this case, the actual power of the solar panel is 77% of the rated value. When cooling with fans, taking into account the power costs for fans (III), the actual power increases to 83% of the rated value. When using evaporative cooling, without taking into account the costs of humidification (IV), the actual power is 94% of the rated value.

5 Conclusions

1. A criterion for rational cooling of a solar panel is proposed, which allows determining the limit of using solar panel cooling.
2. It is advisable to obtain comparative characteristics of solar panel operation under changing operating conditions in laboratory conditions, where it is easier to ensure stability of conditions.
3. Experimental studies of a solar panel have established that at an ambient temperature below 40 °C, the temperature of the front side of the solar panel is lower than the temperature of the back side of this panel by 5–25 °C.
4. Blowing the back surface of the solar panel with an air flow at a speed of 2.5 m/s at an ambient temperature of 40–45 °C provides an increase in the generation power by 10–15%. The efficiency of blowing the back surface of the solar panel depends on the air temperature.

5. The best air cooling effect is observed when blowing on the wetted back side of the solar panel, when intensive evaporative cooling occurs.
6. Evaporative cooling at an ambient temperature of 40 °C provides an increase in the solar panel power up to 94% of the nominal value.

References

1. Rüther, G.: Global Solar Energy: A Turning Point Year. Deutsche Welle (May 29, 2013). https://www.energyplaces.org/news/mirovaya-solnechnaya-energetika-perelomnyj-god. Archived June 19, 2013
2. Gulkov, V. N: Study of the effect of heating solar modules on the efficiency of radiation conversion/Gulkov, V. N., Kolesnichenko, I. D., Korotkov, K. E.//Bulletin of ETU "LETI". 1, 10–16. (2019)
3. Asanov M. M., Bekirov E. A., Voskresenskaya S. N.: Reducing the effect of solar cell surface heating on its operating efficiency//Construction and technogenic safety. **51**, 92–95 (2014).
4. Patel, M.R.: Wind and Solar Power Systems: Design, Analysis, and Operation, p. 448. Taylor & Francis, New York, USA (2006)
5. Lemeshko, M.A.: In: Lemeshko, A.S., Bondareva, D.A. (eds.) Increasing the Efficiency of a Solar Panel by Cooling it with Fans/M.A. Chebotarev//Science Diary (2022)
6. Kirpichnikova I.M. Increasing the energy efficiency of solar modules by reducing surface temperature. In: Kirpichnikova, I.M., Makhsumov, I.B. Technical sciences (2). Electrical complexes and systems, 489–499 (2020)
7. Trembach, S.A.: The problem of overheating of solar panels. A way to solve the problem by electronic tinting. Energy Bulletin magazine No. **26**, 46–51 (2020) https://cyberleninka.ru/article/n/problema-peregreva-solnechnyh-paneley-sposob-resheniya-problemy-putyom-elektronnoy-tonirovki
8. Hybrid solar collector.: http://zvt.abok.ru/articles/106/Gibridnii_solnechnii_kollektor (2022)
9. Patent Russia, RU 2164722 C2, H01L 31/058), Multipurpose solar battery (1918)
10. Patent, C. N.: 201518459U, Y02A20/00 automatic cleaning and cooling system for solar photovoltaic power generation system (2021)
11. Patent, K. R.: 101243175B1, Priority date 2011-10-14, Publication date 2013-03-13. 12. Patent CN 209659240 H02S 40/42. Solar panel with cooling device
12. Patent, R. U.: 2832143 SPK H02S 40/42. Priority (15.06.2023). Method and device for cooling photovoltaic modules

Analysis of Health Mechanisms of the Digital Landscape Environment in Game Media: Illustrated Using the Case of Sky: Children of the Light

Siyu Yu and Yanyan Wang(✉)

School of Journalism and Communication, Nanjing Xiaozhuang University, Nanjing 211171, China
2022101@njxzc.edu.cn

Abstract. This study investigates the health mechanisms of digital environments in gaming media, aiming to provide a theoretical foundation and practical insights for applying gaming media in mental health interventions. The study used questionnaire surveys and participatory observation. Using convenience sampling, 154 valid participants were recruited from among the players of the smartphone game Sky: Children of the Light. The Perceived Stress Scale (PSS-10) and a stress-relief effectiveness scale were administered. Descriptive statistics, reliability and validity testing, correlation analysis, and multiple linear regression were conducted using SPSS 19.0. Players' stress scores significantly decreased after experiencing the game's digital natural landscapes. Visual (color, lighting), auditory (music, natural sound effects), and interactive (social interaction) elements of the landscapes received moderately high ratings (mean = 3.79–3.99). Multiple linear regression revealed that in the interaction dimension, "chatting with friends/strangers or idling" had the strongest predictive power for stress relief ($\beta = 0.245$, $P < 0.05$). Digital natural landscapes in Sky can significantly reduce short-term stress through audiovisual immersion and social interaction. However, these effects are temporary and context-dependent, underscoring the need for rational engagement to prevent overindulgence.

Keywords: Health mechanisms · Digital environments · Gaming media

1 Introduction

With the rapid advancement of mobile internet and digital technologies, smartphone games have become a significant form of leisure and entertainment in modern life. The digital environments they create are increasingly recognized as novel avenues for emotional regulation and stress relief. Launched in 2019, the social adventure smartphone game *Sky: Children of the Light* quickly gained popularity through its unique stranger-socializing mechanism and its aesthetically designed natural landscapes. As of April 2025, it ranked eighth on Weibo's Super Topic Ranking for Smartphone Games, with 5.86 million discussion threads, reflecting a high level of user engagement.

Y. Zhu (Ed.): ISESIE 2025, LNCE 000, pp. 314–323, 2026.
https://doi.org/10.1007/978-981-92-2389-3_25

Currently, adolescents and young adults face mounting academic and professional pressures, leading to increasingly prominent mental health concerns. As a game centered around "healing aesthetics," *Sky: Children of the Light* features virtual natural landscapes (e.g., Daylight Prairie, Hidden Forest, and Valley of Triumph) whose stress-relieving effects remain underexplored. Existing studies have largely focused on the game's social interaction mechanisms, paying insufficient attention to the therapeutic functions of its digital natural landscapes. From the perspectives of communication studies and psychology, this study explores the stress-relief mechanisms of digital natural landscapes in gaming media, thereby providing a theoretical supplement to the field of digital health.

2 Literature Review

2.1 Digital environments and Digital Nature Healing

The integration of digital environments in games has garnered significant attention for their potential to alleviate stress and promote relaxation. Virtual reality (VR) has emerged as a promising tool in this domain, with demonstrating that VR-based relaxation technologies can effectively support emotional recovery and reduce stress, particularly by addressing mental disorders such as anxiety and phobias [1]. This aligns with the broader understanding that immersive digital environments can facilitate psychological well-being. The gamification strategies in health interventions has also been explored [2], and educational applications of digital relaxation tools also demonstrate their versatility [3]. Although focused on health education, this approach underscores the capacity of multimedia and game-based environments to foster positive emotional states. Overall, the literature indicates that digital environments—particularly those involving immersive VR, full-body interaction, games—offer promising avenues for stress relief and relaxation. These technologies leverage engagement, presence, and affective regulation to support mental health, especially in contexts where traditional methods may be less accessible or effective. Continued research into the design and application of such digital tools is essential to optimize their therapeutic potential and accessibility in promoting psychological well-being.

Healing usually focuses on psychological repair, aiming to reduce mental stress, enhance individual self-recovery abilities, and achieve internal self-healing. Recent rapid advancements in virtual reality (VR) technology have significantly enhanced its applicability, offering robust technical support for both research and clinical implementation in rehabilitation contexts [4]. Studies indicate that virtual healing environments (VHEs) demonstrate significant therapeutic potential across domains such as emotional intervention, cognitive training, and sports rehabilitation. Typical healing scenarios often incorporate virtual simulations of natural landscapes and biophilic elements, which have been associated with positive psychological and physiological outcomes [5–8], in addition to the content of healing scenarios, multisensory factors and interaction modalities also serve as critical determinants influencing the efficacy of rehabilitation outcomes [9–12]. These findings provide valuable references for future research and practice in areas such as the restorative benefits of virtual nature and the design of biophilic environments. In

PTSD therapy, a research team utilized virtual reality technology to design virtual healing spaces and games for veterans, as well as an open space surrounded by a jungle. The patients' clinician-rated PTSD decreased by 34%, and self-rated PTSD scores dropped by 45% [13], New media technologies enable therapists to autonomously select perspectives and modulate spatial parameters, thereby significantly enhancing their sense of presence and environmental engagement. When individuals are immersed in such highly participatory naturalistic healing settings, they can derive subjective comfort and harmonious therapeutic experiences through self-created healing scenarios, ultimately fostering self-actualization, accomplishment, and psychological well-being [14] .

2.2 Research on the Game Sky: Children of the Light

As a form of entertainment, video games not only allow players to release emotions through intense and stimulating gameplay, but certain design elements—such as virtual natural environments and social interactions—also possess emotional therapeutic functions. Since it launches on the Apple Store in 2019, Sky: Children of the Light has attracted a large number of players with its exquisite visuals and moving soundtrack. Warm and healing qualities have made it a place for many players to relax and unwind. Zou Yufei (2021) used observational methods to study the mobile game Sky: Children of the Light, and found that unique stranger-socialization mechanics and immersive storytelling provided players with a space for emotional exchange and psychological comfort. As an art form characterized by interactivity and virtual realism, games in the digital age and virtual technology—offers players increasingly authentic and immersive experiences [15]. Gong Jie (2022), employing methods such as literature analysis, participatory observation, and in-depth interviews, explored the subjects, means, scenarios, and functions of interpersonal communication in the social game Sky: Children of the Light, highlighting both the positive effects and potential issues of social mobile games in the context of interpersonal communication [16].

Overall, digital environments are rapidly evolving and demonstrating broad application potential across various fields, particularly in healthcare, education, and gaming. The aforementioned literature also confirms that digital natural environments can provide emotional healing benefits, though their application in games warrants further development. At the same time, academic research on Sky: Children of the Light remains relatively scarce, with limited diversity in research focus—most studies concentrate on online social interaction and interpersonal communication. Research on digital virtual natural environments in games is still in its preliminary stages, leaving considerable room for further exploration. This study identifies the stress-reducing effects and key design elements of digital virtual natural landscapes in Sky: Children of the Light, offering references for game developers to optimize scene design with a focus on mental health while highlighting a novel approach that can supplement mental health interventions.

3 Research Participants and Methods

3.1 Research Participants

The participants of this study were players of Sky: Children of the Light. From April 27 to April 30, 2025, convenience sampling was employed to distribute questionnaires to individuals via Sky-related Weibo super topics, gaming forums, and player communities. Inclusion criteria were as follows: mature players with at least one year of gameplay experience and voluntary participation with completion of a valid questionnaire. A total of 172 questionnaires were collected, of which 154 were deemed valid after excluding incomplete or patterned responses, resulting in an effective response rate of 89.5%. All participants gave informed consent to participate in the study.

3.2 Research Instruments

3.2.1 Perceived Stress Scale

This study employed the Perceived Stress Scale (PSS-10) scale developed by Cohen et al., consisting of 10 items, including 4 reverse-scored items. Responses were rated on a 5-point Likert scale (0 = "never," 4 = "very often"), with a total score ranging from 0 to 40. Higher scores indicated greater stress levels. In this study, the Cronbach's α coefficients for the pre-gameplay and post-gameplay questionnaires were 0.861 and 0.890, respectively, indicating good reliability.

3.2.2 Stress-Relief Effectiveness Scale

Based on the attention restoration theory, a self-designed 7-item scale was developed to assess stress-relief effectiveness from three dimensions: visual (color, lighting), auditory (music, natural sound effects), and interactive (social chat, action interaction). Responses were rated on a 5-point Likert scale (1 = "strongly disagree," 5 = "strongly agree"). The scale demonstrated good reliability, with a Cronbach's α of 0.877.

3.3 Statistical Methods

Data were processed using SPSS 19.0. Quantitative data are presented as mean ± standard deviation ($\bar{x} \pm s$). Descriptive statistics were applied to analyze sample characteristics, pre-and post-gameplay stress scores, and stress-relief effectiveness. Cronbach's α was used to assess the reliability of the scales, while the Kaiser–Meyer–Olkin (KMO) test and Bartlett's test were employed to confirm validity. Pearson correlation analysis was conducted to examine the relationships between each dimension and stress scores, and multiple linear regression analysis was applied to identify predictors of stress relief. A P value of < 0.05 was considered statistically significant.

The PSS (pre-gameplay) demonstrated a KMO value of 0.870 and a significant Bartlett's test of sphericity ($P < 0.001$); the PSS (post-gameplay) showed a KMO of 0.880 and a significant Bartlett's test ($P < 0.001$); and the stress-relief effectiveness scale had a KMO of 0.894 and a significant Bartlett's test ($P < 0.001$). All scales exhibited good construct validity.

4 Results

4.1 Demographic Characteristics of Participants

As shown in Table 1, of the 154 respondents, 115 (74.7%) were female and 39 (25.3%) were male. The majority were aged 19–25 years (75 participants, 48.7%), followed by the 13–18 age group (47 participants, 30.5%). Regarding educational or employment status, 58 (37.7%) were university students, 32 (20.8%) were high school students, and 30 (19.5%) were employed workers, with the student population accounting for a total of 77.3%.

Table 1. Demographic characteristics of participants

Gender	Frequency	Percentage (%)	Valid percentage (%) cumulative percentage (%)	
Male	39	25.3	25.3	25.3
Female	115	74.7	74.7	100.0
Age				
12 or younger	6	3.9	3.9	3.9
13–18 years	47	30.5	30.5	34.4
19–25 years	75	48.7	48.7	83.1
26 or older	26	16.9	16.9	100.0
Educational and employment status				
Middle school	14	9.1	9.1	9.1
High school	32	20.8	20.8	29.9
University	58	37.7	37.7	67.5
Graduate or higher	15	9.7	9.7	77.3
Employed	30	19.5	19.5	96.8
Unemployed	5	3.2	3.2	100.0
Total	154	100.0	100.0	

4.2 Statistical Analysis of Stress Scores and Stress-Relief Effectiveness before and after Gameplay

4.2.1 Changes in Stress Scores before and after Gameplay

The total stress score before gameplay was 21.32 ± 6.71, indicating a moderately high level, while the post-gameplay score was 17.97 ± 6.73, falling within the normal range. This difference was statistically significant ($P < 0.05$). Among specific items, the most notable decrease was observed in the item "feeling upset about things beyond control" (pre-gameplay: 2.63 ± 0.98; post-gameplay: 1.39 ± 0.90), while the item "feeling confident in handling personal problems" showed the most significant increase (pre-gameplay: 1.51 ± 1.02; post-gameplay: 2.51 ± 1.06).

Table 2. Stress levels perceived in daily life (pre-gameplay)

Item	Frequency	Min	Max	Mean	SD
Feeling upset by unexpected events	154	0	4	2.61	0.986
Feeling unable to control important life matters	154	0	4	2.47	0.991
Feeling nervous, anxious, and stressed	154	0	4	2.44	0.949
Feeling confident in handling personal problems	154	0	4	1.51	1.018
Feeling that things are going your way	154	0	4	1.63	1.126
Feeling unable to cope with responsibilities	154	0	4	2.47	1.063
Having control over irritating life matters	154	0	4	1.55	0.991
Often feeling in control of things	154	0	4	1.58	1.036
Feeling upset about things beyond your control	154	0	4	2.63	0.977
Often feeling overwhelmed by challenges and unable to overcome them	154	0	4	2.44	0.963

Table 3. Perceived stress levels after playing in the natural landscapes (post-gameplay)

Item	Frequency	Min	Max	Mean	SD
Feeling upset by unexpected events	154	0	4	1.34	0.924
Feeling unable to control important life matters	154	0	4	1.38	0.825
Feeling nervous, anxious, and stressed	154	0	4	1.34	0.923
Feeling confident in handling personal problems	154	0	4	2.51	1.056
Feeling that things are going your way	154	0	4	2.53	1.061
Feeling unable to cope with responsibilities	154	0	4	1.11	0.852
Having control over irritating life matters	154	0	4	2.49	1.005
Often feeling in control of things	154	0	4	2.52	1.011
Feeling upset about things beyond your control	154	0	4	1.39	0.903
Often feeling overwhelmed by challenges and unable to overcome them	154	0	4	1.36	0.906

After calculating the total mean scores from Tables 2 and 3 (see Table 4), the results indicate that the average stress score before gameplay was approximately 21, reflecting a moderately high level of stress. Following gameplay, the average stress score decreased to around 18, corresponding to a low-to-moderate stress level within the normal range. This decrease of about 3 points suggests that players experienced a notable reduction in stress after engaging with the virtual natural landscapes in Sky: Children of the Light. This finding aligns with Kaplan et al.'s [6] attention restoration theory, which posits that exposure to tranquil natural environments allows individuals to detach from everyday stressors, enhance attentional capacity, and facilitate psychological recovery.

Table 4. Pressure value variation table

	Perceived stress (pre-gameplay)	Perceived stress (post-gameplay)
N valid	154	154
Missing	0	0
Mean	21.3247	17.9675
SD	6.70565	6.72855
Min	5.00	1.00
Max	37.00	34.00

4.2.2 Statistical Analysis of Stress-Relief Effectiveness

As shown in Table 5, the mean scores for stress-relief items ranged from 3.79 to 3.99, indicating a moderately high level of agreement. Among them, "color design of the natural landscapes" received the highest score (3.99 ± 0.76), while "immersive quality of scene music" scored the lowest (3.79 ± 0.91). The items "chatting with friends/strangers or idling" (3.82 ± 0.86) and "natural elements helping escape stress" (3.91 ± 0.82) also received relatively high ratings, reflecting the dual stress-relief effectiveness of social interaction and environmental immersion.

Table 5. Descriptive statistics of stress-relief items

Item	Frequency	Min	Max	Mean	SD
The color design of natural landscapes in *sky: children of the light* gives me a sense of beauty, relaxation, and stress relief	154	1	5	3.99	0.758
Natural elements in *sky* (e.g., grasslands, ocean, starry sky, and rainforest) help me temporarily escape the pressures of daily life and studies	154	1	5	3.91	0.819
The lighting design of natural landscapes in *sky* helps relax my mind	154	1	5	3.81	0.869
The music in different scenes of sky draws me in and makes me feel relaxed	154	1	5	3.79	0.907
The natural sound effects in *Sky* (e.g., bird calls, rain, ocean waves, and wind) help alleviate my stress. Interactions with friends or strangers in *sky*'s natural scenes	154	1	5	3.94	0.802

(continued)

Table 5. (*continued*)

Item	Frequency	Min	Max	Mean	SD
(E.g., hugging, head patting, bear hugs, and kissing) help me relieve stress and feel joy	154	1	5	3.81	0.906
Chatting with friends or strangers or simply idling in *sky*'s natural landscapes helps me reduce stress and fosters a sense of belonging	154	1	5	3.82	0.864

4.3 Regression Analysis of Stress-Relief Mechanisms

A multiple linear regression was conducted using post-gameplay total stress score as the dependent variable and items from the visual, auditory, and interaction dimensions as independent variables. The model yielded an R^2 of 0.063. Among the predictors, the item "chatting with friends/strangers or idling" showed the strongest predictive power for stress relief ($\beta = 0.245$, $t = 2.272$, $P = 0.025$) (see Table 6).

Table 6. Regression coefficients (a)

Model	Unstandardized coefficients standardized coefficients			Sig.	Collinearity statistics	
	Std.					
	B	Error	Trial version t		Tolerance	VIF
1 (Constant)	15.306	3.393	4.511	0.000		
9.1	– 1.501	0.935	– 0.169–1.604	0.111	0.577	1.732
9.2	1.281	0.929	0.156 1.380	0.170	0.502	1.993
9.3	– 0.218	0.860	– 0.028–0.253	0.800	0.520	1.923
9.4	0.283	0.800	0.038.353	0.724	0.552	1.812
9.5	– 0.950	0.927	– 0.113–1.026	0.307	0.526	1.902
9.6	– 0.040	0.815	– 0.005–0.049	0.961	0.534	1.874
9.7	1.907	0.840 a.	0.245 2.272 Dependent variable: post-gameplay total stress score	0.025	0.552	1.812

5 Discussion

5.1 Significant Stress-Relief Effectiveness of Virtual Natural Landscapes

This study found that the virtual natural landscapes in *Sky: Children of the Light* significantly reduce players' short-term stress, with post-gameplay stress scores dropping from a moderately high level to within the normal range. These findings align with those of Zhang et al. [17], who concluded that virtual natural landscapes reduce negative emotions. The underlying mechanisms may include the following: visually, the therapeutic color palette and lighting designs conform to the "biophilia hypothesis," eliciting positive emotions; auditorily, natural sound effects and background music create an immersive atmosphere that regulates mood through sensory relaxation; and interactively, free exploration and low-pressure social interactions provide a sense of control, consistent with the "soft fascination" mechanism in the attention restoration theory [18].

5.2 The Key Role of the Interaction Dimension in Stress Relief

Regression analysis revealed that "chatting with friends/strangers or idling" was the strongest predictor of stress relief, confirming the "progressive emotional engine" social attribute of *Sky: Children of the Light* proposed by Zou [15]. The high proportion of female players (74.7%) may be attributed to women's stronger emphasis on social connections [16]. The game's non-competitive social interactions satisfy players' need for belonging, thereby enhancing stress-relief outcomes.

5.3 Study Limitations

The study found that the virtual natural landscapes in *Sky: Children of the Light* reduce players' short-term stress through audiovisual immersion and social interaction mechanisms. However, the strongest predictor was "social chatting and sense of belonging," indicating that while virtual natural landscapes provide immediate stress relief, their effects are temporary and context-dependent. They cannot replace professional psychotherapy or real-world nature therapy (Zhang et al., 2020). Given that 93.5% of players engage in gameplay on four or more days per week, it is important to guard against excessive dependence in order to prevent a shift from its use as a "stress-relief tool" to a potential "addiction risk."

Author Contributions. Conceptualization, Y.W. and S.Y.; methodology, Y.W. and S.Y.; writing—original draft preparation, S.Y.; writing—review and editing, Y.W.; supervision, S.Y.; funding acquisition, Y.W. All authors have read and agreed to the published version of the manuscript.

Funding. This paper forms part of the phased research outcomes of the Major Project of Philosophy and Social Science Research in Jiangsu Universities (Project No. 2024SJZD145).

Data Availability Statement. The data supporting the findings of this study were collected through questionnaires from Chinese companies and are available upon request from the authors. Due to privacy concerns, the raw data are not publicly available.

Acknowledgments. The authors would like to thank the editor and anonymous reviewers for helpful comments and suggestions, which helped to improve the manuscript.

Conflicts of Interest. The authors declare no conflict of interest.

Abbreviations

KMO Kaiser–Meyer–Olkin
PSS Perceived Stress Scale

References

1. Gu, L., Sun, B., Liu, L., Li, Y., Yang, J.: Research and verification of a new relaxation technology based on virtual reality for human psychological security and psychosomatic recovery. In: 2021 IEEE International Conference on Intelligence and Safety for Robotics (ISR), pp. 33–37. IEEE (2021)
2. Amer, N.A., Shohieb, S.M., Eladrosy, W.M., Elbakry, H.M., Elrazek, S.M.A.: Sokoon: a gamification-based cognitive behavioral therapy application—an application for depression, stress, and anxiety. Int. J. Gaming Comput.-Mediat. Simul.S (IJGCMS). **15**(1), 1–26 (2023)
3. Vozza, I., Guerra, F., Marchionne, M., Bove, E., Ottolenghi, L.: A multimedia oral health promoting project in primary schools in Central Italy. Ann. Stomatol. **5**(3), 87 (2014)
4. Xiao, Q.: Research on the art design of group psychological healing installation based on user needs. In: International Conference on Human-Computer Interaction, pp. 80–95. Springer Nature Switzerland, Cham (2024)
5. Reese, G., Kohler, E., Menzel, C.: Restore or get restored: the effect of control on stress reduction and restoration in virtual nature settings. Sustainability. **13**(4), 1995 (2021)
6. Andringa, G., Nuijten, P.L., Macville, M.M., Mertens, E.G.A., Roos, A.G.: Feasibility of virtual reality in elderly with dementia. In: Dementia Lab Conference, pp. 146–149. Springer International Publishing, Cham (2019)
7. Cheng-Shih, L., Mei-Yuan, J., Tsu-Ming, Y.: The elderly perceived meanings and values of virtual reality leisure activities: a means-end chain approach. Int. J. Environ. Res. Public Health. **15**(4), 663 (2018)
8. Yu, C.-P., Lee, H.-Y., Luo, X.-Y.: The effect of virtual reality forest and urban environments on physiological and psychological responses. Urban For. Urban Green. **35**, 106–114 (2018)
9. Rockstroh, C., Blum, J., Göritz, A.S.: Virtual reality in the application of heart rate variability biofeedback. Int. J. Hum.-Comput. Stud. **130**, 209–220 (2019)
10. Jeff, T., Jeremy, V., Hannah, C.: Virtual reality for anxiety reduction demonstrated by quantitative eeg: a pilot study. Front. Psychol. **9**, 368656 (2018)
11. Hedblom, M., Gunnarsson, B., Iravani, B., Knez, I., Schaefer, M., Thorsson, P., et al.: Reduction of physiological stress by urban green space in a multisensory virtual experiment. Sci. Rep. **9**(1), 10113 (2019)
12. Park, S.H., Lee, P.J., Jung, T., Swenson, A.: Effects of the aural and visual experience on psycho-physiological recovery in urban and rural environments. Appl. Acoust. **169**, 107486 (2020)
13. Rothbaum, B.O. et al.: Virtual reality exposure therapy for PTSD Vietnam veterans: a case study. J. Trauma. Stress.: Off. Publ. Int. Soc. Trauma. Stress. Stud. **12**(2), 263–271 (1999)
14. Chai, Q.X.: Immersive experiences in digital media interactive art. Decoration. **02**, 73–75 (2012)
15. Zou, Y.F.: The social attributes of the mobile game sky: children of the light: a progressive emotional engine. J. Hubei Norm. Univ. (Philos. Soc. Sci.). **41**, 76–81 (2021)
16. Gong, J.: Interpersonal communication in the social mobile game sky: children of the light. Doctoral dissertation, Nanjing University of Aeronautics and Astronautics (2022)
17. Zhang, Z., Xu, L.Q.: The healing benefits and application of the virtual natural environment. South Archit., 34–40 (2020)
18. Kaplan, R., Kaplan, S.: The Experience of Nature: a Psychological Perspective. Cambridge University Press, Cambridge (1989)

Evaluation of Smart and Sustainable Alternatives: AHP and GIS Applications

Fuad Mohsin Berjess[1], Hamid Ahmed Awad[1], Thaer Shaker Mahmood[1], and Firas Alrawi[2,3](✉)

[1] College of Engineering, University of Anbar, Ramadi, Iraq
{fua23e1003,hamid.awad,drthaersh}@uoanbar.edu.iq
[2] School of Geographical Sciences and Urban Planning, Arizona State University, Tempe, AZ, USA
dr.firas@uobaghdad.edu.iq
[3] Urban and Regional Planning Center, University of Baghdad, Baghdad, Iraq

Abstract. In Ramadi, urban mobility is hindered by major hurdles stemming from fast urban expansion, poor infrastructure, and insufficient strategic planning efforts. The principal aim of this article is to assess (36) regions within Ramadi for their appropriateness for intelligent and sustainable transportation solutions and to ascertain the most suitable alternative. We developed five core categories of criteria, complete with sub-criteria, to systematically evaluate intelligent and sustainable transport systems. The AHP, or analytical hierarchy process, was used to assess these criteria along with alternatives, and its results were merged with a geographic information system (GIS) for analysis in a spatial context. This study proposes four principal alternatives for intelligent and sustainable transportation in Ramadi: Bus Rapid Transit (AL1), bicycle and pedestrian pathways (AL2), smart solutions (AL3), and electric vehicle infrastructure (AL4). The findings revealed that (AL3) (smart solutions) attained the highest overall performance evaluation. A sensitivity analysis demonstrated that various transportation alternatives are unevenly influenced by alterations in specific criteria, with environmental considerations being paramount for electric vehicle infrastructure, traffic considerations for smart solutions, and road characteristics for bicycle and pedestrian pathways. The academic article showcases maps that illustrate how suitable each choice is throughout Ramadi, with regions categorized accordingly.

Keywords: Urban mobility · Hierarchy process · GIS · Ramadi City · Transportation index

1 Introduction

Urban advancement demands a strategic framework that interweaves technological innovations, systemic organization, and practices that promote sustainability. Core aspects that boost intelligent advancement encompass the adoption of progressive tech solutions, principles of sustainable urban architecture, strong governance systems, and community

Y. Zhu (Ed.): ISESIE 2025, LNCE 000, pp. 324–333, 2026.
https://doi.org/10.1007/978-981-92-2389-3_26

involvement. These ingredients synergize to combat the hurdles connected to urbanization, covering resource depletion, ecological balance, and social inequities [1–3]. The alignment of smart transportation with sustainable urban development is evident in several dimensions, including environmental, social, and economic aspects. This alignment is crucial for the transition from traditional to smart cities, particularly in developing regions. Below, the criteria of smart transportation and their alignment with sustainable urban development goals are explored in detail [4–6]:

- Environmental Sustainability: Smart transportation systems contribute to minimizing emissions and improving air quality. Sophisticated transport methodologies harness data evaluation and AI technologies to better traffic routing schemes. The incorporation of electric and driverless cars into innovative transit choices remarkably raises environmental sustainability through the reduction of carbon emissions.
- Social Sustainability: Smart transportation enhances social inclusion by improving accessibility and connectivity within urban areas. The implementation of smart mobility solutions provides seamless travel experiences and encourages the use of public and shared transport options, which can improve the quality of life.
- Economic Sustainability: Intelligent transportation systems are positioned to refine the operational efficacy of transport networks. The creation of state-of-the-art transport infrastructure, which integrates advanced traffic control mechanisms and automated parking solutions, may generate fiscal benefits for municipal governments and travelers alike.

The primary aim of this article is to evaluate thirty-six districts within Ramadi with respect to smart and sustainable transportation and identify the most suitable alternative for analysis. The analysis is carried out on selected criteria, considering social, economic, environmental, traffic, and roadway criteria, employing the MCDM (Analytic Hierarchy Process) framework and GIS applications.

2 Smart and Sustainable Transportation Criteria

The criteria for evaluating smart and sustainable transportation systems are multi-faceted, encompassing several dimensions [14, 15]. The principal criteria for smart and sustainable transportation are delineated as follows:

- Social Equity: Environmentally sustainable transportation alternatives must be accessible for all at just costs. The requirements of marginalized communities should be prioritized.
- Economic Viability: Achieving recognition as sustainable transport networks requires them to display financial feasibility, which concerns their proficiency in balancing costs and benefits across their whole operational timeline. Moreover, such networks need to alleviate the effects of externalities and promote job creation while enticing investors. These structures should entail policies and design aspects that lessen the odds of unfortunate events, unlawful conduct, and aggressive confrontations.
- Sustainable transportation systems: must guarantee seamless connectivity across various transport modalities and establish accessible routes to essential destinations such as workplaces, educational institutions, healthcare facilities, and recreational areas. Accessibility must remain a core focus of these systems.

- Land Use Integration: It is imperative to integrate sustainable transportation strategies with land use planning methodologies. The promotion of environments conducive to pedestrian navigation and bicycle accessibility ought to be prioritized, in conjunction with an emphasis on transit-oriented development principles.
- Health and Well-Being: Active transportation choices like walking and biking should be at the forefront of sustainable transport initiatives because they yield many health perks. This program is expected to substantially boost public health and overall well-ness. Furthermore, creating spaces that are friendly to pedestrians should take prece-dence to motivate individuals to engage in physical activity as a critical aspect of their daily routines.

3 Study Area

Ramadi is situated in the western region of Iraq, namely at the convergence of longitudes (43°20′) and (43°12′) east, as well as latitudes (33°27′) and (33°23′) north. The urban settlement is predominantly situated in the southern quadrant. The Euphrates River delineates the geographical boundary of the Al-Anbar Governorate, with the eastern quadrant positioned on one side of the river. As per the data provided by the Anbar Statistics Directorate, Ramadi encompasses an extensive area of 138,088 ha, with a developed expanse of 7,375 ha. It operates as a vital transportation hub where numerous transit routes, including principal highways and international thoroughfares such as Routes 10 and 12, converge. Consequently, it is perceived as an essential conduit for movement between the eastern and western sectors of Iraq. This endows the city with regional importance that significantly enhances its progress and growth.

4 Data Collection

The initial phase of this study encompassed an extensive array of attributes within 36 districts in Ramadi, which included indices of the most important aspects. It compiled pertinent data from a diverse array of sources, which included governmental agencies, libraries, as well as studies and theses, Documents, official statistics, both published and unpublished data, cartographic materials, and satellite imagery. Additionally, it executed field surveys and investigative studies (Table 1).

Table 1. Social, economic, environmental, traffic, and road criteria

Criteria	Sub criteria	Code	Definition
Social (m1)	District area index	C1	Total area in each district (m^2)
	District population index	C2	Number of population in each district (in thousands)
	Service building index	C3	Number of schools, healthcare centers in each district
Economic (M2)	Car ownership index	C4	Percentage of population that have at least one vehicle in each district

(continued)

Table 1. (*continued*)

Criteria	Sub criteria	Code	Definition
	Land price index	C5	Average price of one meter in each district (in million)
Environment (M3)	Green area index	C6	The proportion of land allocated for green area (m^2)
	Air pollution index	C7	Amount of co gas emission in each district (ppm)
	Noise index	C8	Level of noise in each district (Db)
Traffic (M4)	Traffic volume index	C9	Average traffic volume in each district (Veh/h)
	Traffic accidents index	C10	Number of traffic accidents in each district (for two years 2021–2023)
Road (M5)	Parking index	C11	Proportion of land allocated for park area (m^2)
	Side walk index	C12	Proportion of land allocated for sidewalk area (m^2)
	Road density index	C13	Area of road network in each district (m^2)
	Road complex index	C14	Average of urban indices in each district

5 Smart and Sustainable Transportation Alternatives

- Alternative 1: Bus Rapid Transit (BRT): BRT is a high-quality, bus-based transit system that distributes fast, easy, and cost-effective urban mobility.
- Alternative 2: Bicycle and Pedestrian-Friendly Corridors: This alternative helps non-motorized transport by creating safe, accessible routes for cyclists and pedestrians within urban areas.
- Alternative 3: Smart Solutions (e.g., Smart Traffic Signals and Smart Parking): This alternative uses advanced technology and real-time data to manage urban.
- Alternative 4: Electric Vehicle (EV) Infrastructure: Development of a network for supporting electric vehicles, including public charging stations, battery-swapping facilities.

6 Analytic Hierarchy Process (AHP)

This approach leverages numerical weights to distinguish critical criteria from ancillary ones, grounded in their relative importance within the evaluative framework. This research implemented a three-stage AHP: (1) Identification of the problem and clarification of the objective of the decision-making. (2) Development of the hierarchical model.

The hierarchical structure was systematically organized into three discrete tiers: the apex level, the intermediary level, and the foundational level. (3) Execution of pairwise comparisons among the criteria and alternatives utilizing a scale ranging from 1 to 9. This scale was employed to assess the relative importance of the evaluation. In the nine-point evaluative framework, the degrees of relative significance were articulated through the numerical representations of 1, 3, 5, 7, and 9. These numerical values signified the categories of equal, moderate, strong, very strong, and extreme levels of thematic relevance in relation to the gravity of the circumstances. The values of two, four, six, and eight were classified as intermediate. The execution of Consistency Ratio (CR) assessments was undertaken for the purpose of ensuring control, with the objective of averting any discrepancies within the evaluative matrix. The CR is ascertained through Eq. (1):

$$CR = \frac{CI}{RI} \tag{1}$$

The parameters employed within this framework consist of CR representing the consistency ratio, CI denoting the consistency index, and RI signifying the random index. The confidence interval (CI) is ascertained through the Eq. (2):

$$CI = \frac{\lambda_{\max} - n}{n - 1} \tag{2}$$

The value of the maximum is denoted as λ_{max}, and n represents the total number of elements, the outcomes are deemed inconsistent and then disregarded. The computation of a weighting score for each of the potential options encompasses both alternatives and criteria. The relative weights are ascertained through the principal eigenvector (w) linked to the maximal eigenvalue (λ_{max}), articulated as:

$$AW = \lambda_{max}{}^{*}W \tag{3}$$

The normalization of the vector culminated in the creation of a vector that encapsulates the prioritization of the evaluation themes pertinent to the objective. The weighting coefficient can be efficiently ascertained utilizing software applications such as (Super Decision) or (Expert Choice), or alternatively, (Microsoft Office Excel).

7 Results

The matrix for pairwise comparison based on expert opinion for the primary criteria can be found in Tables 2, 3, 4 and 5.

Table 2. Main criteria weight for alternative 1 (bus rapid transit)

Criteria	SC	EC	EN	TR	RO	Weight (%)
Social	1.00	1.50	2.00	0.75	0.50	16
Economic	0.75	1.00	1.50	0.50	0.33	14

(continued)

Table 2. (*continued*)

Criteria	SC	EC	EN	TR	RO	Weight (%)
Environment	0.50	0.75	1.00	0.33	0.20	10
Traffic	1.50	2.00	3.00	1.00	0.67	18
Road	2.00	3.00	5.00	1.50	1.00	41
CR = 0.010						

Table 3. Main criteria weight for alternative 2 (bicycle and pedestrian-corridors)

Criteria	SC	EC	EN	TR	RO	Weight (%)
Social	1.00	1.00	1.75	1.00	0.50	16
Economic	1.00	1.00	1.50	0.75	0.33	14
Environment	0.75	0.75	1.00	0.50	0.25	10
Traffic	1.00	1.50	1.75	1.00	0.50	18
Road	2.50	3.00	4.00	2.33	1.00	41
CR = 0.010						

Table 4. Main criteria weight for alternative 3 (smart solutions)

Criteria	SC	EC	EN	TR	RO	Weight (%)
Social	1.00	1.75	1.00	0.50	0.50	16
Economic	0.50	1.00	0.50	0.33	0.33	9
Environment	1.00	1.75	1.00	0.50	0.50	16
Traffic	2.00	3.00	2.00	1.00	1.00	30
Road	2.00	3.00	2.00	1.00	1.00	30
CR = 0.010						

Table 5. Main criteria weight for alternative 4 (electric vehicle (EV) infrastructure)

Criteria	SC	EC	EN	TR	RO	Weight (%)
Social	1.00	1.00	0.33	1.00	1.00	15
Economic	1.00	1.00	0.50	1.00	1.00	16
Environment	3.00	2.00	1.00	2.00	2.00	36
Traffic	1.00	1.00	0.50	1.00	1.00	16
Road	1.00	1.00	0.50	1.00	1.00	16
CR = 0.010						

- Alternative 1: Bus Rapid Transit (BRT)

1. Road Criteria Dominance: For the Bus Rapid Transit alternative, the 'Road' criterion holds the highest weight at 41%. This suggests that factors related to road infrastructure and its capacity are considered most critical for BRT implementation.
2. Traffic and Social Importance: 'Traffic' and 'Social' criteria follow with weights of 18% and 16% respectively, indicating their significant roles in the evaluation of BRT systems.
3. Economic and Environmental Contributions: 'Economic' and 'Environment' criteria have lesser, but still notable, weights of 14% and 10% respectively.

- Alternative 2: Bicycle and Pedestrian Corridors

1. Road Criteria Dominance: Similar to BRT, the 'Road' criterion is the most heavily weighted for Bicycle and Pedestrian Corridors, also at 41%. This expectation reflects the importance of dedicated and well-maintained road space for these modes of transport.
2. Traffic and Social Importance: 'Traffic' and 'Social' criteria are weighted at 18% and 16% respectively, mirroring their importance in the BRT alternative.
3. Economic and Environmental Importance are weighted at 14% and 10% respectively, for the BRT alternative.

- Alternative 3: Smart Solutions

1. Traffic and Road criteria have the highest weight, both at 30%. This indicates that these solutions are heavily reliant on and significantly impact traffic flow and road network efficiency.
2. Social and Environmental Importance: 'Social' and 'Environment' criteria are equally weighted at 16%.
3. Economic Contribution: The 'Economic' criterion has the lowest weight at 9% for this alternative.

- Alternative 4: Electric Vehicle (EV) Infrastructure

1. Environmental Dominance: The 'Environment' criterion is the most significant for Electric Vehicle Infrastructure, with a weight of 36%. This highlights the primary environmental benefits and considerations associated with EV adoption (Table 6).
2. Balanced Weights for Other Criteria: 'Economic', 'Traffic', and 'Road' criteria each hold a weight of 16%, indicating a balanced importance across these factors.
3. Social Contribution: The 'Social' criterion has a slightly lower weight at 15%.

Table 6. Alternatives score (smart and sustainable transportation alternatives)

No.	District	AL1	AL2	AL3	AL4
1	AL-SOFIYA	0.508	0.460	0.534	0.465
2	AL-TAQADUM	0.350	0.328	0.376	0.377
3	AL-QUDUS	0.560	0.538	0.562	0.527
4	AL-KATANA	0.328	0.287	0.377	0.327
5	AL-AZIZIYA	0.347	0.318	0.375	0.356
6	AL-NASR	0.255	0.217	0.293	0.253
7	AL-AMJAD	0.372	0.341	0.400	0.340
8	AL-RIYAD	0.387	0.384	0.390	0.393
9	AL-JUMHURIU	0.333	0.299	0.373	0.316
10	AL-MUAALMIN	0.450	0.433	0.465	0.431
11	AL-SALAM	0.379	0.361	0.402	0.337
12	AL-TAAMIM	0.521	0.518	0.489	0.420
13	AL-NOOR	0.403	0.381	0.423	0.358
14	AL-TAHRIR	0.351	0.339	0.350	0.309
15	AL-HUSAYN	0.279	0.247	0.305	0.249
16	AL-NAKHIL	0.432	0.412	0.461	0.449
17	AL-SHAFEE	0.349	0.348	0.350	0.375
18	AL-EKHA	0.301	0.295	0.291	0.250
19	AL-WARAR	0.460	0.454	0.517	0.550
20	AL-KHADRA	0.323	0.295	0.352	0.309
21	AL-ZUHUR	0.415	0.403	0.420	0.394
22	MUJMAE DAWR ALATIBA	0.320	0.324	0.320	0.334
23	AL-ZAYTUN	0.281	0.265	0.300	0.279
24	OTHMAN BIN AFAAN	0.479	0.477	0.477	0.465
25	AL-JAMIEA	0.507	0.511	0.551	0.574
26	AL-FIRDAWS	0.409	0.411	0.449	0.529
27	AL-WAHDAH	0.252	0.230	0.270	0.236
28	AL-MUSTAQBAL	0.545	0.548	0.556	0.604
29	AL-WAAHAH	0.362	0.362	0.365	0.375
30	AL-EADIL	0.150	0.143	0.158	0.150
31	AL-DUBAAT ALTHANNIH	0.169	0.161	0.166	0.129
32	AL-SIKAK ALSHARQIA	0.149	0.132	0.156	0.108
33	AL-BAKAR	0.459	0.439	0.481	0.452

(continued)

Table 6. *(continued)*

No.	District	AL1	AL2	AL3	AL4
34	AL-MALAB	0.179	0.168	0.183	0.153
35	AL-ARAMIL	0.053	0.044	0.059	0.042
36	AL-ZANKURA	0.098	0.089	0.089	0.084

8 Conclusions

- The alternatives that achieved the most elevated scores are considered the most advantageous options for selection. The overall sum of scores for all 36 districts indicates that AL3 (Smart Solutions) has the highest total score (13.085), followed by AL1 (12.515), AL4 (12.299), and AL2 (11.962).
- The social criteria (M1) show relatively consistent sensitivity across all alternatives, with AL1 generally exhibiting the highest sensitivity in this category. Economic criteria (M2) and environmental criteria (M3) demonstrate significantly higher sensitivity for AL4 (EV Infrastructure), indicating their critical importance for EV adoption. Traffic criteria (M4) show the highest sensitivity for AL3 (Smart Solutions), implying that traffic-related factors are particularly influential for these interventions. Road criteria (M5) exhibit the highest sensitivity for AL2 (Bicycle and Pedestrian-Friendly Corridors), underscoring the relevance of road infrastructure for such corridors.
- Overall, the sensitivity analysis highlights that different transportation alternatives are affected disproportionately by changes in specific criteria. Environmental factors are most critical for EV infrastructure, traffic factors are most impactful for smart solutions, and road characteristics are most influential for bicycle and pedestrian corridors.
- The suitability analysis, visually represented through maps, categorizes areas into 'Very Low' to 'Very High' suitability levels for each alternative. This spatial understanding provides a guide for targeted implementation, suggesting that BRT systems and bicycle/pedestrian corridors would be most effective if concentrated in specific, well-suited zones. Similarly, smart city technologies and EV infrastructure should be strategically placed in areas with high suitability to maximize usage and benefit.

References

1. Shiva, C.K., Vedik, B., Manoharan, G.: Sustainable smart transportation: technologies, benefits, challenges to urbanization concept. In: Secure and Intelligent IoT-Enabled Smart Cities, pp. 355–370. IGI Global (2024)
2. Kumar, S.S., Usha, P., Balakrishnan, P., Kannan, V.K., Manjula, M., Vijayakumar, M.: Deep learning driven engineering innovations: advancing sustainable green building and smart infrastructure for a resilient future. In: 2024 First International Conference for Women in Computing (InCoWoCo), pp. 1–5. IEEE (2024)

3. Tiwari, S., Bhushan, A., Singh, A.K., Yadav, R.K.: Unleashing the potential of IOT integration for energy optimization in smart homes. In: 2024 3rd International Conference on Power Electronics and IoT Applications in Renewable Energy and its Control (PARC), pp. 82–85. IEEE (2024)
4. Braga, I.F., Ferreira, F.A., Ferreira, J.J., Correia, R.J., Pereira, L.F., Falcão, P.F.: A DEMATEL analysis of smart city determinants. Technol. Soc. **66**, 101687 (2021)
5. Oladimeji, D., Gupta, K., Kose, N.A., Gundogan, K., Ge, L., Liang, F.: Smart transportation: an overview of technologies and applications. Sensors. **23**(8), 3880 (2023)
6. Hummels, H., Argyrou, A.: Planetary demands: redefining sustainable development and sustainable entrepreneurship. J. Clean. Prod. **278**, 123804 (2021)
7. Luna-Nemecio, J., Tobón, S., Juárez-Hernández, L.G.: Sustainability-based on socioformation and complex thought or sustainable social development. Resour., Environ. Sustain. **2**, 100007 (2020)
8. Bradley, P.: An institutional economics framework to explore sustainable production and consumption. Sustain. Prod. Consum. **27**, 1317–1339 (2021)
9. Ruggerio, C.A.: Sustainability and sustainable development: a review of principles and definitions. Sci. Total Environ. **786**, 147481 (2021)
10. Filippi, F.: A paradigm shift for a transition to sustainable urban transport. Sustainability. **14**(5), 2853 (2022)
11. Kraus, L., Proff, H.: Sustainable urban transportation criteria and measurement—a systematic literature review. Sustainability. **13**(13), 7113 (2021)
12. Hotak, M., Turcios, E., Dayrit, D.G., Arora, A.: The road to sustainable transportation in supply chains: a stage-level approach. Sustain. Innov.: Trends Mark. Manag., 59–69 (2020)
13. Litman, T., Burwell, D.: Issues in sustainable transportation. Int. J. Glob. Environ. Issues. **6**(4), 331–347 (2006)

Author Index

Y. Zhu (Ed.): ISESIE 2025, LNCE 000, pp. 335–336, 2026.
https://doi.org/10.1007/978-981-92-2389-3

MIX
Papier aus verantwortungsvollen Quellen
Paper from responsible sources
FSC® C105338

FSC
www.fsc.org

If you have any concerns about our products,
you can contact us on
ProductSafety@springernature.com

In case Publisher is established outside the EU,
the EU authorized representative is:
Springer Nature Customer Service Center GmbH
Europaplatz 3, 69115 Heidelberg, Germany

Printed by Libri Plureos GmbH
in Hamburg, Germany